VLSI Signal Processing

OTHER IEEE PRESS BOOKS

VLSI: Technology and Design, *Edited by Otto G. Folberth and Warren D. Grobman*
General and Industrial Management, *By Henri Fayol, revised by Irwin Gray*
MOS Switched-Capacitor Filters: Analysis and Design, *Edited by G.S. Moschtyz*
Distributed Computing: Concepts and Implementations, *Edited by P.L. McEntire, J.G. O'Reilly, and R.E. Larson*
Engineers & Electrons, *By J.D. Ryder and D.G. Fink*
Electronic Displays, *Edited by H.I. Refioglu*
Advanced Microprocessors, *Edited by A. Gupta and H.D. Toong*
A Guide for Writing Better Technical Papers, *Edited by C. Harkins and D.L. Plung*
Digital MOS Integrated Circuits, *Edited by M.I. Elmasry*
Modern Active Filter Design, *Edited by R. Schaumann, M.A. Soderstrand, and K.R. Laker*
Analog MOS Integrated Circuits, *Edited by P.R. Gray, D.A. Hodges, and R.W. Brodersen*
Integrated Injection Logic, *Edited by J.E. Smith*
Sensory Aids for the Hearing Impaired, *Edited by H. Levitt, J.M. Pickett, and R.A. Houde*
Programs for Digital Signal Processing, *Edited by the Digital Signal Processing Committee, IEEE*
Automatic Speech & Speaker Recognition, *Edited by N.R. Dixon and T.B. Martin*
Speech Analysis, *Edited by R.W. Schafer and J.D. Markel*
The Engineer in Transition to Management, *I. Gray*
Analog Integrated Circuits, *Edited by A.B. Grebene*
Integrated-Circuit Operational Amplifiers, *Edited by R.G. Meyer*
Digital Signal Computers and Processors, *Edited by A.C. Salazar*
Charge-Coupled Devices: Technology and Applications, *Edited by R. Melen and D. Buss*
Waveform Quantization and Coding, *Edited by N.S. Jayant*

VLSI Signal Processing

Edited by
Peter R. Cappello, University of California
Chi-Foon Chan, Intel Corporation
Bijoy Chatterjee, ESL
Gwyn Edwards, Racal-Vadic
Ed Greenwood, CHI Systems
Leah Jamieson, Purdue University
S.Y. Kung, University of Southern California
Richard Lyon, Fairchild Research Center
J. Greg Nash, Hughes Research Laboratories
Graham R. Nudd, Hughes Research Laboratories
Robert Owen, Consultant
Earl E. Swartzlander, Jr., TRW

The chapters in this book are based on papers presented at the Workshop on VLSI Signal Processing held November 27–29, 1984, at the University of Southern California. The Workshop is a biennial activity of the VLSI Technical Committee of the IEEE Acoustics, Speech & Signal Processing Society.

Published under the sponsorship of the
IEEE Acoustics, Speech & Signal Processing Society.

The Institute of Electrical and Electronics Engineers, Inc., New York.

IEEE PRESS
1984 Editorial Board
M. E. Van Valkenburg, *Editor in Chief*
M. G. Morgan, *Editor, Selected Reprint Series*
Glen Wade, *Editor, Special Issue Series*

J.M. Aein	Thelma Estrin	R.C. Jaeger
J.K. Aggarwal	L.H. Fink	E.A. Marcatili
J.E. Brittain	S.K. Gandhi	J.S. Meditch
R.W. Brodersen	Irwin Gray	W.R. Perkins
R.F. Cotellessa	H.A. Haus	A.C. Schell
M.S. Dresselhaus	E. W. Herold	Herbert Sherman

W. R. Crone, *Managing Editor*
Teresa Abiuso, *Administrative Assistant*

Copyright © 1984 by
THE INSTITUTE OF ELECTRICAL AND ELECTRONICS ENGINEERS, INC.
345 East 47th Street, New York, NY 10017
All rights reserved.

PRINTED IN THE UNITED STATES OF AMERICA

IEEE Order Number: PC01800

Library of Congress Catalog Card No.: 84-82077

ISBN 0-87942-186-X

AUTHORS

J.G. Ackenhusen, Bell Telephone Laboratories, 316
H.M. Ahmed, Codex Corporation, 52
J. Annevelink, Delft University, 294
L. Auslander, City University of New York, 172
A.Z. Baraniecki, M/A-Com Research Center, 88
M.R. Baraniecki, M/A-Com Research Center, 88
A. Benveniste, IRISA, 282
R. Bianchini, ESL, 39
P. Bournai, IRISA, 282
R.W. Brodersen, University of California, 239
D.S. Broomhead, Royal Signals and Radar Establishment, England, 375
A.M. Brown, TMC, Ltd., 112
C.F. Chan, Intel Corporation, 328
M. Chase, NEC Electronics, Inc., 133
J.W. Cooley, Cleveland State University, 172
A. Delaruelle, Philips Laboratories and Philips KommuniKations Industrie, 70
P.B. Denyer, University of Edinburgh, 252
P. Dewilde, Delft University, 294
R.J. Dunki-Jacobs, General Electric Co., 329
S. Eidson, Advanced Micro Devices, 99
J.A. Eldon, TRW Systems, 16
J.A. Eldon, TRW LSI Products, 124
R.D. Etchells, Hughes Research Laboratories, 2
R.S. Freedman, Hazeltine Corporation, 266
T. Gautier, IRISA, 282
M. Glesner, Technische Hochschule Darmstadt, 216
J.H. Gray, ESL, 36
J. Grinberg, Hughes Research Laboratories, 2
G. Hallnor, TRW Systems, 27
K. Hanson, Texas Instruments, 98
Y.A. Haque, American Microsystems, 117
R.M. Hardy, General Electric Co., 329
J.G. Harp, Royal Signals and Radar Establishment, England, 375
K. Iwano, Princeton University, 387
H. Joepen, Technische Hochschule Darmstadt, 216
M. Kahrs, University of Rochester, 228
S.Y. Kung, Delft University, 294
C.J. Kuo, Massachusetts Institute of Technology, 271
P. LeGuernic, IRISA, 282
S. Levitan, University of Massachusetts, 15
B.C. Levy, Massachusetts Institute of Technology, 271

M.J. Little, Research Laboratories, 2
R.F. Lyon, Fairchild Research Center, 64
J.G. McWhirter, Royal Signals and Radar Establishment, England, 375
D.I. Moldovan, University of Southern California, 350
C.T. Mullis, University of Colorado, 158
A.F. Murray, University of Edinburgh, 252
B.R. Musicus, Massachusetts Institute of Technology, 271
G.R. Nudd, Hughes Research Laboratories, 2
Y.H. Oh, Bell Telephone Laboratories, 316
R. Owen, Consultant, 308
K.J. Palmer, Royal Signals and Radar Establishment, England, 375
R.M. Perlman, Advanced Micro Devices, 155
S. Pope, University of California, 239
W.J. Premerlani, General Electric Co., 329
S.U.H. Qureshi, Codex Corporation, 52
J. Rabaey, University of California, 239
I.V. Ramakrishnan, University of Maryland, 363
P. Ramesh, Intel Corporation, 328
J.T. Rayfield, Brown University, 162
D. Renshaw, University of Edinburgh, 252
K. Rinner, Philips Laboratories and Philips KommuniKations Industrie, 70
E. Riseman, University of Massachusetts, 15
J.G.B. Roberts, Royal Signals and Radar Establishment, England, 375
R.A. Roberts, University of Colorado, 158
M.S. Schlansky, ESL, 39
J. Schmid, Philips Laboratories and Philips KommuniKations Industrie, 70
R.O. Schmidt, Guiltech Research Co., 38
J. Schunk, Technische Hochschule Darmstadt, 216
J.D. Seals, AT&T Bell Laboratories, 37
R.R. Shively, AT&T Bell Laboratories, 37
A.J. Silberger, Cleveland State University, 172
H.F. Silverman, Brown University, 162
K. Steiglitz, Princeton University, 387
A.A. Stroll, TRW Sysstems, 16
E.E. Swartzlander, Jr., TRW Systems, 27
D. Taylor, Advanced Micro Devices, 99
P.M. Toldalagi, Analog Devices, 145
J.L. van Meerbergen, Phililps Laboratories and Philips KommuniKations Industrie, 70
P.J. Varman, University of Maryland, 363
C. Weems, University of Massachusetts, 15
F.P. Welten, Philips Laboratories and Philips KommuniKations Industrie, 70
J.E. Wheeler, General Electric Co., 329

PREFACE

There is a special interrelationship between VLSI technology and signal processing. It is described succinctly in the June 1984 issue of VLSI Design:
"VLSI and Digital Signal Processing: A Symbiotic Relationship."
Based on this belief, the IEEE Workshop on VLSI Signal Processing has been organized to provide a forum for this research, and to promote interaction among researchers in these disciplines.
Each Chapter of this book is based on a paper that was presented at the workshop, though not all such presentations have a corresponding chapter. The chapters are organized into ten Parts which correspond to the workshop's ten sessions.

Part I: Image Analysis

This part was organized by Leah Jamieson of Purdue University. Its chapters apply VLSI to several problems associated especially with image analysis, such as its high throughput requirement, image rotation, and fast Fourier transformation.

Part II: Algorithmic Processors

This part was organized by Bijoy Chatterjee of ESL. Its chapters deal with the rapidly emerging field of algorithmically specialized architectures for high performance signal processing. Two of the presentations are not included in this book in order to legally protect patent applications.

Part III: VLSI Architectures I

This part was organized by Ed Greenwood of CHI Systems. The chapters in this part, and also in VLSI Architectures II, disclose the chip architectures of several digital signal processors that are being developed currently.

Part IV: Telecommunications

This part was organized by Gwyn Edwards of Racal-Vadic. The chapters in this part deal with VLSI chip (sets) that are being designed specifically for use in telecommunications.

Part V: VLSI Architectures II

This part which was organized by Robert Owen, Consultant in DSP & VLSI, is a companion to VLSI Architectures I.

Part VI: Advances in Algorithms

This part was organized by Earl E. Swartzlander, Jr. of TRW. Its chapters emphasize the interrelationship between VLSI technology and signal processing algorithms with respect to numerical characteristics.

Part VII: Digital Signal Processing CAD

This part was organized by Richard Lyon of Fairchild Research Center. These chapters present research on computer-aided design tools for VLSI signal processing applications.

Part VIII: Languages and Software

This part was organized by Sun-Yuan Kung of the University of Southern California. These chapters are concerned with languages and software for specifying, transforming, and verifying signal processing algorithms and array architectures.

Part IX: Real-Time Speech

This part was organized by Chi-Foon Chan of Intel. These chapters disclose VLSI chip architectures that are intended specifically for use in real-time speech processing. One of the presentations is not included in this volume for legal reasons.

Part X: Systolic Architectures

This part was organized by J. Greg Nash of Hughes Research Laboratories. The chapters in this part deal with several aspects of systolic arrays, such as partitioning problems for computation on arrays, fault-tolerant array algorithms, global vs. local array synchronization, and optimization of array elements.

The workshop was held in the Davidson Conference Center, University of Southern California, Los Angeles, California, November 27–29, 1984, and was sponsored by the IEEE Acoustics, Speech, and Signal Processing Society.

I wish to thank all of those who submitted extended summaries, and full papers, for consideration. The goal of having a small workshop precluded the inclusion of some papers of quality and merit. It is expected that these papers will appear in conferences and journals.

This workshop has benefited from the encouragement of many individuals within the IEEE ASSP Society, particularly from Ron Crochiere, Charles Teacher, Tom Crystal, and Ken Steiglitz, as well as from Ted Gerlach of IEEE Conference Services and Reed Crone of the IEEE Press. It has been a pleasure to work with J. Greg Nash and Barbara Cory. They have given greatly of their time and expertise, improving the workshop in myriad ways. The ten Session chairpersons and two Panel Moderators, Jonathan Allen of MIT and Harper Whitehouse of NOSC, are to be credited largely for determining the basic quality of the Workshop's technical program. Graham Nudd, Earl Swartzlander, Gwyn Edwards, Dick Lyon, and Rob Owen, however, contributed crucially in its finalization. I wish, finally, to thank Sun-Yuan Kung and Earl Swartzlander for the support and counsel they have so graciously given me.

Peter R. Cappello

TABLE OF CONTENTS

Part I: IMAGE ANALYSIS
L. Jamieson, Purdue University 1

1. Three-Dimensional Computing Structures for High Throughput Information Processing, *J. Grinberg, R.D. Etchells, G.R. Nudd, and M.J. Little, Hughes Research Laboratories* 2

2. Signal to Symbols: Unblocking the Vision Communication/Control Bottleneck, *S. Levitan, C. Weems, and E. Riseman, University of Massachusetts* 15

3. VLSI for Image Rotation, *Z.Z. Stroll, J. Eldon, and E.E. Swartzlander, TRW Systems* 16

4. High Speed FFT Processor Implementation, *E.E. Swartzlander and G. Hallnor, TRW Systems* 27

Part II: ALGORITHMIC PROCESSORS
B. Chatterjee, ESL. 35

5. A VLSI FFT System, *J.H. Gray, ESL* 36

6. EMSP: A Data-Flow Computer for Signal Processing Applications, *J.D. Seals and R.R. Shively, AT&T Bell Laboratories* 37

7. Digital/Optical Computing and Linear Algebra, *R.O. Schmidt, Guiltech Research Co.* 38

8. A High Performance Interconnect for Concurrent Signal Processing, *M.S. Schlansky and R. Bianchini, Jr., ESL* 39

Part III: VLSI ARCHITECTURES I
E. Greenwood, Motorola 51

9. A Custom Chip Set for Digital Signal Processing, *S.U.H. Qureshi and H.M. Ahmed, Codex Corporation* 52

10. MSSP: A Bit-Serial Multiprocessor for Signal Processing, *R.F. Lyon, Fairchild Research Center* 64

11. A 2-micron CMOS, 10-MHz, Microprogrammable Vector Processing Unit with On-Chip 3-port Registerfile for Incorporation in Single Chip General Purpose Digital Signal Processors, *F.P. Welten, A. Delaruelle, F.J. van Wijk, J.L. van Meerbergen, J. Schmid, and K. Rinner, Philips Laboratories and Philips Kommunikations Industrie* 76

12. VLSI Multiprocessor Architecture for Signal Processing, *M.R. and A.Z. Baraniecki, M/A-COM Research Center* 88

Part IV: TELECOMMUNICATIONS
G. Edwards, Racal-Vadic 97

13. An Integrated Coherent Demodulation Technique, *K. Hanson, Texas Instruments* 98

14. Design of a Single-Chip Bell 212A Compatible Modem Employing Digital Signal Processing, *D. Taylor and S. Eidson, Advanced Micro Devices* 99

15. The Integration of a 144 Kb/s Line Transmission System, *A.M. Brown, TMC, Ltd.* 112

16. An Echo Canceller, *Y.A. Haque, American Microsystems* 117

Part V: VLSI ARCHITECTURES II
R. Owen 123

17. New Modular CMOS Correlator Chip, *J.A. Eldon, TRW LSI Products* 124

18. A Pipelined Data Flow Architecture for Digital Signal Processing: The NEC uPD7281, *M. Chase, NEC Electronics, Inc.* 133

19. New Word-Slice Components Architecturally Optimized for High-Performance DSP, *P.M. Toldalagi, Analog Devices* 145

20. A New Approach to Floating-point DSP, *R.M. Perlman, Advanced Micro Devices* 155

Part VI: ADVANCES IN ALGORITHMS
E.E. Swartzander, Jr., TRW 157

21. Digital Processing Structures for VLSI Implementation, *C.T. Mullis and R.A. Roberts, University of Colorado* 158

22. Some VLSI Design Implications for DFT Calculations Based on a WFTA/M68000 Implementation, *J.T. Rayfield and H.F. Silverman, Brown University* 162

23. Numerical Stability of Fast Convolution Algorithms for Digital Filtering,
    L. Auslander, City University of New York, J. Cooley, IBM,
    T.J. Watson, Research Center, and A. Silberger, Cleveland
    State University 172

Part VII: DIGITAL SIGNAL PROCESSING, CAD
R. Lyon, Fairchild Research Center 215

24. ALGIC—A Flexible Silicon Compiler System for Digital Signal
    Processing Circuits, J. Schuck, M. Glesner, and H. Joepen,
    Technische Hochschule Darmstadt 216

25. Silicon Compilation of a Very High Level Signal Processing
    Specification Language, M. Kahrs, University of Rochester 228

26. Automated Design of Signal Processors Using Macrocells, S. Pope,
    J. Rabaey, and R.W. Brodersen, University of California 239

27. FIRST: Prospect and Retrospect, P.B. Denyer, A.F. Murray,
    and D. Renshaw, University of Edinburgh 252

Part VIII: LANGUAGES AND SOFTWARE
S.Y. Kung, University of Southern California 265

28. Logic Programming of Systolic Arrays, R.S. Freedman,
    Hazeltine Corporation 266

29. The Specification and Verification of Systolic Wave Algorithms,
    C.J. Kuo, B.C. Levy, and B.R. Musicus, Massachusetts Institute
    of Technology 271

30. SIGNAL: A Data Flow Oriented Language for Signal Processing,
    P. LeGuernic, A. Benveniste, P. Bournai, and T. Gautier, IRISA 282

31. Hierarchical Iterative Flowgraph Integration for VLSI Array Processors,
    S.Y. Kung, University of Southern California, and J. Annevelink and
    D. DeWilde, Delft University 294

Session IX: REAL TIME SPEECH PROCESSING
C.F. Chan, Intel Corporation 307

32. Designing a Single-Chip Dynamic-Programming Processor for
    Connected and Isolated Word Speech Recognition, R. Owen 308

33. Single-Chip Implementation of Feature Measurement for LPC-based
    Speech Recognition, *J.G. Ackenhusen and Y.H. Oh,
    Bell Telephone Laboratories*... 316

34. Applications in Speech Recognition Using Speech-specific VLSI,
    P. Ramesh and C.F. Chan, Intel Corporation...................... 328

35. A Programmable VLSI Digital Signal Processor, *R.J. Dunki-Jacobs,
    R.M. Hardy, W.J. Premerlani, and J.E. Wheeler, General Electric Co.* ... 329

Part X: SYSTOLIC ARCHITECTURE
J.G. Nash, Hughes Research Laboratories 249

36. Partitioned Q. R. Algorithm for Systolic Array, *D.I. Moldovan,
    University of Southern California* 350

37. Fault-Tolerant Array Algorithms for Signal Processing, *P.J. Varman,
    Rice University, and I.V. Ramakrishman, University of Maryland* 363

38. A Practical Comparison of the Systolic and Wavefront Array
    Processing Techniques, *D.S. Broomhead, J.G. Harp, J.G. McWhirter,
    K.J. Palmer, and J.G.B. Roberts, Royal Signals
    and Radar Establishment, England*................................ 375

39. Some Experiments in VLSI Leaf-Cell Optimization, *K. Iwano and K. Steiglitz,
    Princeton University* 387

Part I
IMAGE ANALYSIS
Edited by:
Prof. Leah Jamieson
Purdue University

1

3-D COMPUTING STRUCTURES
FOR HIGH THROUGHPUT INFORMATION PROCESSING

J. Grinberg, R. D. Etchells, M. J. Little, G. R. Nudd
Hughes Research Laboratories
Malibu, California

ABSTRACT

Current trends in advanced VLSI development and system design are spurring an interest in "three-dimensional" (3-D) integration techniques. Three-dimensional integration technologies have a number of inherent advantages over more conventional 2-D approaches, but carry certain limitations as well. System using these new technologies must be designed with both the strengths and weaknesses of the hardware and packaging in mind. Our intent here is to present a brief overview of the field, followed by a somewhat more detailed discussion of the particular 3-D approach that we have developed. This discussion will also cover a particular computer architecture that we have developed to be uniquely suited to 3-D implementation.

INTRODUCTION

If a single goal could be said to have motivated the development of advanced electronic systems over the past thirty years, it would probably be the search for an ever higher processing capability per unit volume. From the vacuum tube through the transistor, to the microelectronic era, and through a multitude of accompanying advances in physical packaging, the drive has consistently been to increase the amount of functional capability that could be contained within a given physical envelope. In applications where limited space was a dominant design parameter, the increase in compactness has been the primary motivating factor. In other circumstances, the increased processing speed and reduced power consumption that accompany reduced physical size have been the driving constraints. In virtually all instances though, the end goal has been the same: increase the packing density of the active electronic elements.

These factors have led in recent years to an increased interest in ever greater levels of system integration at the microelectronic level. System performance has increased in almost direct proportion to the degree of integration that can be achieved in the system design. Several constraints have traditionally limited the maximum level of integration attainable. Fabrication and yield considerations have for the most part limited the maximum practical size of microelectronic chips. Additionally, packaging and signal-routing constraints have further limited the maximum densities in the overall systems.

In an attempt to sidestep these constraints, researchers are showing increasing interest in a variety of "three-dimensional" integration and packaging techniques. In what follows, we will survey several classes of 3-D techniques, and examine more closely one that we have developed at Hughes [1,2]. As we shall see, while solving many problems associated with two-dimensional designs, 3-D technologies impose new constraints of their own. We will present a specific architecture that has been developed with these constraints in mind, and discuss some of its capabilities and characteristics.

TRENDS IN SYSTEM DEVELOPMENT

As background for the following discussion of 3-D techniques, it would be appropriate to first discuss in greater depth the underlying issues motivating this research. Several specific trends are important.

Figure 1 plots the cycle time of various state-of-the-art commercial computer systems against the year of their introduction. While there is a good deal of scatter in the data points, the overall asymptotic nature of the trend-line is fairly clear. In the last two decades the cycle time has decreased only by less than one and a half orders of magnitude. In the absence of some as yet undiscovered circuit technology, processor cycle times are approaching fundamental limits. Forseeable advances in present technologies could result in only another one or at the most two orders of magnitude decrease in state-of-the-art CPU cycle times. Apart from the basic limitations of device physics, other factors contribute to this asymptotic behavior. A basic consequence of decreased clock periods is that the time-alignment of signals travelling different paths within the processor becomes much more important, and significantly more difficult to achieve. Current work in LSI gallium arsenide circuitry shows that, even at the chip level, signal propagation delays must be very carefully balanced in order to obtain optimum performance. Also, logic signals with frequency components in the range of multiple gigahertz require low-impedance transmission lines to ensure their fidelity. This results in power dissipation significantly beyond that associated with would result from the simple toggling of the logic elements themselves at those frequencies.

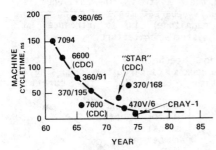

Fig. 1. Historic CPU Cycle Time Improvement

Thus we see that there are conflicting constraints involved in the design of extremely high-speed circuitry. On the one hand, the need to reduce clock and signal skews to the absolute minimum calls for the circuitry to be compacted to the highest degree possible.

At the same time though, the very high system clock rates produce a very high degree of cross talk and the requirement for low-impedance signal lines results in high system power dissipation. In practice, the radiative or conductive area required to remove dissipated power from a system often determines its minimum practical size, and thereby its maximum functional clock rate.

Memory access delays also limit processor cycle time. Throughout the history of computer evolution, memory access speed has consistently lagged that of arithmetic logic elements by one to one and a half orders of magnitude. This condition has been somewhat ameliorated by the use of multitiered memory hierarchies, with very fast cache memories closest to the logic circuitry, but the fundamental discrepancy still remains. This is at least partially a result of the large physical separation between the two types of circuitry that exists in most computer architectures. One approach to overcoming this physical separation between logic and memory will be discussed later.

An important conclusion may be drawn from the preceding. If we take the change of state of a logic gate as the fundamental unit of computational activity, the total number of such state changes in a system per unit time would be a measure of the total computational power of that system. Accepting this admittedly simplistic metric (gate cycles/sec) without criticism for the moment, we can see two ways by which it may be maximized: either the number of active gates or the clock rate of the system may be increased. Ideally, of course, we would like to increase both, but considerations of power dissipation, heat removal, and overall system size mentioned earlier generally preclude that.

The arguments above indicated that we are approaching the limits to which we may extend system clock rates. In the long term, the path available to system designers for further performance increase involves increasing the amount of active hardware in their systems. The challenge is to do so in a meaningful fashion. The only apparent means of doing so to a significant extent (several orders of magnitude) is to introduce a great degree of parallelism into the system architecture.

Advancing microelectronic fabrication technology has permitted system designers to integrate ever larger amounts of circuitry onto single chips. Figure 2 shows this trend over time, including the current limit reaching 10^6 components per chip. Wafer scale integration techniques may push this limit to the range of 10^7 within the next five years. As chip sizes have increased, so has the design effort associated with such chips. Computer-aided design and design methodologies taking advantage of it (ie, Mead-Conway), have done much to ease the task of system design at this level. It has been shown [3] that, in random, two-dimensionally arranged logic arrays, the average interconnect length between active nodes of the network is proportional to the number of nodes connected. Furthermore, decreasing microelectronic design rule dimensions result in devices capable of driving shorter and shorter interconnect lengths.[4] Thus, although our ability to integrate high-density systems is increasing, our ability to effectively interconnect the elements of these systems may in fact be approaching basic limits as well.

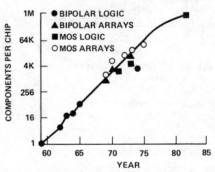

Fig. 2. Historic Circuit Density Increase

THREE-DIMENSIONAL INTEGRATION

It is at this point that three-dimensional integration techniques become important, for they offer several means by which to bypass the limitations that we have just discussed. The most obvious of these is the added compactness that the various approaches provide. This carries a number of advantages in itself. First, it somewhat reduces problems of timing skew between signals, permitting larger networks to be effectively integrated. Furthermore, the greater degree of integration thus afforded encourages a more intimate mingling of logic and memory circuitry than is otherwise possible, thereby alleviating some of the problems with memory access/logic cycle time discrepancies.

The simple availability of a third dimension in which to route signal paths permits significant additional improvements in circuit density, by reducing the "lead-length" problem discussed in the last section. Rosenberg [3,5] has shown that average interconnect lengths in random logic systems increase proportional to the square root of the number of nodes connected when the third dimension is available for signal routing and node placement. This contrasts with a direct proportionality to the number of nodes in strictly two-dimensional configurations. Fully three-dimensional implementations of very large systems then, should show average interconnect lengths significantly smaller than those of strictly two-dimensional ones. This of course would result in much lower parasitic impedances to be driven by the active elements, allowing the effective use of much smaller devices.

Signal-routing and active component placement in the third dimension and the topological flexibility thus provided also allows system designers to achieve much better structural mappings between their hardware systems and the data sets upon which they are intended to operate. This can improve performance by greatly reducing the physical and electrical distances that data must be moved in the course of computations. (This could be considered to be just another expression of the reduction in average signal path length in three-dimensional systems). Systems characterized by structural homology between hardware and algorithms are also much easier to program than those lacking such conformity.

A Historical Perspective

Since we do, in fact, live in a world characterized by three geometric dimensions, it can successfully be argued that all of our electronic systems are three-dimensional in nature. Any system containing even a simple card-rack and backplane could be thought of as a "3-D" implementation. Through the years though, there has been a steady progression of technology, leading to the present state-of-the-art in three-dimensional microelectronics.

Table 1. EVOLUTION OF 3-D TECHNOLOGIES

Technology	Number of Z-axis Interconnects	Horizontal λ (cm)	Z-axis λ (cm)	Active Components distributed along Z-axis
Backplane/PCB	10^2-10^3	10^{-1}	1	y
Multilayer PCB	10^3-10^4	10^{-2}	3×10^{-2}	n
IBM "TCM"	10^4-10^5	5×10^{-3}	10^{-2}	n
Hughes 3-D	10^4-10^7	2×10^{-4}	10^{-2}	y
SOI	10^4-10^7	2×10^{-4}	5×10^{-4}	y

Table 1 details some of the stages in the development of three-dimensional integration techniques. The technologies are listed in approximate order of sophistication and date of introduction. The parameters shown are as follows: The number of z-axis interconnects is taken to mean the number of connections which pass through circuit or interconnect levels in the "z" direction. The horizontal λ (lambda) column shows the minimum spacing of signal traces in the X-Y plane, while the Z-

VLSI Signal Processing

axis λ shows the minimum z-axis spacing between adjacent circuit or interconnect planes that can be obtained with each of the technologies. Finally, the last column of the table indicates whether the third dimension is used only for interconnect routing, or whether it may contain active circuit elements as well.

The insertion of multilayer printed circuit boards into backplane assemblies is representative of early use of the third dimension in packaging. This technology is characterized by limited z-axis interconnection capability, as well as large lambdas in both the horizontal dimension, and along the z-axis. Modern multilayer printed circuit boards increase the number of z-axis interconnections possible, while slightly reducing the horizontal lambda and greatly reducing the z-axis lambda. Note, though, that multilayer printed circuits do not permit the three-dimensional arrangement of active circuitry throughout the wiring matrix. The state of the art in multilayer techniques is represented by the IBM "thermal conduction module" (TCM) technology. This is basically an extension of thick-film hybrid technology to printed-circuit dimensions. This advance has again increased the number of z-axis interconnections that are feasible, while further reducing both horizontal and z-axis minimum dimensions. As with multilayer printed circuits, the third dimension is used primarily for signal routing. While some limited provision is made for the distribution of passive circuitry across the third dimension, no active circuitry may be so incorporated. The Hughes 3-D approach, which makes use of stacked silicon wafers again increases the maximum density of z-axis interconnections, while at the same time further increasing the circuit density at given design-rules (reducing the area taken by metallization). approach also permits the distribution of active circuitry throughout the third dimension, on the same planes as the horizontal signal lines. Finally, so-called "silicon-on-insulator" (SOI) technology makes use of stacked thin-film and epitaxial layers on a single silicon substrate to reduce the minimum z-axis spacing to microelectronic dimensions as well.[6]

The current status of these technologies is that all up to and including the IBM "TCM" approach are currently in commercial use. The Hughes 3-D system is currently under development: A demonstration machine will be operational by late 1986, and large systems using the technology should be seen by the early 1990s. Silicon-on-insulator 3-D techniques are still in the preliminary laboratory stage, and will not likely see widespread application before the mid- to late-1990s.

The Hughes 3-D Approach

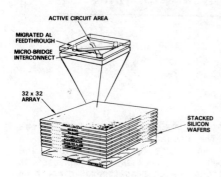

Fig. 3. 3-D Computer Concept

As mentioned above, the Hughes 3-D approach involves the stacking of multiple silicon wafers to assemble completed systems (Figure 3). Two underlying technologies are important [2]: Thermomigrated feedthroughs provide highly conductive, electrically isolated signal paths through the silicon wafers. Signals are carried between adjacent wafers in the stack by "microbridge" interconnects, thin metallic bands 15 mils long, connected to circuit pads at their ends, and arching above the silicon substrate to a height of 1-2 mils. Bridges on adjacent wafers make right angles to make electrical contact. The combination of these technologies allows the creation of highly conductive vertical signal paths through the stacks of silicon wafers, with relatively low parasitic impedances. This approach yields as many as several thousand z-axis interconnection paths per square inch of horizontal area. Future refinements could extend this capability to the realm of two orders of magnitude increase in such paths per square inch.

The architecture that has been developed to take advantage of these technologies is of the type known as a "cellular array." The configuration was derived from the structure of the image-processing algorithms that we want to implement on this new machine. The excellent mapping of data and algorithms onto this type of hardware will (1) make programming very easy, and (2) make algorithm execution much more efficient than otherwise possible. This architecture has further advantages for 3-D implementation in that its data routing and circuitry are very regular, simplifying the task of physical layout (the replicated to designed ratio of the circuitry could be very high). Also, using the 3-D approach we can provide the enormous amount of hardware associated with cellular designs. The massive parallelism also permits a further tradeoff against device speed; it allows sufficient margin to use serial logic in the individual cells, rather than parallel logic. Serial cellular logic greatly simplifies the cell circuitry, permitting much larger arrays to be integrated. Additionally, serial cellular logic greatly reduces the amount of interconnection required between levels of the 3-D structure.

While the 3-D technologies under development represent an unprecedented level of three-dimensional integration, the ability they provide to communicate vertically, between wafers, is still nearly two orders of magnitude lower than standard 2-D communication. Accordingly, the partitioning of hardware across the various levels of the stack of wafers is a very important consideration.

The overall structure of a cellular array is that of an N x N array of identical computing elements, which work in lockstep, executing a common program. Our task was to decide how to distribute the circuitry of such an array across a stack of wafers. Since our choice of serial logic for the cell circuitry meant that there would be only a single primary data line associated with each computing element in the array, it was a natural decision to run these (relatively few) data lines vertically through the stack, spreading the functional units of each computing element vertically across multiple wafers.

The detailed partitioning of the processing elements across the multiple wafers of the stack is determined by a variety of factors. A primary consideration was that the cells on all the wafers needed to be approximately the same size. If they are not, the largest cell would determine the cell areas on other levels, which would result in wasting of Si real estate. The cell circuitry on all wafers should occupy as near to the same amount of area as possible.

Overall constraints on the cell size are imposed by (1) the size of the array that we wished to integrate, (2) the minimum feature size of the circuit technology we were using, and (3) the overall size of the wafers upon which we would construct the array. In general, the tradeoff is to use more layers in the stack, as opposed to larger cells in each layer. Current and projected application requirements call for the ultimate fabrication of processing arrays of dimension 256 x 256 or even 512 x 512. Since the horizontal dimensions of the array are limited by the size of the wafers available for fabrication, the cell circuitry on each

VLSI Signal Processing

wafer should be made as simple as possible. On the other hand, several factors argue for larger cells. An absolute lower limit is set by the requirement that the circuitry on each level possess some minimum meaningful level of functionality. Beyond that, though, there is a certain amount of overhead associated with the cells themselves, and that of getting information off-chip. One component of this overhead is the area consumed by the 3-D feedthroughs themselves, presently about 2 x 2 mils each. In addition to this, something on the order of 5-10 transistors would be required to implement the necessary interface between the cell circuitry and the data bus. Since this overhead would be the same regardless of the size of the rest of the cell, efficient use of silicon indicates the use of larger cells.

The balance between these constraints that we have arrived at is to have cells of roughly 100 transistors each on each of the functional planes of the 3-D computer. Our projections indicate that this design choice will permit the integration of 256 x 256 arrays on 4-inch diameter wafers in the near term (1988), and 512 x 512 arrays on 6-inch wafers in the forseeable future (1993).

The Architecture

The hardware of the Hughes 3-D processor consists of an array of N x N identical processors. The array itself is distributed horizontally, and the elements of each processor in the array are connected vertically. Signals travel horizontally across each wafer of the stack through conventional aluminum and polysilicon conductive layers. Signals are passed vertically through the stack by way of wire-like bus lines composed of feedthroughs through the wafers and "microbridge" interconnects between the wafers. Each wafer in the stack contains a complete N x N array of one particular type of functional element (such as an n x n array of memory registers, an n x n array of accumulators, etc). All processors in an array are identical, being composed of the same combination of "unit" processing elements.

Figure 4 is a more schematic illustration of the same structure as Figure 3, this time viewing the stack edge-on, and showing the control processor used to exercise the array hardware as a separate unit. In this illustration, the silicon wafers of the previous figure are represented by the vertical rectangles. The horizontal lines running through the stack represent the data and control buses. The resulting machine may be thought of as "word-parallel, bit-serial," in that

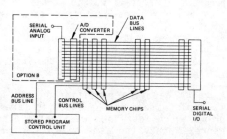

Fig. 4. "Side" view of 3-D Computer

while the logic of the individual processors in an array employs bit-serial arithmetic, all processors operate simultaneously, in word-parallel fashion. The massive parallelism at the processor level more than compensates for the relatively minor loss of speed incurred by the use of serial arithmetic.

The overall architecture of the machine is "single instruction, multiple data" (SIMD), in that all processors in the array work in lock-step, under the direction of the control unit. A capability for data-dependent operations does allow independent treatment of any definable subset of the data contained in the array.

The control processor, shown as the "stored program control unit" in Figure 4, handles all program storage and sequencing, as well as the direct control of the array elements. This is a conventional Von Neumann processor, which communicates with the array hardware via address lines and a control bus, which are connected to every wafer of the stack. Whenever the address line associated with a given wafer in the stack becomes active, data present on the control bus is strobed into an on-chip configuration data latch. This action moves that wafer from the quiescent state, in which it had previously existed, into an active state. Prior to the execution of any instruction by the array, all wafers are in a neutral state. In this state they do not communicate with the data buses, and ignore the system clock signals. In preparation for an operation, the control processor sequentially configures each wafer that is to be involved by enabling the first wafer and then transmitting the appropriate configuration codes over the control bus, enabling the second wafer, and so on. Once the necessary wafers have been configured, the control processor gates through the appropriate number of system clock pulses required for that operation. At the conclusion of the operation, the stack elements are returned to their "neutral" state by toggling a reset line common to all wafers.

The Array Hardware

For image processing, we have found that only five elemental wafer types are required to perform virtually all of the algorithms that we have studied to date. These element types are listed in Table 2, along with a brief enumeration of their functions. Of the five types shown, only the first two are absolutely necessary to the functioning of the machine. The other two were developed to enhance performance on certain common operations.

Table 2. 3-D WAFER TYPES

WAFER TYPE	FUNCTION
Memory	Store, shift, invert/noninvert, OR, full word/MSB only, destructive/nondestructive readout
Accumulator	Store, add, full word/MSB only, destructive/nondestructive readout
Replicator Plane	I/O, X/XY short, stack/control unit communication
Counter	Count in/shift out
Comparator	Store (reference), greater/equal/lower

By far the majority of the processing occurring in the 3-D computer takes place in the memory and accumulator wafers. Each cell of these wafers has a 16-bit serial memory register for data storage, and CMOS circuitry to provide the required logic functions. A schematic diagram of a memory cell is shown in Figure 5. The cells of the memory wafer are used to store data, but also perform the important auxiliary function of lateral data transfer between adjacent processors in the array. Each memory cell has nearest-neighbor communication with others on the same plane, and may pass data values in any of the four compass directions on the array (north, south, east, or west). Data within each memory cell may therefore be passed to any one of its four nearest neighbors in the array, to the input of the register itself (allowing non-destructive readout), or through the serial data bus connection to other wafers, all under software control.

The accumulator circuitry is similar to that of the memory, differing mainly in that there is no nearest-neighbor communication, and a one-bit serial adder is included. Twos-complement subtraction can easily be accomplished by allowing a "carry" to be introduced into the least-significant bit of the word prior to execution of the arithmetic operation. (The bit-wise inversion of the subtrahend

VLSI Signal Processing

required by twos-complement subtraction is provided by the memory cell circuitry.) A schematic diagram of typical accumulator cell circuitry is shown in Figure 6.

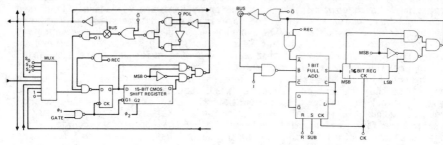

Fig. 5. Memory Cell Schematic   Fig. 6. Accumulator Cell Schematic

The other three wafer types listed in Table 2 are not absolutely essential to the operation of the 3-D computer, serving primarily to speed up the execution of various algorithms. Of these, by far the most important is the "Replicator Plane" wafer. The replicator plane wafer provides a means of rapidly propagating data values across the entire area of the array. One example of its use would be the transmission of a threshold or constant value to all elements of the array, prior to a comparison operation. A second special purpose type of wafer is an N x N array of counters. These wafers are used in situations where it is necessary to quickly count the number of occurrences of a single-bit piece of data. A normal accumulator cell could be used for this purpose, but it would require a full 16 clock cycles to tally each single-bit occurrence. The counter cell requires but a single clock cycle to perform the same action. This capability is very useful in calculating histogram distributions of image data, an operation important in vision processing for image segmentation.

The third special purpose type of wafer is an N x N array of comparators. As with the counter cells of the previous paragraph, comparator hardware exists mainly to speed up operations that could be performed, albeit more slowly, by the accumulator circuits. Each cell of a comparator wafer contains a 16-bit register into which a reference value may be loaded, together with circuitry for performing serial magnitude comparisons.

Yield Policy

As mentioned earlier, the primary factor which determines the maximum practical size of microelectronic circuits is the decline of yield with increasing chip area. The very large array sizes contemplated for the 3-D computer therefore make yield a primary concern.

It is generally assumed that yield falls off exponentially with increasing chip area. This is the case if defect densities obey a Poisson distribution. In actual practice, the fall-off is somewhat less abrupt [7], but still very significant. Merchant semiconductor manufacturers, particularly those making memory devices, have long since turned to the use of redundant circuitry on their chips to minimize this effect.[8,9] The success of these methods is evidenced by their popularity in the commercial marketplace.

A complete discussion of microelectronic failure modes and yield statistics is considerably beyond the scope of the current paper, but a few brief observations would be worthwhile. The basic concept of yield improvement through the addition of redundant circuitry is quite simple. One merely includes a number of "spare" circuit elements in the system, and provides some means of effecting their substitution for other elements that may fail. A naive view of this process might lead one to suppose that it can be extended indefinitely, at some point virtually guaranteeing perfect yield. This turns out not to be the case.

As has been pointed out by Mangir and Avizienis [10], both the redundant elements themselves and the means by which they are substituted for defective ones are affected by defect processes in the same fashion as the original circuitry. The importance of this is that there is an unavoidable overhead associated with interconnecting the substituted redundant elements into the primary system. The more redundant circuitry is added to a system, the greater this overhead becomes. Eventually, the yield of the primary-to-redundant interconnect wiring becomes the dominant factor in the yield of the system as a whole.

Another parameter that must be considered in evaluating redundancy schemes is the resolution that they permit in selecting "good" material. This may be restated in terms of the "module size" of the substituted components. Very large modules require that a great deal of otherwise functional componentry be discarded as the result of a single malfunctioning element. This is obviously inefficient, in that much greater amounts of circuitry will be required to obtain a given yield level than would be called for in a system with smaller modules. On the other hand, systems with larger modules generally require less circuitry and wiring to be devoted to the primary-redundant interconnect function.

From the preceding then, we can see that there is a tradeoff involved in the design of a redundancy scheme. On the one hand, it is desirable to choose a small module size for primary/redundant substitution, in order to minimize the amount of good circuitry that must be thrown away with the bad. At the same time though, it is important to minimize the amount of interconnect wiring required to effect the substitution of redundant circuitry into the system, in order to avoid yield loss from that wiring. Both of these factors were considered in designing the yield policy for the 3-D computer. The small size and relative independence of the circuitry in the functional cells make it a natural choice to introduce redundancy at the cell level. Furthermore, the topological constraints imposed by the physical structure of the machine require that the redundant elements be located close to the units that they were intended to replace. Finally, the large size of the array calls for a substantial amount of redundancy to be used. These factors led us to adopt a 2:1 redundancy approach, in which every functional cell on each wafer of the completed stack contains two identical circuits. If both circuits in any given cell are found to be bad, then that entire wafer would have to be discarded. A second level of yield policy implementation will permit a cell with both circuits defective to "borrow" a working circuit from a neighbor having both circuits functional.

3-D COMPUTER SOFTWARE PERFORMANCE

Space limitations preclude a detailed discussion here of software implementation on the 3-D computer. Instead, we will present only a brief discussion of the performance of the machine in two basic problem areas of image and signal processing. For a more detailed treatment, the reader is

VLSI Signal Processing

directed to reference [1].

Many applications in image processing involve operations on local areas of the image, either forming weighted averages, convolutions, or spatial filters of some type. Such operations usually involve the computation of a value for each pixel in the original image. This is most often done by linearly combining the values of the surrounding pixels, usually multiplying each value by a "weight" factor first. A simple example of such an operation is the 3 x 3 "Gaussian" function, a weighted, local-area averaging function, which acts as a low-pass spatial filter.

Because the topology of the 3-D machine is so well matched to two-dimensional problems, address calculation within picture arrays is considerably simpler than is the case with more conventional serial machines. Here, rather than specifying the location of the data by nested address displacements, the nearest-neighbor communication of the memory planes is used to "slide" the data across the processing array so as to align the required operands over each other in the stack of computing elements. This is highly efficient in the 3-D architecture, since the data movement often can occur simultaneously to the computation.

Recall that the elements of the array hardware appear as functional units attached to serial data lines. Thus, operands presented to a data bus line are available to other units anywhere along that line, in either direction. Therefore, while programmers must pay attention to the relative displacements of data-sets on various functional planes of the machine, they need not concern themselves with just what plane those data reside on. In a machine with a 16-bit word size and a 10-Hz clock rate, the gaussian calculation would take roughly 21.5 microseconds to complete. In that amount of time, the gaussian mean function would have been computed for every pixel in the image. In the case of a 256 x 256 image, this corresponds to a processing throughput of 1.8×10^{10} additions/second.

Matrix inversion is a very important function in many two-dimensional signal processing applications such as maximum entropy filtering, modern spectral analysis, and beam forming. It is a computationally intensive operation typically requiring $O(N^3)$ operations. Matrix inversion is accomplished on the 3-D computer through the use of the "Faddeev" algorithm.[11,1] This is one of a general class of algorithms useful for solving linear systems of equations. These algorithms are particularly suited for implementation of the 3-D hardware, in that they allow matrix inversion without back-substitution. Furthermore, they may be performed "in-place" on cellular array computers.

This algorithm requires one multiplication, one division, and 10 data transfers in its inner loop, which must execute a total of only "n" times for an n x n matrix. Assuming 16-bit registers and a 10-MHz clock rate, the time required to invert a 256 x 256 matrix is $\sim$23 msec. As mentioned earlier, the execution time on the 3-D computer is proportional to n, the order of the matrix, while a serial machine would require time proportional to n^3. This is an advantage of n^2 for the 3D computer. In the case n = 256, the serial machine would be required to perform its operations at 65536 times the speed of the 3D computer to attain the same execution time.

General Advantages of the 3-D Architecture

One of the most striking characteristics of this 3-D architecture is the extent to which memory and logic circuitry have been merged. Indeed, every functional element of the 3-D computer contains at least some memory. Likewise, even the dedicated memory cells are

VLSI Signal Processing

capable of performing a good many logic operations. In addition, due to the use of serial arithmetic, the memory access and logical operations fully overlap, so that there is no access delay between memory and processing circuitry. These factors, combined with the close topological mapping between the architecture and two-dimensionally structured data sets virtually eliminate the overhead normally associated with data address calculation and memory access. Finally, the intimate intermingling of logic and memory circuitry reduces the communication overhead of the architecture as compared to other designs. This helps to greatly reduce the power dissipation of the 3-D structure.

The 3-D computer architecture is also a very general one, adaptable to problems covering a wide range of disciplines. This is attributable to two factors. First, the extreme simplicity of the basic computational circuitry, coupled with software control virtually to the gate level, provides much broader and more comprehensive control of the logic configuration than is possible in more conventional machines. Furthermore, the highly modular structure of the design provides for a diversity of resources within a common architectural framework. That is, wafers of various types may be added to or subtracted from a baseline configuration, without altering the overall nature of the machine. Programmers or compilers need only know the amount of each resource type present: Any variation in this parameter does not significantly affect the data flow within the machine. Too, as mentioned earlier, the topology of the machine greatly eases the programming task in applications involving two-dimensional data sets.

Finally, 3-D computers of the type described here will be both extremely compact, and very inexpensive to manufacture. The source of the compactness becomes obvious when one compares the ratio of active silicon to packaging materials in a 3-D machine to that of a more ordinary computer. A conventional machine has chips, in packages, on circuit boards, in backplanes. As much as 99% of the total volume of the machine is occupied with materials other than silicon circuitry. A 3-D machine, on the other hand contains as much as 90% active silicon. As to the costs of manufacture associated with this 3-D approach, an industry rule of thumb is that each level of packaging employed in a product increases the cost of that product by roughly a factor of 10. In a conventional machine, the chips must be tested, diced, packaged, tested again, mounted onto circuit boards, soldered, the boards tested, a backplane wired, the boards inserted into the backplane connectors, and the final system tested. With a 3-D computer, on the other hand, one need only test the chips, batch fabricate the interconnects, stack the wafers on top of one another, and test again. Many of these processes are fundamentally parallel in nature, such as the attachment of the interconnects and the actual "wiring" of the system which is accomplished by stacking the wafers. Each wafer, when added to the stack, makes thousands of interconnections simultaneously. By contrast, connections in a conventional machine are for the most part made a few at a time to as many as a hundred at a time with some of the new pin grid packages.

Summary

We have discussed the development of a computer architecture, uniquely tailored to implementation with a "three-dimensional" circuit integration technology. A brief overview of three-dimensional circuit technologies was presented as historical background. The strengths and limitations of the particular integration technology chosen were considered, and their impact on the design process discussed. Finally, the desirable characteristics of the resulting architecture were presented. Notable among these are: ease of

VLSI Signal Processing

programming, low power consumption, inexpensive construction, and an inherent speed advantage of n^2 over conventional monoprocessor architectures on two-dimensional data sets. We believe that this architecture and others like it will play an increasingly important role in image and signal processing in the years to come.

REFERENCES

1. Grinberg, J. et al., "A Cellular VLSI Architecture," IEEE Computer, Volume 17, No. 1, January 1984, pp 69-81.

2. Etchells, R. D. et al, "Development of a Three-Dimensional Circuit Integration Technology and Computer Architecture," SPIE Vol. 282, pp 64-72, 1981.

3. Rosenberg, A. L., "Three-dimensional Integrated Circuitry," VLSI Systems & Computations, H. T. Kung, R. Sproull, and G. Steele, Eds., Computer Science Press, Rockville, Md., 1981, pp 69-80.

4. Bloch, E. and Galage, D. J., "Component Progress: Its Effect on High Speed Computer Architecture and Machine Organization," High Speed Computer and Algorithm Organization, D. T. Kuck, D. H. Lourie, A. H. Sameh, Eds., Academic Press, 1977.

5. Rosenberg, A. L., "Three-Dimensional VLSI: A Case Study," Journal of the ACM, V 30, No. 3, July, 1983, pp. 397-416.

6. Gibbons, J. F., and Lee, K. F., "IEEE-Electron Devices Letters, EDL-1, 117 (1980).

7. Bernard, J., "The IC Yield Problem: A Tentative Analysis for MOS/SOS Circuits," IEEE Trans. Electron Devices, Vol. ED-25, pp. 939-944, Aug. 1978.

8. Cenker, R. P., et al, "A Fault-Tolerant 64k Dynamic RAM," in Dig. ISSCC, Vol. 22, Feb. 1979, pp. 150-151.

9. Smith, R. J. et al, "32k and 16k Static RAMs Using Laser Redundancy Techniques," in Dig. ISSCC, Vol. 25, Feb. 1982.

10. Mangir, T. E. and Avizienis, A., "Fault-Tolerant Design for VLSI: Effect of Interconnect Requirements on Yield Improvement of VLSI Designs," IEEE Trans. on Computers, Vol. C-31, No. 7, July, 1982, pp. 609-616.

11. Faddeev, V. N., Computational Methods of Linear Algebra, C. D. Benster, Trans., Dover Publications, New York, 1959, pp. 90-98.

SIGNAL TO SYMBOLS:
UNBLOCKING THE VISION COMMUNICATION/CONTROL BOTTLENECK

S. Levitan, C. Weems and E. Riseman

University of Massachusetts

THE COMPLETE TEXT OF THIS MANUSCRIPT WILL BE FOUND BEGINNING ON PAGE 411.

VLSI Signal Processing   3

VLSI FOR IMAGE ROTATION

Zoltan Z. Stroll*, Earl E. Swartzlander, Jr.*, and John Eldon**

* TRW Defense Systems Group, One Space Park, Redondo Beach, CA 90278
** LSI Products Division, 4243 Campus Point Court, San Diego, CA 92121

ABSTRACT

   This paper introduces an Image Rotation Controller Chip that performs real time two dimensional digital image rotation in conjunction with a Multiplier Accumulator Chip (MAC). The functions performed include image resampling (rotation, scaling) and two dimensional convolution (image filtering). These operations are at the heart of the rapidly emerging "digital optics" techniques used in the television industry for the preparation of special effects footage. Related applications include the geometric correction of satellite images for NASA and "electronic darkroom" type operations involved in commercial graphics and composition.

INTRODUCTION

   The requirement for manipulating imagery exists in both DoD and commercial arenas. Currently most image manipulation is performed optically, although there is a growing trend to the use of electronic techniques.

   In the DoD area, there are a number of image sources including radar, visible and IR data from stationary, airborne, or spaceborne sensors. The data is generally in the form of electronic samples which undergo a variety of corrective processes prior to classification and other analyses. The corrective processes include geometric and radio-metric rectification to compensate for sensor-system-induced errors. Sensor data as well as map data from a variety of sources are routinely converted to a variety of coordinate systems. All of these geometric operations involve electronic resampling of the original sampled data to a new coordinate system. The data handling requirements are met through the use of system products that include microprogrammed special-purpose processors for real-time image processing and specialized image display processors for operator/analyst interaction with the system. At the heart of both these subsystems is the resampling electronics. One of our goals is to bring VLSI solutions to this area, enabling increased productivity and decreased system acquisition and operational costs.

   The commercial applications area can be split into two segments: the generation of special effects as used in the TV industry and page make-up activities in the publishing industry.

   Until recently, the process of assembling a variety of specially rotated, scaled and cropped images for special effects footage was performed optically in a film studio using photographic techniques. This approach is both labor intensive and expensive. It frequently requires up to a week to prepare a one minute commercial. Digital electronic systems such as the ADO (Ampex Digital Optics) and Grass Valley's Digital Video Effects equipment are replacing photographic techniques. The results are impressive:

VLSI Signal Processing

- Production schedules of a day or less
- More capabilities at the disposal of the director/editor
- Instant feedback
- Lower equipment and operational costs.

In short, commercial television systems benefit greatly from the capability to perform rotation of digital image data in real time or near real time at standard video rates. With the development of a VLSI image rotator circuit the equipment cost and size will be significantly reduced which will encourage the conversion to electronic processing in the television industry.

Related to the special effects generation application is the image display/processor workstation employed in the analysis of LANDSAT or other image data. The typical operation involves manipulation of a 512 x 512 image on a workstation with response time adequate for satisfactory human interaction. Image rotation, scaling, and warping is performed in special purpose hardware [e.g., the Warper in International Imaging Systems' Model 75 workstation (Ref. 1)]. The Warper is a bit sliced technology based board implementing up to second order warping using bilinear interpolation with a response time of 1.8 sec to process a 512 x 512 image. It is our goal in developing the VLSI image rotator circuit to increase the performance level and decrease the circuit complexity by at least an order of magnitude.

The publishing industry represents additional applications for image rotation. Although this industry is undergoing tremendous transformation as it becomes more automated, it is still characterized by highly labor-intensive procedures. For instance, text and graphics are generated increasingly by electronic means; but photographs still require a significant amount of optical processing: reduction/magnification, rotation, cropping, tonal modification, etc. The text, graphics, and photographs are all created separately, so that the creation of camera-ready copy must be performed by a layout artist. The VLSI image rotator circuit represents an important step towards the creation of an "electronic darkroom" that integrates and automates the composition process, resulting in higher productivity.

In the following sections of this paper we examine the algorithms, architecture, and VLSI implementation of the VLSI image rotator circuit.

<u>IMAGE ROTATION ALGORITHMS</u>

Conceptually, a general image transformation or resampling capability requires two steps: a coordinate system transformation followed by pixel interpolation. Interpolation is necessary when the transformed pixel positions do not coincide with the original pixel positions. New pixel values are obtained by interpolating original pixels in the neighborhood of the transformed pixel position.

Interpolation is an image reconstruction operation, involving a discrete two-dimensional convolution of a function derived from the image data with a given weighting function known as an interpolation kernel. If $p(x,y)$ is the original sampled m x n image, then the interpolated image $g(x,y)$ can be written:

$$g(x,y) = \sum_{i=0}^{m-1} \sum_{j=0}^{n-1} c_{i,j} \, W(x-x_i, y-y_j) \qquad (1)$$

Where: x_i and y_j are the interpolation nodes, W is the interpolation kernel and g is the interpolated function. The c's are parameters which depend upon the sampled data. A fundamental property of interpolating functions is that they must coincide with the sampled data at the interpolation nodes, or sample points. If an interpolation kernel is chosen such that $W(0,0) = 1$ and $W(m,n) = 0$ when m and n are nonzero integers, then the c's are just equal to the sampled image points. Theoretically, convolution of the image data with the sinc function provides ideal resampling (Ref. 2); but this requires processing an infinite number of pixels. Systems, however, use a finite length interpolation kernel (usually up to 4 by 4 points).

Popular interpolation functions include cubic convolution, bilinear, and nearest neighbor. The one-dimensional versions of the kernels associated with these functions are depicted in Figure 1. The cubic convolution is a four by four point kernel, the bilinear is a two by two point kernel, and the nearest neighbor is a one point kernel. There is a direct cost-performance tradeoff to be considered when choosing among these four interpolation kernels. Of the kernels mentioned above, the nearest neighbor kernel requires the least computation to implement. No arithmetic is needed. The interpolated output data point is assigned the value of the nearest sample point from the original data. Next in simplicity is the bilinear kernel. This kernel satisfies the condition that $W(0,0) = 1$ and $W(m,n) = 0$ for m,n nonzero integers; thus the c coefficients are equal to the original image points. For this kernel four sample points (two in both the x and y directions) are involved in the interpolation of each new point. Four multiplications and three additions are required for each interpolated point. With the cubic convolution kernel the c's are also equal to the original image points. This four by four point kernel requires 16 multiplications and 15 additions for each sample point.

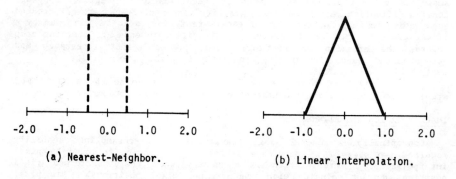

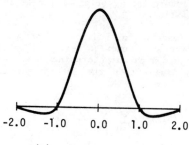

Figure 1. Interpolation Kernels

An important consideration when comparing interpolation methods is their accuracy, i.e., the exactness with which the interpolated function reconstructs the sampled function. Some indication of the accuracy of the method is given by the type of function which can be exactly reconstructed. The cubic convolution, bilinear and nearest neighbor functions exactly reproduce any second-, first-, and zero-degree polynomial, respectively (Ref. 3).

The relative accuracy of different interpolation methods can be determined from their convergence rates, or by how fast the approximation error goes to zero as the sampling increment decreases. For cubic convolution the approximation error consists of terms proportional to h^3, where h is the sampling increment. In this case, the approximation error is said to be cubic since it goes to zero at least as fast as h^3 goes to zero. Similarly, the convergence rates for bilinear and nearest neighbor functions are quadratic and linear (i.e., the error decreases in proportion to h^2 and h respectively).

The accuracy of the various interpolation techniques was evaluated in (Ref. 4). A LANDSAT MSS scene of Baltimore was resampled to a grid half-sample space offset in the along line direction. Interpolation with cubic convolution, bilinear, and nearest neighbor algorithms yielded RMS errors of 1.2%, 1.7%, and 4.1%, respectively with maximum errors of 7.8%, 14.1%, and 40.6%. Cubic convolution exhibits the best performance. In terms of spatial frequency fidelity, interpolation with the cubic convolution is clearly superior to both bilinear as well as nearest neighbor. Bilinear interpolation tends to wash out fine detail as a result of its low-pass filter characteristics. In contrast, nearest neighbor interpolation exhibits jagged edges as a consequence of high-frequency aliasing.

ARCHITECTURE CONSIDERATIONS

The interpolation functions examined have different system design implications. A system capable of input-to-output address mapping as required by the coordinate system transformation would compose a simple and efficient nearest neighbor image resampler. For the other kernels, at least one multiplier accumulator and kernel memory are required in addition to the basic input-to-output address mapping hardware.

The coordinate transformation determines the location of the resampled pixel in the original image space. An affine transform which allows image rotation, scaling, and translation can be described in matrix notation as follows (Ref. 5): let the original coordinates be x,y and the resampled coordinates be u,v. Then [u v 1] = [x y 1]M and post-multiplication with M^{-1} yields [x y 1] = [u v 1]M^{-1}, where

$$M = \begin{bmatrix} a & d & 0 \\ b & e & 0 \\ c & f & 1 \end{bmatrix} \text{ and } M^{-1} = \begin{bmatrix} Dxu & Dxv & 0 \\ Dyu & Dyv & 0 \\ x0 & y0 & 1 \end{bmatrix} \quad (2)$$

are the forward and reverse coordinate transformation matrices, respectively, and Dxu = e/D, Dxv = -d/D, Dyu = -b/D, Dyv = a/D, x0 = (bf - ce)/D, y0 = (cd - af)/D and D = ae - bd. The geometrical significance of these parameters is evident in Figure 2.

The affine transformation outlined above allows scaling, translation, and rotation. However, some applications such as LANDSAT and video/display oriented processing require nonlinear image warping. For these applications

VLSI Signal Processing

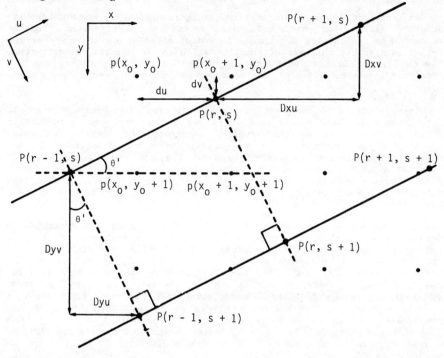

Figure 2. Geometry of Single Pass Resampling

a two-dimensional second-order function could be used to approximate the desired geometric warping:

$$x(u,v) = a_1 u^2 + a_2 u + a_3 uv + a_4 v^2 + a_5 v + a_6 \quad (3)$$
$$y(u,v) = b_1 u^2 + b_2 u + b_3 uv + b_4 v^2 + b_5 v + b_6.$$

The functions above are evaluated using the following recursive difference equations:

and
$$x(u,v) = x(u-1,v) + DxuDu(u-1) + DxuDv(v) + Dxu0$$
$$x(u,v) = x(u,v-1) + DxvDu(u) + DxvDv(v-1) + Dxv0$$
and
$$y(u,v) = y(u-1,v) + DyuDu(u-1) + DyuDv(v) + Dyu0$$
$$y(u,v) = y(u,v-1) + DyvDu(u) + DyvDv(v-1) + Dyv0; \quad (4)$$

with the following initial values:

$$x0 = a_6, \; y0 = b_6 \quad (5)$$
$$Dxu0 = a_1+a_2, \; Dxv0 = a_4+a_5, \; Dyu0 = b_1+b_2, \; Dyv0 = b_4+b_5$$
$$DxuDu = 2a_1, \; DxvDu = DxuDv = a_3, \; DxvDv = 2a_4$$
$$DyuDu = 2b_1, \; DyvDu = DyuDv = b_3, \; DyvDv = 2b_4.$$

VLSI Signal Processing

Mechanization of the above computation involves the use of adders and registers only; expensive multipliers are not required.

Let an array of m x n pixels in the [x y] coordinate system with pixel values at (i,j) denoted by p(i,j) be defined for the integers $0 \leq i \leq m-1$, $0 \leq j \leq n-1$ to represent the original image. (The p(i,j) must be preprocessed if $W(m,n) \neq 0$ for m,n nonzero integers.) The pixel value p(x,y) at (x,y) is obtained by convolution with the two-dimensional interpolation kernel function W(x,y).

$$p(x,y) = \sum_{i=0}^{m-1} \sum_{j=0}^{n-1} p(i,j) W(x-i, y-j) \qquad (6)$$

This can be simplified by interpolating only k neighbors. That is,

$$p(x,y) = \sum_{i=0}^{k-1} \sum_{j=0}^{k-1} p(x_i, y_j) W(x-x_i, y-y_j) \qquad (7)$$

where x_i and y_j are one of the k nearest integers of x and y, respectively; p(x,y) has a constant background value for the pixels undefined in the original pixel array. Then the resampled pixel at the grid point (r,s) in the [u v] coordinate system is $P(r,s) = p(x_r, y_s)$ where $[x_r \; y_s \; 1] = [r \; s \; 1]M^{-1}$, $x_r = x_o + du$, $y_r = y_o + dv$, and if $k = 2$ then

$$\begin{aligned} p(x_r, y_s) &= p(x_o, y_o) W(du, dv) \\ &+ p(x_o+1, y_o) W(du-1, dv) \\ &+ p(x_o, y_o+1) W(du, dv-1) \\ &+ p(x_o+1, y_o+1) W(du-1, dv-1). \end{aligned} \qquad (8)$$

The transformation and interpolation shown above form the basis for the one-pass resampler. In the one-pass implementation, the original data scanlines are retrieved from disk (unless the entire image is available in a frame buffer), buffered in main memory, and segments along the new scan direction are then shipped to the resampling hardware. Here the interpolator resamples the data into complete new scanlines and stores them. Convolution with a k x k kernel requires k^2 multiply/accumulate operations.

The two-pass resampler implementation is so named because the image data passes through the resampling hardware twice, once for along-line resampling and a second time for across-line resampling, with intermediate storage on disk (or in a frame buffer) (Ref. 5,6). The original data is read from the input frame buffer, buffered through main memory and transferred to the resampler hardware where along-line resampling is performed by interpolating between neighboring pixels on the scanline. Subsequently, the data is returned in along-line segments with each segment internally transposed to the across-line direction. The buffered data is stored in a format that allows efficient access to it in the across-line segment order for the second pass. In the second pass, the segments are collected in across-line sequence, and passed to the resampler for across-line resampling in a fashion similar to the along-line process. Finally the segments are returned to be buffered through main memory and stored. Unlike the one-pass resampler, the two-pass resampler does not restrict rotation while decoding. The separate processing along the x and y coordinates of this algorithm requires a separable interpolation function, i.e., $W(x,y) = W(x)W(y)$. Convolution with a k x k kernel requires 2k multiply/accumulate operations.

VLSI Signal Processing

As with interpolation functions, the one- and two-pass implementations have efficiency, accuracy, and complexity tradeoffs. In terms of computational complexity, the single-pass implementation exhibits k-squared growth and the two-pass grows linearly with k, where k is the number of pixels to be interpolated along a scanline. From the performance viewpoint, the single-pass implementation handles each data only once with computation in parallel as opposed to a two-stage computation in the two-pass implementation. The two implementations exhibit different limitations. The single-pass approach is characterized by a static error in the desired versus actual angle of rotation and scale factor (Ref. 5). The two-pass approach exhibits unequal static along/across line rotation angle and scale factor errors, resulting in the inability to create a perfectly square resampled grid. These static errors are under a fraction of a percent, quite adequate for most applications.

The input-to-output address mapping of both the one- and two-pass algorithm requires primarily registers and adders. Adders may be used instead of multipliers to generate the addresses since the x and y values are integral. Registers are required to store the transform parameters and intermediate results. In addition, an x,y position comparator is needed to disable the interpolation accumulation when the requisite input position is out of range. In the two-pass algorithm, the interpolation accumulation is performed in only one dimension. Thus, for a k by k point kernel the multiplier/accumulator is only cycled 2k times rather than k^2 times as in the one-pass implementation. For the two-pass approach, additional logic is required to transpose the input and/or output pixel addresses.

VLSI IMPLEMENTATION

As illustrated in Figure 3, the image rotator generates read and write addresses and arithmetic control signals for a subsystem comprising two frame-size RAMS, up to two interpolation coefficient lookup tables, and a multiplier accumulator. The image rotator extracts pixels as needed from the source frame store RAM, accesses the necessary signal processing coefficients, and writes the transformed data into the output frame store RAM.

In the discussion which follows, "x" and "y" denote coordinates in the original pixel space; "u" and "v" are the coordinates of the resampled pixel space.

Figure 3, a simplified block diagram of the image rotator, summarizes the chip's major internal components. The user loads the necessary resampling parameters and the image rotator generates pairs of addresses for:

1. The original data in the source RAM;
2. The coefficients of the interpolation lookup table; and
3. The writing of the final results into the output RAM.

The image rotator also generates instructions for the external MAC and lookup tables. In addition to the input (parameter) and output (address) registers, the image rotator includes two sets of position generators and a set of position comparators. The image rotator has 20 user set parameters: x starting point (x0); y starting point (y0); initial increments along the new u and old x and y axes (Dxu0 and Dyu0, respectively); corresponding initial increments along the new v axis (Dxv0 and Dyv0); second-order increments DxuDu, DxuDv, DyuDv, DyuDu, DxvDv, and DyvDv; the x and y axis limits of the original picture xmin, xmax, ymin, and ymax; and the descriptors for the final image: upper left hand corner starting point (S), scanline width (M), active pixels for scanline (W), and number of scanlines (H). All registers

VLSI Signal Processing

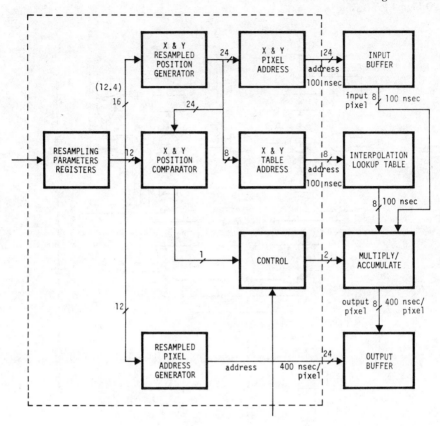

Figure 3. Image Rotator Circuit Functions

are strobed by the common chip clock and fed data from the 24-bit Parameter In/Final Address Out port (R23:0), but each has its own load enable signal, decoded by the image rotator's internal controller. To initialize the image rotator, the user activates the "load instructions" signal, then presents the 20 parameters to the in/out port in sequence on 14 successive rising edges of the chip clock. The 14th load completes the initialization cycle. In piecewise-linear work, the user can update individual resampling parameters at will. In single-pass and two-pass applications, the chip is limited to input and output frame sizes of 4096 lines of 4096 pixels each, i.e., a 12-bit x 12-bit field.

Parameters x0 and y0 are 17-bit, two's complement values, as are the computed resampling coordinates x and y. The first-order incrementors (the D's) are 13-bit two's complement values, whereas the second-order incrementors are 8-bit two's complement numbers. The x, y, u, and v limit parameters are 12-bit unsigned magnitude values.

Figure 4 depicts the four multiplexers and three accumulators that comprise the x portion of the Resampled Position Generator. The y portion (not shown) is identical. Inputs to this module include resampling parameters x0, y0, Dxu0, Dyu0, Dxv0, Dyv0, DxuDu, DyuDu, DxuDv, DxvDv, DyuDv,

VLSI Signal Processing

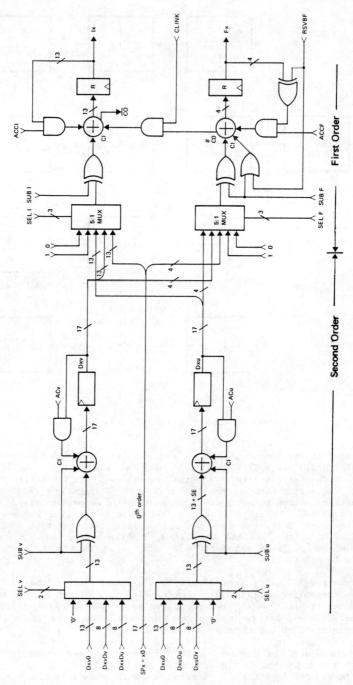

Figure 4. Resampled Position Generator (x coordinate)

and DyvDv. The outputs are the 17-bit coordinates, each of which is partitioned into a 4-bit fraction (between pixels) and a 12-bit integer (even pixel positions). The 17th (sign) bit flags negative positions which are out of range. The two integer portions, Ix and Iy, comprise the address of a pixel to be fetched from the source frame store RAM. The two fractions, Fx and Fy, jointly select the interpolation coefficient to be multiplied by that pixel value. In filtering mode, the "fractions" select the convolutional coefficients.

The x-position generator's 4:1 multiplexer routes either 0, the x coordinate of the starting position (x0), the rate of change of x in the u direction (Dxu), or the rate of change of x in the v direction (Dxv) into the input of the accumulator. The XOR gate and carry in implement subtraction, and the AND gate in the accumulator path facilitates the initialization (LOAD x0) function. The four controls (accumulate enable, subtract, and the two multiplexer addresses) are controlled by the image rotator internal instruction generator.

The extra accumulator in each half of the position generator facilitates second-order warping, in which the distance between resampling points changes at a constant rate. (In simple translation and rotation, the horizontal and vertical distances between resampled points are both constant.) The second-order incrementing terms are accumulated as necessary to compute the appropriate interpixel spacings for use in the first accumulators.

A position comparator is used to flag all attempts to retrieve data from outside the original pixel space. The user preloads the x and y limits imposed by the boundaries of the original picture (or the desired portion of it). Then, as scanning progresses, this circuit continuously compares the integer portion of each new x and y coordinate against the appropriate limits. The control flag goes high if either x or y falls outside its permissible range. In a typical system, this control forces the immediate coefficient to zero. Arithmetic is carried to 17 bits, where the MSB is the two's complement sign bit. Only positive (or zero) limits may be set, since we are dealing with a single quadrant coordinate system.

The addresses generated are output through heavy-duty three-state buffers to pads R23:0, the resampled address output port. The internal tristate control disables these buffers, allowing R to double as the input port for the resampling parameter registers.

All controls and addresses generated within the image rotator are registered and are output via high-current output buffers to their respective ports.

The image rotator supports static image filtering, as well as resampling. In the filtering mode, the fraction and integer portions of the resampled position generators are separated (CLINK=0) to avoid undesired carries. The fractional adders are stepped through the coefficient table in synchronism with the "walk" of the integer adders through the pixel window. The "reverse subtract" controls, used to step through the coefficient table in nearest neighbor interpolation, are not used in the filtering mode.

The MSI implementation of the image rotator chip requires approximately 150 IC's which consume about 25 watts of power. A board based on this implementation would sell for $4000 to $6000 in quantity. By comparison, the VLSI implementation based on 1-micrometer CMOS design rules would fit in an 88-pin package and consume a fraction of a watt.

VLSI Signal Processing

CONCLUSIONS

This paper outlines the design of a chip to rotate video image data at real time rates. The chip is targeted mainly at the television industry's special effects market, but it also has applications in related markets such as DoD systems, publishing, robotics, etc. The chip design is based on user-selectable functions and flexible parameters, allowing chip usage in a wide variety of applications.

REFERENCES

1. J. Adams, C. Patton, C. Reader, and D. Zamora, "Hardware for Geometric Warping," Electronic Imaging, Vol. 3, No. 4, pp. 50-55, April 1984.

2. W. K. Pratt, Digital Image Processing, New York: Wiley, 1978.

3. R. G. Keys, "Cubic Convolution Interpolation for Digital Image Processing," IEEE Transactions on Acoustics, Speech, and Signal Processing, Vol. ASSP-29, pp. 1153-1160, December 1981.

4. K. W. Simon, "Digital Image Reconstruction and Resampling for Geometric Manipulation," Proceedings of the Purdue Conference on Machine Processing of Remotely Sensed Data, pp. 3A-1 - 3A-11, June 2-5, 1975.

5. Z. Z. Stroll and S. C. Kang, "VLSI Based Image Resampling for Electronic Publishing," K.-S. Fu, ed., in VLSI for Pattern Recognition and Image Processing, New York: Springer-Verlag, 1984.

6. Z. Z. Stroll and S. C. Kang, "System Design of Image Resamplers for Electronic Publishing," Proceedings of 1983 IEEE International Conference on Computer Design.

… VLSI Signal Processing

HIGH SPEED FFT PROCESSOR IMPLEMENTATION

Earl E. Swartzlander, Jr.
TRW Defense Systems Group
Redondo Beach, California

George Hallnor
TRW Defense Systems Group
McLean, Virginia

ABSTRACT

This paper describes recent progress in the implementation of high speed spectrum analysis systems with state-of-the-art commercial and semi-custom VLSI circuits. Initial efforts are producing Fast Fourier Transform (FFT) and inverse FFT processors that operate at data rates of up to 40 MHz (complex). The current implementation computes transforms of up to 16,384 points in length by means of the McClellan and Purdy radix 4 pipeline FFT algorithm. The arithmetic is performed by commercial single chip 22 bit floating point adders and multipliers, while the interstage reordering is performed by delay commutators implemented with semi-custom VLSI. This paper explains the pipeline FFT implementation and focuses attention on our current activity which involves developing a fixed point arithmetic version for improved performance in applications where the flexibility of floating point arithmetic is not necessary.

INTRODUCTION

Although the Cooley-Tukey FFT algorithm developed in 1965 has made it possible to apply digital signal analysis techniques to many applications, many others (e.g., radar and sonar beam forming, adaptive filtering, communications spectrum analysis, etc.) require combinations of flexibility and speed necessitating computational performance that exceeds the present state of the art. Currently, there are three approaches for signal processing: software implementation on general purpose computers, software implementation on a general purpose computer augmented with a Programmable Signal Processor (PSP), and custom hardware development. Software only and Software-PSP implementations are adequate when the spectral bandwidth is under 10 MHz. Custom processors achieve analysis bandwidths of 10-50 MHz but most are optimized for a specific application and require extensive (and expensive) redesign to modify them to suit other applications. Thus general purpose computers with or without PSP augmentation are too slow while custom processors are too expensive and lack the required flexibility.

Current signal processing systems require many diverse functions: transform processors, time and frequency domain vector processors, and general purpose computers. Work is underway to produce a growing family of building block modules to facilitate the development of such systems on a semi-custom basis. The result is the ability to quickly develop high performance signal processing systems for a wide variety of algorithms. The use of predesigned and precharacterized modules reduces cost, development time, and most importantly, risk.

VLSI Signal Processing

SIGNAL PROCESSING MODULES

The initial set of signal processing modules includes a data acquisition module, building block elements that are replicated to realize pipeline FFT and inverse FFT modules, a frequency domain filter module, a power spectral density computational module, and an output interface module (Ref. 1,2). The modules all have separate data and control interfaces. All data interfaces satisfy a common interface protocol so that modules can be connected together to form architectures that match the data flow of each specific system. The separation of the data and control is analogous to the Harvard mainframe computer architecture which uses separate data and instruction memories to eliminate the "von Neumann bottleneck." In signal processing the separation of data and control allows the simple data interfaces to operate at high speed while the more flexible (and complex) control interfaces operate at a slower rate.

Due to its importance in most signal processing applications, the FFT module was selected for initial development. The FFT processor uses the radix 4 pipeline algorithm developed a decade ago by McClellan and Purdy (Ref. 3) as an extension of the radix 2 pipeline FFT algorithm (Ref. 4). With the radix 4 pipeline algorithm, 4 data pass in parallel through a pipeline network comprised of computational elements and delay commutators as shown on Figure 1. An important feature of this algorithm and architecture is that only two types of elements are used: computational elements and delay commutators. Only minor changes are required to implement forward and inverse transforms of lengths that are powers of 4. The changes involve varying the number of stages connected in series, changing the counter sequence and step size on the computational elements, and changing the length of the delays on the delay commutator. The computational element performs a four point discrete Fourier transform. In our initial implementation 22 bit floating point arithmetic is performed with single chip adders, subtractors, and multipliers (Ref. 5). The delay commutator reorders the data between computational stages as required for the FFT algorithm. Data rates of 40 MHz are achieved using 10 MHz clock rates since the radix 4 architecture processes four data streams concurrently.

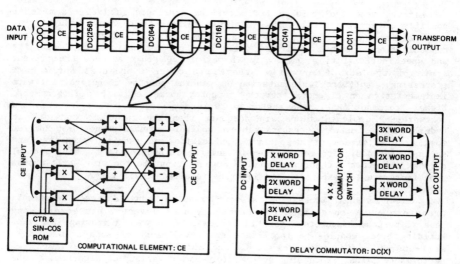

Figure 1. Pipeline FFT Architecture

THE DELAY COMMUTATOR CIRCUIT

Careful examination of the FFT module design revealed that much of the complexity was due to the delay commutator element. Initial complexity estimates are 80 commercial integrated circuits for the computational element and 180 circuits for the delay commutator. The disparity in complexity arises because of the commercial inavailability of shift registers of lengths 1, 2, 3, 4, 8, 12, 16, 32, 48, 64, 128, 192, 256, 512, and 768 as needed to implement the delays in the delay commutator. An alternative approach is to develop a delay that can be programmed to an arbitrary length. The most efficient approach (shown on Figure 2) involves simulating a delay line by using a RAM as a circular buffer with write and read addresses displaced by a constant (i.e., the length of the simulated delay line). Given the high complexity of the commercial implementation of the delay commutator, alternative approaches were examined. A B bit wide data slice of a delay commutator that can be programmed for X = 1, 4, 16, 64, and 256 requires approximately 400 (B+1) logic gates and 3072·B shift register stages. Since a shift register stage is comparable in complexity to 3 random logic gates, this reduces to 400 + 9616·B gates. Table 1 compares VLSI versions based on

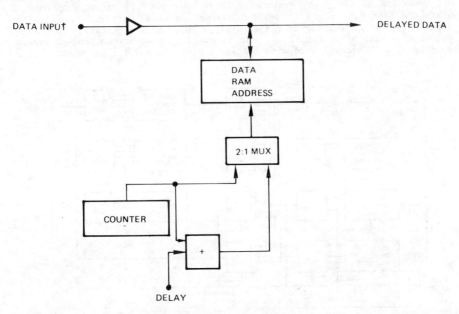

Figure 2. Variable Length Delay Implementation with Circular Buffer

TABLE 1. DELAY COMMUTATOR IMPLEMENTATION OPTIONS

	Gates/Chip	Max B	Chips/Stage	Relative Development Cost
Commercial	–	–	179	1
Gate Array	10K	1	44	2
Standard Cell	40K	4	11	3
Custom	100K	10	5	10

VLSI Signal Processing

gate arrays, standard cells, and custom technology. For these technologies (as of 1984), maximum achievable data bit slice widths are limited to 1, 4, and 10, respectively. For these widths 44, 11, or 5 delay commutator circuits would be required per computational stage of the pipeline FFT. Given the desire to minimize system complexity and to avoid an expensive custom VLSI development activity, we selected the standard cell approach. The resulting delay commutator circuit is a 4 bit wide slice that uses programmable length shift registers and a 4 x 4 switch as shown on Figure 3. Data enters through shift registers with taps and multiplexers to set the delay at 1, 4, 16, 64, or 256 (=X) in the uppermost input register and multiples of 2X and 3X in the middle and lower registers, respectively. Four 4:1 multiplexers implement the commutator function under the control of the programmable rate counter. The final 2 bit counter/decoder that controls the multiplexer settings can be reset and held to disable the commutator switch function. In this mode the chip provides fixed length registers with delays of 256, 2X 768, and 1280 which are used to expand the delay commutator for transform lengths greater than 4096 points. Data from the 4:1 multiplexers is output through programmable length shift registers that are similar to the input registers.

Operation of the delay commutator to reorder data is shown on Figure 4 where the data flow for a 64 point transform is shown (after Ref. 6). The input data (with a spacing of 16) is applied to a radix 4 butterfly producing output data with a spacing of 16. The reordering necessary to produce a data spacing of 4 is accomplished with delays of 0, 4, 8, and 12; commutation at a

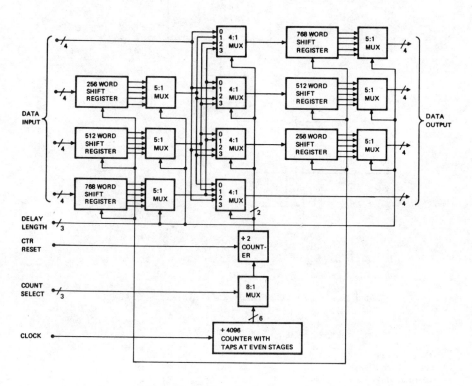

Figure 3. Delay Commutator Circuit

VLSI Signal Processing

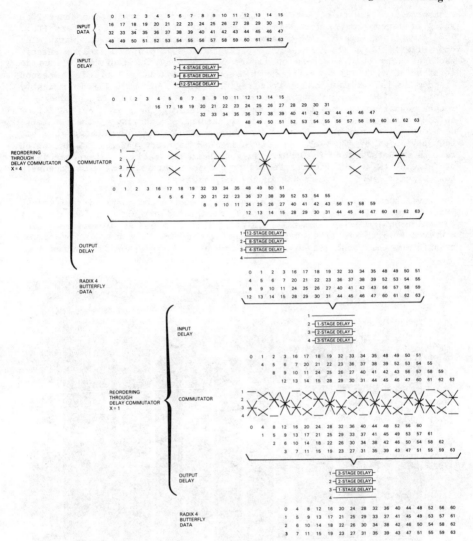

Figure 4. 64 Point FFT Data Flow Through the Delay Commutator (after Ref. 6)

rate of one fourth the data rate; and delays of 12, 8, 4, and 0. This data (with a spacing of 4) is applied to a radix 4 butterfly. The resulting data is reordered by use of delays of 0, 1, 2, and 3; commutation at a rate equal to the data rate; and delays of 3, 2, 1, and 0. The final data has a spacing of 1 as required for the final radix 4 butterfly.

This circuit was designed and implemented with Bell Laboratory's polycell (standard cell) CMOS technology. This technology was selected because it is well suited to the development of VLSI with high density shift registers and

VLSI Signal Processing

random logic. The chip contains 12,288 shift register stages and about 2000 gates of random logic, for a total complexity of 108,000 transistors (Ref. 7). At a clock rate of 10 MHz, the power dissipation is under 1/2 watt. The chip size is 340 x 376 mils. It is packaged in a 48 pin dual-in-line ceramic package. The chip is shown on Figure 5. Each of the four bit slices is constructed with input registers in a column, switching logic in a second "random logic" column, and output register in a column. The four nearly identical slices are about four times as tall as they are wide producing a roughly square chip when they are properly stacked. There is minor variation in the random logic of each bit slice to account for sharing of the counters, decoders, clock drivers, etc. Although a radix 2 delay commutator chip has been developed by NEC (Ref. 8), it is limited to clock rates of 5 MHz which results in data rates of 10 MHz. In contrast, the radix 4 delay commutator operates at twice the clock rate and processes 4 data streams simultaneously to achieve a four fold increase in the data rates.

Development of the delay commutator chip reduces the complexity of the 40 MHz 4096 point FFT from 1375 commercial integrated circuits to 546 circuits (of which 66 are delay commutator chips). This is a 60% complexity reduction achieved through the development of a single semi-custom chip. For larger transforms, the complexity of a 16,384 point FFT processor is reduced from

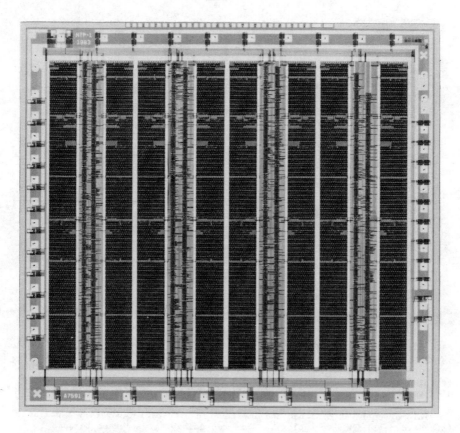

Figure 5. Delay Commutator Integrated Circuit Photograph

1634 integrated circuits to 670 circuits with the VLSI delay commutator circuit. Such a reduction greatly improves system reliability since connections between circuits represent the dominant failure mechanism in modern systems (Ref. 9). With 60% fewer circuits (and a corresponding reduction in the number of interconnections) the reliability is greatly improved.

As shown on Table 2 the 22 bit floating point arithmetic implemented in the FFT module provides 96 dB of precision (i.e., equivalent to 16 bit fixed point arithmetic) over a dynamic range of 476 dB (Ref. 5). Although this dynamic range is less than that of 32 bit floating point arithmetic, it is more than adequate for most high speed signal processing applications. 32 bit arithmetic is useful in scientific computation when inverting matrices, evaluating eigenvectors, etc. but these operations are usually performed at much lower rates than those required for signal processing where input data is often limited in precision to 8 bits or less.

CURRENT DESIGN DIRECTIONS

The arithmetic to implement our initial system is performed by single chip computational elements which are realized with a bipolar technology. This provides high circuit density, producibility, and tolerance to severe environments (i.e., temperatures and radiation). Unfortunately, the power consumption of these circuits is rather high at 3 watts typical. Accordingly, we are currently developing a redesigned pipeline FFT module which uses CMOS arithmetic elements. We are implementing 16 bit fixed point arithmetic which allows use of the recently announced CMOS 16 bit multipliers (Ref. 10).

Part of the sacrifice in using fixed point arithmetic is the loss of flexibility in data scaling. Consider a pathological case where there is one strong component in the spectrum and it is desired to make an accurate measurement of its intensity. Here the radix 4 computational element should downscale all data by 2 bits to avoid arithmetic overflow due to the wordsize growth that normally occurs in the computation of sums. On the other hand, if low level signal extraction is of paramount importance, it may be best to avoid downscaling the intermediate results. In system development it is useful to use floating point arithmetic in the early stages of a project until the data scaling is well understood. At that point it is often possible to change to fixed point arithmetic. This is especially true if the fixed point scaling can be easily modified to accommodate changing conditions (i.e., decreasing signal strengths, the presence of jammers or other interference sources, etc.).

In developing the low power version of the pipeline FFT processor, suitable flexible low power CMOS adders are simply not available. Our desire is for a 16 bit wide parallel adder (cascadable for higher precision) with overflow/underflow detection and selectable fixed or data dependent scaling

TABLE 2. ARITHMETIC COMPARISON

Arithmetic System	Dynamic Range	Precision
12 Bit Fixed Point	72 dB	72 dB
16 Bit Fixed Point	96 dB	96 dB
22 Bit Fixed Point	132 dB	132 dB
22 Bit Floating Point	476 dB	96 dB
32 Bit Floating Point	1686 dB	144 dB

VLSI Signal Processing

or saturation as required. Like the delay commutator we are currently examining a variety of implementation approaches to select the optimum design technology. It represents a fixed point element that can be electrically programmed to provide performance approaching that of floating point arithmetic at a fraction of the complexity (at both the chip and board levels). An initial projection is that the typical power consumption of the 40 MHz 4K FFT processor will decrease by about 75%.

CONCLUSIONS

This paper shows the high payoff of semi-custom integrated circuit development. Specifically the complexity of a 40 MHz pipeline FFT processor was decreased by 60% through the development of the delay commutator. Current emphasis is focusing on reducing power consumption by the development of a flexible 16 bit fixed point adder. The resulting highly integrated processor is simpler and correspondingly lower in power, size, and cost than previous designs. Such improvements can be achieved in a wide variety of signal processing systems by carefully tailoring the algorithms, processor architecture, and technology selection.

REFERENCES

1. E. E. Swartzlander, Jr. and G. Hallnor, "Frequency-Domain Digital Filtering with VLSI," in VLSI and Modern Signal Processing, Englewood Cliffs, NJ: Prentice-Hall, Inc., 1984, Chapter 19.

2. E. E. Swartzlander, Jr., L. S. Lome, and G. Hallnor, "Digital Signal Processing with VLSI Technology," Proceedings IEEE International Conference on Acoustics, Speech, and Signal Processing, pp. 951-954, 1983.

3. J. H. McClellan and R. J. Purdy, "Applications of Digital Signal Processing to Radar," in A. V. Oppenheim, Editor, Applications of Digital Signal Processing, Prentice-Hall, Englewood Cliffs, Chapter 5, 1978.

4. H. L. Groginsky and G. A. Works, "A Pipeline Fast Fourier Transform," IEEE Transactions on Computers, Vol. C-19, pp. 1015-1019, 1970.

5. J. A. Eldon and C. Robertson, "A Floating Point Format for Signal Processing," Proceedings IEEE International Conference on Acoustics, Speech, and Signal Processing, pp. 717-720, 1982.

6. L. R. Rabiner and B. Gold, Theory and Applications of Digital Signal Processing, Englewood Cliffs, Prentice-Hall, Inc., p. 611, 1975.

7. E. E. Swartzlander, Jr., W. K. W. Young, and S. J. Joseph, "A VLSI Delay Commutator for FFT Implementation," Proceedings International Solid State Circuits Conference, pp. 266-267, and 351, 1984.

8. A. Kanemasa, R. Maruta, K. Nakayama, Y. Sakamura, and S. Tanaka, "An LSI Chip Set for DSP Hardware Implementation," Proceedings IEEE International Conference on Acoustics, Speech, and Signal Processing, pp. 644-647, 1981.

9. G. W. Preston, "The Very Large Scale Integrated Circuit," American Scientist, Vol. 71, pp. 466-472, 1983.

10. Y. Kaji, N. Sugiyama, Y. Kitamura, S. Ohya, and M. Kikuchi, "A 45ns 16 x 16 CMOS Multiplier," Proceedings International Solid State Circuits Conference, pp. 84-85, 1984.

Part II
ALGORITHMIC PROCESSORS

Edited by:
Dr. Bijoy Chatterjee
ESL

5
A VLSI FFT SYSTEM

J. H. Gray

ESL

THIS MANUSCRIPT UNAVAILABLE FOR PUBLICATION

EMSP: A DATA-FLOW COMPUTER FOR SIGNAL PROCESSING APPLICATIONS

J. D. Seals and R. R. Shively
AT&T Bell Laboratories

THE COMPLETE TEXT OF THIS MANUSCRIPT WILL BE FOUND BEGINNING ON PAGE 421.

DIGITAL/OPTICAL COMPUTING AND LINEAR ALGEBRA

R. O. Schmidt

Guiltech Research Co.

THIS MANUSCRIPT UNAVAILABLE FOR PUBLICATION

8

VLSI Signal Processing

A HIGH PERFORMANCE INTERCONNECT FOR CONCURRENT SIGNAL PROCESSING

Michael S. Schlansker and Ronald Bianchini Jr.
ESL, a subsidiary of TRW
Sunnyvale, CA 94088-3510

ABSTRACT

This work investigates the design of an interconnect to support the execution of a parallel program on a heterogeneous set of processors. Static inter-process data traffic requirements are established at compile time. Data traffic requirements are input to a scheduler which is used to generate very high performance parallel data transfer microcodes. The removal of network control from the processing function and the provision of an autonomous network function allows very efficient network control for large scale systems.

The architecture defines parallel digital systems which heavily exploit VLSI area. Network programmability allows systematic design methodologies leading to reduced design cost. The network and associated processors can be reprogrammed to support alternate applications, system performance enhancements, and degraded failure modes.

INTRODUCTION

Digital signal processing typically requires processing of many data sets through the same mathematical procedure. Classical operations such as the FFT or matrix inverse are reapplied to a vast number of data sets at a processing rate determined by system requirements. Such special purpose requirements have been addressed by research in the area of special purpose processor design.

Research has shown that special purpose processors, while less flexible, perform appropriate calculations like the FFT much faster [4]. Systolic processor research [1],[2],[6],[7] has shown how scalable hardware arrays solve classical algorithms of linear algebra (matrix multiply, L/U factorization, etc.).

A new signal processing architecture will be presented. This architecture exploits static program behavior to automatically produce a near optimal data routing policy. Microprogrammable switches within the network are used to implement this policy. Since data routing is done at compile time, the problem of scheduling traffic flow through blocking networks of arbitrary topology can be solved in a very efficient manner.

The approach greatly reduces system connectivity which is a critical component of system cost. The interconnect can be viewed as a programmable backplane into which special purpose and general purpose processors are inserted. The network is programmed to autonomously support application requirements. This is achieved with a traffic scheduler which automatically emits network microcode.

NETWORK ARCHITECTURE

The principal objective of this work is to achieve very high computation rates through the efficient networking of special purpose processors. A programmable networking concept which is efficient even when expanded to a large number of processors will support signal processing systems which are extensible to very high performance levels.

Many parallel processing architectures are based on the full crossbar interconnection network. The basic difficulty with this approach is the $O(N^2)$ cost which becomes prohibitive for a large number of processors. If we consider the expansion of a 16 PE system to 18 PE's the number of switches expands from 256 to 324. An additional 68 switches are required to support 2 additional PE's.

Therefore, the full crossbar precludes the use of truly large scale systems. Designers using the crossbar will hesitate to provide spare processors for fault tolerance or to augment existing systems with new processing capabilities.

VLSI Signal Processing

With the use of VLSI technology, the additional cost of a transistor or gate is very small while the cost of interconnect is large. This heavily favors the use of a small number of complex switches over a large number of simple switches. The silicon area at each switching site is exploited to provide distributed network programmability and data buffering. An $O(N)$ mesh network supporting a heterogeneous collection of processors is shown in Figure 1. The identical switch cells run distinct programs describing each switches role in a well defined data transfer operation.

This statically generated multi-switch program is constructed by a compile time network scheduler and is downloaded at load time thus "training the network" to execute a specific data transfer pattern serving the program requirements. The actual operations carried out by each cell are very simple switching and buffering operations and can be readily executed at the full interchip data transfer rate.

The definition of an efficient cell structure is of critical importance in the determination of an appropriate network architecture. We have selected a four nearest neighbor interconnect cell with an additional link for connection to an associated processor shown in Figure 2. The cell serves as the standard interface between the network and the connected processors. Thus, it is used as the standard bulk data connector for large scale signal processing systems.

The use of $O(N)$ networks for the processor interconnect requires scheduling techniques which allow the efficient sharing of network resources among competing processing requirements. The scheduling techniques which have been developed treat networks of arbitrary topology under arbitrary static traffic load. The use of irregular networks is efficiently supported and is a valuable tool for system design. In many cases, the system designer will build networks whose topology is tailored to the interprocess communication requirements of the application.

FLUID FLOW MODEL

The scheduler generates a network routing policy R^* using the fluid flow model of bulk data transfer. Nodes within the network are defined as:

P_i - A network source node,
P_j - An adjacent network node,
P_k - A network destination node.

Inputs to this scheduler are:

L - network topology matrix, such that L_{ij} describes the link capacity between nodes P_i and P_j,
T - traffic matrix, such that T_{ij} represents the bandwidth Requirements between nodes P_i and P_j.

Either unidirectional or bidirectional traffic can be modeled. All traffic is assumed to be steady-state minimizing network buffering difficulties. If the processor produces or consumes data in a burst, it is responsible for buffering its own data to properly interface with network protocol. This is not a fundamental restriction of this work but serves as a solid starting point.

The traffic routing policy is described by a policy matrix R which assigns a probability R_{ijk} that a unit of data in processor P_i destined for processor P_k is routed through neighboring processor P_j.

Two networks $L \propto L^1$ are similar if they are identical in topology and have link bandwidth functions which are in proportion. Thus, the product of a scalar times a network defines a similar network:

$$L^1 = c \times L.$$

Given an input network L and a traffic matrix T the fluid flow scheduler determines an optimal throughput routing policy R^* and a scalar utility u which measures the effectiveness of the policy. This policy will just support the input traffic on a similar network L^* of minimized bandwidth:

$$L^* = (1/u) \times L.$$

VLSI Signal Processing

This procedure can alternatively be viewed as identifying maximal traffic $T^* = u \times T$ which just saturates the input network L.

The fluid flow scheduler uses dynamic programming techniques to determine a policy R^* which is optimal under a specific cost structure. Let U_{ij} be the utilization of the link connecting two adjacent nodes P_i to P_j. The cost of crossing a single network link from adjacent processors P_i to P_j is defined to be:

$$C_{ij} = 1/(1 - U_{ij})^2.$$

This function penalizes heavy traffic on a link and forces the identification of alternate more lightly loaded paths to support traffic requirements. Other cost functions like $1/(1 - U_{ij})$ have been used but will not be discussed here.

The solution procedure follows the general form of a policy iteration as presented in Howard [5]. The policy iteration selects increasingly efficient policies arbitrarily close to the optimal policy R^* which minimizes the average cost under the input workload. The policy R^* concurrently minimizes for each source node P_i and each destination node P_k the cost C_{ik} which can be expressed as a sum over neighboring nodes P_j:

$$C_{ik} = \sum_j R_{ijk}(C_{ij} + C_{jk}).$$

The scheduler cycle is a policy iteration which has been augmented to repeatedly increase traffic flow levels as follows:

step 1) A small initial amount of traffic is scheduled on the network.

step 2) A policy iteration cycle is executed to reduce the cost of each C_{ik}. The traffic matrix is adjusted to move traffic from costly paths to lower cost paths. The cost penalty on heavily utilized links forces traffic smoothing.

step 3) Traffic is scaled to a higher level as determined by the saturation of the most heavily utilized link.

step 4) Repeat steps 2&3 until additional traffic cannot be supported.

NETWORK MICROCODE GENERATOR

The output of the "Fluid Flow Scheduler" is a policy R^* which gives a probabilistic description of traffic flow direction and traffic splitting. The optimized schedule R^* is discretized and used to generate network microcode.

The continuous probability assignments for all flows must be mapped into discrete network clock cycles such that only one message unit is on any link at any given time. To accomplish this, every switch in the system network will contain M steps of microcode. Once started, each switch will execute the microcode in a continuous loop. This creates an M word "Micro-cycle". The value of M is selected to be large enough to allow sufficient accuracy when discretizing the continuous solution produced by the fluid flow scheduler. All traffic will be scheduled in fractions which are multiples of $1/M$ times the link bandwidth. In examples thus far, satisfactory results have been obtained with M on the order of 100.

A preliminary "switch" design is showed is Figure 2. It contains four network ports (N, E, S, W), a connection to the local processing unit, and a connection network internal to the switch (labeled: C). Each network port contains a small FIFO (first in, first out) queue on the output side of the port. The processor connection has a microcode addressable reordering queue. This queue is used to reorder data whose order has been altered due to traffic splitting within the network. The connection network internal to the switch is controlled by the microcode as well, and can connect any port to any other port. The internal connection network is four times as fast as the external ports and allows the movement of four words of data on a single cycle.

The discrete approximation to steady-state traffic is not truly steady state. Instantaneous traffic flow may briefly rise and fall throughout the cycle. Flow discretization requires buffering on each output port shown as a queue of depth d_1. Presently, using a primitive discretizer, queue depths of six are sufficient. A highly efficient flow discretizer can hope to be very efficient with four words or fewer per port.

VLSI Signal Processing

The processor queue is significantly more complex. It is addressable so that data arriving at the switch can be reordered. Data is ordered so that it enters the local processor in the same order that it left the source processor. Using this constraint we can calculate d_2:

$$d_2 = d_1 + l_{net}$$

In this equation, l_{net} is the difference between the latency of the shortest and longest path from source to destination. Again, d_1 represents the non-steady-state effect of the discretizer. The processor queue will remain static as long as the lowest element of the queue is empty. Added data will simply be loaded in from the connection network. Once the lowest queue element is filled, data can be shifted down to the processor.

SYMMETRIC NETWORK PERFORMANCE

This section analyzes the performance of four symmetric networks to illustrate the effects of increasing the number of processors N on network cost. The data traffic workload used in this analysis requires that the traffic from processor P_i to processor P_j is equal for all $i \neq j$ while the traffic from a processor to itself is 0. Thus:

$$T_{ij} = B/(N-1) \quad (i \neq j)$$

and

$$T_{ii} = 0$$

so that

$$\sum_j T_{ij} = B.$$

We will call this the uniform workload. This workload allows the use of a simple analytic calculation for the prediction of the performance of the scheduler on any of the symmetric networks. Data locality is properly exploited, uniform traffic may be a pessimistic model of actual system traffic.

The fluid flow scheduler when confronted with uniform traffic on a symmetric network converges in one iteration. First it calculates minimal paths to support each traffic requirement T_{ij}. After identical shortest path traffic routes are selected from each source node the scheduler loads all links equally with the uniform traffic input. The scheduler then scales traffic flow to uniformly saturate all network links and terminates with an optimal schedule. Any asymmetry in either the network or the traffic will result in multiple iteration convergence.

Consider the mesh connected network of Figure 1. Each switch has a single link to its connected processor, while four nearest neighbor links are connected to an adjacent switch. The nearest neighbor links have bandwidth L. The total number of links in the mesh plane is $2 \times N$ where N represents the number of processors. Messages entering the mesh plane may travel through a number of links to reach the destination switch.

Let D_{ij} represent the minimum distance from processor i to processor j. The average distance traveled by each message from any source node P_i can be calculated for the case of the uniform workload:

$$R_i = \sum_{j \neq i} D_{i,j} / (N-1).$$

For symmetric networks, the average radius is source node independent and may be called R.

Each word of traffic entering the mesh plane must move an average distance of R units. The volume of traffic in the mesh plane can be no greater than the total bandwidth available in the mesh plane. Let the bandwidth requirement of each of the N processors be B. The average radius establishes an upper bound on network performance as follows:

$$N \times B \times R \leq N \times 2 \times L$$

or

$$B \leq 2 \times L / R.$$

This becomes an equality when a symmetric network is saturated with uniform traffic. Solving for the link bandwidth L:

$$L = R \times B / 2.$$

For any given processor I/O requirement B there is an appropriate link bandwidth L such that uniform traffic at rate B can be supported within the mesh plane.

Consider the interconnection of 49 processors. A 7×7 mesh network has an average radius $R=3.5$. The scheduler described above will produce microcode for a mesh network with link bandwidth $L=(3.5\times1)/2$ equivalent in performance to the crossbar under uniform load. A mesh 1.75 times as wide as a crossbar can match the crossbar in performance under uniform load. Figure 3 illustrates the average radius as a function of N the number of processors for four different symmetric network types.

Using data pin count as a cost function, each full crossbar switch has two 1 bit wide connections. The full crossbar network has 49×49×2=4802 pins. Each switch in a two bit wide mesh network has 5 connections. The mesh connected network has 49×5×2=490 pins. The mesh network is 87.5% saturated and the spare bandwidth and can be used to support higher radius traffic or performance extensibility.

Figure 4 shows pin count for a crossbar of size $N \times N$ and each of three other symmetric network types which have been sized to yield equivalent performance under uniform workload. These results have been calculated using the average radius results of Figure 3. Any rounding to integral bit widths has been neglected.

DIAMOND MESH

Networks of minimal radius are very attractive in that they provide maximal extensibility. A careful examination of the mesh network reveals that a smaller radius interconnect exists for the interconnection of mesh nearest neighbor type networks. The "diamond network" is illustrated in Figure 5.

A diamond network of radius R is constructed by first placing a center element in a gridded plane. Then, all elements of radius $\leq R$ are placed surrounding the center. The wrap around arcs are determined by replicating the diamond to tile the infinite plane. After identically labeling the perimeter of a single diamond and its neighbors, the interconnect is determined for wrap around arcs. Both the simple mesh and the diamond mesh have a radius which grows as $O(\sqrt{N})$.

PERFECT SHUFFLE

The perfect shuffle is a well known network described in [9]. The networks illustrated are multi-stage shuffles with $2^{(n-1)}$ switches per column and n columns connected in a cycle. Such a network is shown for n=3 in Figure 6.

The shuffle network can be shown to have an average radius which grows as $O(\log N)$. Of the four networks shown in Figures 3 & 4, the shuffle has the lowest asymptotic growth in cost when supporting uniform traffic. A network like the shuffle which is constructed for minimal large network radius is not necessarily optimal when supporting traffic which exhibits locality.

PROGRAMMING METHODOLOGY

To facilitate programming of the network, a graphical program language is desired. The graphical language will contain nodes and arcs. Each node will represent a process, and each arc will represent steady-state input or output data flows from various processes. The language is similar to other languages with statically connected processes with queued data links[12].

Each process node may be expanded to several processes using sub graphs. At the lowest level each terminal process node is mapped onto a processor. Network arcs are parameterized by steady-state bandwidths. In the example program of Figure 7, each program arc has a fixed data transfer requirement. For example, the arc connecting the sensor input to the preprocessing unit will consist of a vector of 4096 real numbers per processing cycle. The arc bandwidths can then be calculated for input to the network scheduler once the number of cycles per second has been determined. The network could also support switch processes. These would be used when one output sends data to two inputs. To simply eliminate unneeded data, the switch process could be mapped to a switch node (not processor) near the two destination processors.

VLSI Signal Processing

Once all process to process flows have been calculated, the process to processor mapping must be specified. Currently this mapping is left to the programmer. Research similar to [11] provides for a fully automated process mapping capability. With a fully automated mapping capability, fault tolerance could be supported. Process flows are rescheduled on a working subset of the hardware configuration after a failure. This approach will lead to graceful system degradation with modest down time.

The primary factors involved in choosing the process to processor mapping are the relative distances from data sources to destination, and the relative efficiencies in which a process runs on the various processors.

APPLICATION

A digital signal processing example has been derived from work done on the Chimera project at ESL. This project uses digital circuitry to perform acoustic beamforming. This system achieves approximately one third of a billion floating point operations per second. The system consists of an interconnected collection of special purpose processors. The interconnect is hand engineered and scheduled for high performance. This approach is engineering labor intensive and not flexible enough for future applications.

A more systematic approach is required to reduce engineering cost, risk, and schedule and to develop systems which are inherently much more flexible. The very high data rate requirements of this class of systems necessitates very efficient network data transfer mechanisms. An sample beamformer program selected from Chimera can be seen in Figure 7.

We will describe a systematic implementation of this program on an interconnected network of special purpose processors exploiting static scheduling techniques. Table 1 specifies the list of processors. If this networking technology were mature, these processors would be "off the shelf" units due to the standard nature of the processor interface into the network. A network of sufficient size and appropriate topology is constructed. This network can be thought of as an empty backplane as shown in Figure 8. All processors must be assigned positions in the backplane.

The location of the processors should be assigned according to application requirements. Such a careful assignment of processors to switches will lead to traffic with very low mean radius. In the Chimera project, all processing is done in the frequency domain. For this reason, the FFT processor module is located nearby the sensor input unit. In the final orientation these two units are assigned locations one and two. See Table 1 for the processor / network assignments.

Once the processor assignments have been specified, the network is ready to be programmed. A program as shown in Figure 7 is used to both specify the code running on each processor as well as the network communication requirements. Program sub graphs are not shown for brevity. At the lowest level of the program, terminal nodes must be mapped onto a specific network processor. For example the 8K FFT which is applied to all input samples is mapped onto the FFT processor of location two. All data traffic volumes must be specified either directly by the programmer or may be indirectly generated through compile time support software (e.g. system simulation). The mapping of all program processes onto the network of processors is shown in Table 2.

A complete program specification and process mapping allows the creation of a simple source destination list as shown in Table 3. This list of messages is input to the fluid flow compiler which determines an optimal schedule with associated link utilizations, and a network utility.

The output of the fluid flow compiler is shown in Figure 9. Assuming a link bandwidth of 10 Mhz for all network links, the network utility for this problem is: $u = .2012$. This utility represents the fraction of the required data transfer supported by a one bit wide network. A $5 = \lceil 1/u \rceil$ bit wide network is sufficient to support required traffic. A very high performance beamformer system realized using standard eight bit wide links would have a utilization of $<5/8$ on each link.

Each of the network links are utilized by the percentages given in figure 9 (assuming a 4.97 (1/u) bit wide bus). Using a cut-set analysis, the results of the scheduler are provably optimal. In this example, a single unit, the FFT unit forms the cut-set. If a circle were drawn around the cut-set (the FFT unit), all links entering and leaving the circle are completely utilized. This reflects the fact that in this example, the FFT unit requires more data than any of the other units. The fluid flow scheduler has routed data so that it enters and leaves the FFT unit through all possible paths. Superfluous data which does not need to

VLSI Signal Processing

enter or depart the FFT unit has been routed around the cut-set. This cut-set analysis illustrates that it is impossible to re-route data to the FFT unit to accomodate additional traffic.

CONCLUSION

The use of statically scheduled control which is distributed throughout a synchronous network allows very efficient use of network bandwidth. Where program data rate requirements are predictable, this approach can be used in a systematic manner to construct very large scale parallel processors with networks of modest cost. This approach has advantages with respect to engineering cost, reprogrammability, and extensibility to future requirements while exploiting a very efficient network technology.

REFERENCES

[1] H. M. Ahmed, J. M. Delosme, and M. Morf, "Highly Concurrent Computing Structures for Matrix Arithmetic and Signal Processing", Computer, pp65-82, Jan, 1982.

[2] J. H. Avila, and P. J. Kuekes, "A One Gigaflop VLSI Systolic Processor", Proc. 27th annual SPIE Conf., Aug, 1983.

[3] N. H. Brown, "The EMSP Dataflow Computer", Proc. of the Seventeenth Annual Hawaii International Conference on System Sciences, pp 39-48, 1984.

[4] J. Gray, and M. Greenstreet, "A VLSI FFT System Design", IEEE Workshop on VLSI Signal Processing, Nov 27, 1984.

[5] R. Howard, Dynamic Probabilistic Systems, VOLS 1 & 2, Wiley, 1971.

[6] H. T. Kung, , C. E. Leiserson, "Why Systolic Architectures?", Computer Magazine, Vol. 15, NO. 1, pp36-46, Jan 1982.

[7] H. T. Kung, L. M. Ruane, and D. W. L. Yen, "A Two-Level Pipelined Systolic Array for Multi-Dimensional Convolution".

[8] B. R. Rau, et. al. "A Statically scheduled VLSI Interconnect for Parallel Processors," VLSI Systems and Computations, Computer Science Press, Vol. 14, No. 9, Oct 1981.

[9] M. S. Schlansker, "A VLSI Based Architecture for Scientific Computation", HISCC-17, Honolulu, Hawaii, January 1984.

[10] Stone, H. S. "Parallel Processing with the Perfect Shuffle", IEEE Trans. on Computing, Vol. C-20, No. 2 :pp 153-61, February 1971.

[11] D. C. Vogel, "Real-Time Mapping of Signal Processing Jobs onto Multiprocessor Networks", M. S. thesis, M. I. T., 1982.

[12] Y. S. Wu, "A Common Operational Software (ACOS) Approach to a signal Processing Development System," Proc. 1983 IEEE International Conference on Acoustics, Speech and Signal Processing, Boston, pp 1172-1175.

VLSI Signal Processing

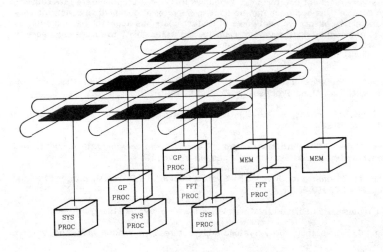

Figure 1: System with Mesh Interconnect

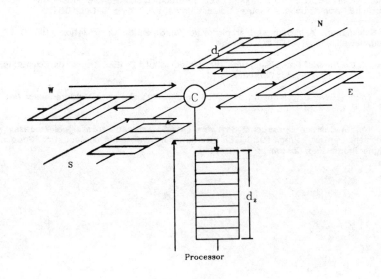

Figure 2: Switch Architecture

VLSI Signal Processing

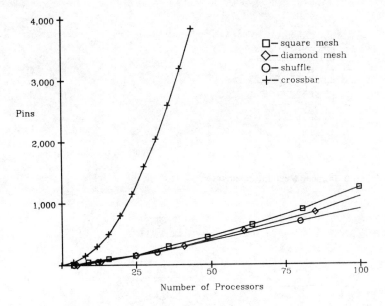

Figure 3: Average Network Radius

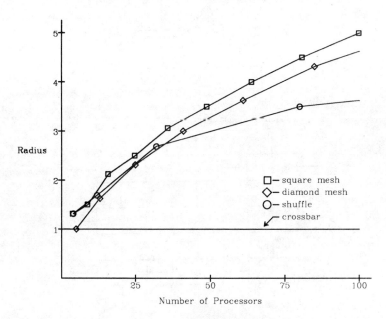

Figure 4: Network Pin Count

VLSI Signal Processing

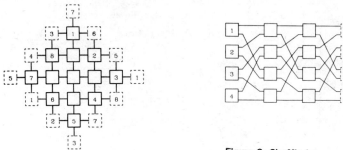

Figure 5: Diamond Mesh Interconnect

Figure 6: Shuffle Interconnect

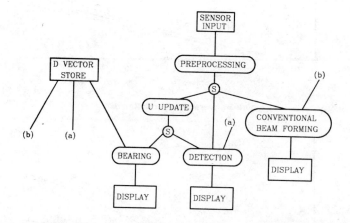

Figure 7: Beamformer Program Graph

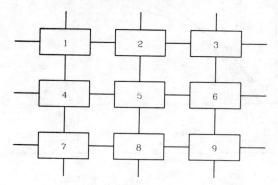

Figure 8: Empty Backplane

VLSI Signal Processing

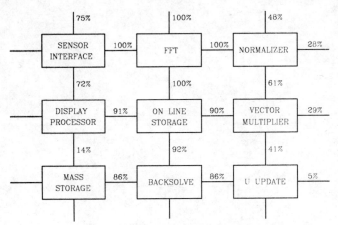

Figure 9: Scheduled Beamformer Traffic

PROCESSOR / SWITCH MAPPING

Processor	Switch
Sensor Interface	1
FFT	2
Normalizer	3
Display Processor	4
On Line Storage (Disk)	5
Vector Multiply	6
Mass Storage (Tape)	7
Matrix Backsolve	8
U Update	9

Table 1: Processor to Backplane Mapping

PROCESS / SWITCH MAPPING

Process	Switch
DVectorStore	5
S1	6
UUpdate	9
S2	8
preproc.SensorInterface	1
preproc.FFT	2
preproc.Normalizer	3
bearing.DBacksolve	8
bearing.IO	4
detect.S1	8
detect.S2	8
detect.S3	2
detect.DBacksolve	8
detect.ZBacksolve	8
detect.Multiplier	6
detect.FFT	2
detect.IO	4
detect.Tape	7
cbeamform.Multiplier	6
cbeamform.IO	4
cbeamform.Tape	7

Table 2: Process to Backplane Mapping

SOURCE / DESTINATION TRAFFIC LIST

SOURCE	DEST.	TRAFFIC K bits/s	SOURCE	DEST.	TRAFFIC K bits/s
1	2	136320	2	3	42000
3	6	32000	3	9	10240
9	8	10240	3	2	3200
8	6	1440	8	6	10240
8	2	2880	2	4	14080
2	7	200	8	4	180
9	8	10240	6	7	3200
6	4	9600	5	8	32000
5	8	16000	5	6	32000

Table 3: Traffic Specification of Beamformer

Part III
VLSI ARCHITECTURES I

Edited by:
Dr. Ed Greenwood
Motorola

A CUSTOM CHIP SET FOR DIGITAL SIGNAL PROCESSING

Shahid U. H. Qureshi and Hassan M. Ahmed
Codex Corporation
20 Cabot Blvd.
Mansfield, MA 02048

ABSTRACT

A novel programmable digital signal processor based on two custom integrated circuits is described. This processor has novel arithmetic addressing and program flow control capabilities that result in a high utilization of the arithmetic unit and a low program overhead for 'housekeeping' tasks.

INTRODUCTION

Numerous programmable VLSI circuits for digital signal processing (DSP) have recently been introduced [1-4]. Recognizing the severe computational burden associated with filtering algorithms common to digital signal processing [5,6], all of these chips provide for very fast multiplication and accumulation. However, commercially available DSP chips generally lack good address generation capability, a flexible and efficient microprocessor interface and the ability to connect sufficient external program and data memories. This paper describes a custom DSP chip set, called the signal processing element (SPE), that offers these features in addition to a fast arithmetic capability.

Fully featured (or 'premium end user') modems can effectively utilize the above mentioned capabilities to provide multiple modes of operation, built-in network management, circuit quality monitoring and multiplexing functions. The Codex premium end user modem uses an SPE with a Motorola MC68000 in a loosely coupled architecture (Figure 1) supported by the flexible microprocessor interface of the SPE. The microprocessor performs control and management functions while the SPE executes all the signal processing tasks. Such a dual processor architecture is extremely powerful since each processor is assigned only those tasks that are best suited to its capabilities.

The remainder of this paper is devoted to describing the custom chips forming the SPE.

THE SIGNAL PROCESSING ELEMENT (SPE)

The SPE is a custom Digital Signal Processor having its own instruction set and operating its own program. It periodically communicates with the host processor to receive new signal processing tasks and data as well as to deliver the results of processed data. Interprocessor communications is generally initiated by the host through a unique host interface.

Two custom integrated circuits are the basis of the SPE. The first of these is the MAR or 'multiplier-accumulator-RAM' circuit. It is the arithmetic engine of the SPE having a two cycle pipelined multiply-accumulator and 256 words of data memory. The MAR is capable of performing 32 functions and has a very flexible data movement facility.

The SPC or signal processing controller circuit generates the necessary signals to control the MAR, external data memory and input-output devices which may be connected to the SPE data bus. Two MARs can be simultaneously controlled for complex arithmetic operation. In fact, although the MARs actually perform the calculations of the DSP algorithms, the SPC generates addresses and orchestrates the execution of the SPE program.

The basic SPE operations fall into essentially four categories: arithmetic, data structure control, program structure control and host operations. Program control and host operations lie in the domain of the SPC while arithmetic is performed by the MAR. Data structure control is a shared activity in which the SPC maintains addressing parameters while the data structure resides in the MAR memory.

We will now focus on the architectures of the MAR and SPC chips.

THE MULTIPLIER - ACCUMULATOR-RAM CIRCUIT

The MAR, depicted functionally in figure 2, is a single-bus arithmetic engine operating at a 300 ns bus transfer cycle rate. Offering both integer and fixed point fractional calculations, it embodies a two cycle pipelined 16 X 16 bit multiplier with 40 bits of accumulation. Arithmetic operations are specified by a five bit function field while data transfers are determined by a five-bit control field. A 256 X 16 RAM provides on chip data storage. Although primarily intended to function synchronously with the SPC circuit, the MAR may also be used as an asynchronous peripheral to a MC68000 type microprocessor.

During each instruction cycle, the MAR decodes the control and function fields and performs a data transfer (e.g. loading a register) and a function. The function can either set one of the many MAR modes or initiate an ALU operation such as multiplication or addition. ALU operations are completed in subsequent instruction cycles during which new functions are being decoded and performed. The structured calculations of most signal processing tasks lend themselves well to such a pipelined arrangement.

The MAR arithmetic logic unit (ALU) consists of three data registers, X, Y and A, one mode control register, an eight bit status register, and a 16 X 16 multiplier-adder. The X, Y registers contain 16 bit operands for arithmetic operations. A forty bit accumulator is segmented into three parts A, AH, AL. Both AH, AL are 16 bits wide and

VLSI Signal Processing

together form a 32 bit accumulator. When operating in fraction mode the binary point is to the right of the most significant bit of AH, X and Y. An eight bit extension, AX, of the most significant part of the accumulator, provides overflow protection. The accumulator status information is stored in a STATUS register. In addition to the usual status flags, the MAR gives separate overflow indications for AH and AL thus facilitating multiple precision calculations.

The MAR has a two cycle, radix four Booth multiplier [7] whose operation is initiated by loading an operand into the X register (the other operand having been loaded earlier into Y). The calculation is completed during the following two cycles. Multiplication is pipelined, allowing the next operation to be set up during the execution of the present one without affecting the ongoing calculation. While a single cycle multiplier could have been employed, our architecture did not warrant this. The single bus architecture of figure 2 was purposely chosen to limit the die size of the MAR. Dual operand instructions such as multiplication require two data transfer cycles to set up X and Y. In well structured calculations such as with filtering where multiplication is performed repeatedly, the two data transfers can occur concurrently with the completion of the ongoing multiply thus neccessitating only a two cycle multiplier to maintain a full pipeline. Figure 3 illustrates this point with the list of operations necessary to realize a tapped-delay line or finite impulse response digital filter.

Space does not permit a description of the utility of each of the MAR operations in DSP tasks. We will mention some of the unique features of the MAR.

A 'delayed' multiply operation is provided to maintain a full multiplier pipeline during array multiplication applications in which the intermediate results must be saved. Delayed multiplies are not executed at the time of the instruction but are later triggered by other functions. In particular, delayed multiplication is triggered by accumulator read operations while delayed multiply and accumulate is triggered by writing to the accumulator. Consider the Schur product of two 'n' component vectors A, B i.e.

$$P = \underline{A} \ O \ \underline{B} = (A_0 B_0 \quad A_1 B_1 \quad \ldots \quad A_{n-1} B_{n-1})$$

in which the result of each product A_i, B_i must be saved. Figure 4 provides a convenient implementation using delayed multiplications which requires only 3 cycles/component compared to the 4 cycles/component which would otherwise be required. The delayed multiply-accumulate operation is of similar value in calculating.

$$Z_{i+1} = Z_i + X_i Y_i \qquad i=0,1,2,\ldots,n-1$$

The MAR can also efficiently calculate metrics such as $(X-Y)^2$ and $|X-Y|$ using its 'DIFF' functions. Furthermore, MAX and MIN functions are provided to simplify the task of determining the largest or smallest values in an array. These functions identify the extreme values as well as their indices.

Two MAR's can operate simultaneously under the control of a single SPC as shown in figure 1. Notice that the F,C,A fields are common to each MAR i.e. both devices basically perform the same operation. If internal memory is used, the operands required by each MAR must occupy the same address. (The SPC can selectively idle one of the MARs on each instruction cycle.)

The dual MAR structure can be used to simultaneously perform two streams of real valued calcuations but is especially valuable for complex calculations. Generally, the real parts of a set of complex quantities are stored in one MAR while the imaginary parts are stored in the other MAR at the corresponding memory location. Therefore complex quantities can be readily added with a single ADD instruction performed concurrently in the two MARs. Complex multiplication is also readily handled as shown in figure 5. Notice that

$$(a_r + ia_i)(b_r + ib_i) = (a_r b_r - a_i b_i) + i(a_r b_i + a_i b_r)$$

The imaginary component of the product can be calculated as a multiply and add but the real component involves subtraction which is difficult to execute concurrently with addition since each MAR receives the same function code. This problem is obviated with a rich complement of instructions for loading the Y register. A single instruction allows loading the Y registers of the two MARs; however, one may be tagged with a sign reversal. Hence executing MAC in the two MARs will actually yield a 'multiply and subtract' in one of them.

The complex multiplication example also serves to demonstrate the data movement that is possible with an MAR. Inter MAR transfers are required to compute the cross terms that form the imaginary part of the product. Over 30 data transfer possibilities involving the external bus, internal bus, internal memory and accumulators may be selected using the control field.

THE SIGNAL PROCESSING CONTROLLER (SPC) CIRCUIT

The SPC, the second custom chip of the SPE, is a controller that executes a program written in its own powerful instruction set. Depicted in figure 6, the SPC, essentially has three external interfaces. First, the program memory interface is a dedicated parallel path to the SPE program store. Secondly, the SPE interface provides data and control information within the SPE. Sixteen bit data words are transferred on DB0-DB15 while function, control and address information is passed to the MARs and external devices via the F, C, A lines. Two edge triggered interrupts provide for exception processing. Finally, a unique host interface provides a low overhead communications link with the host processor. The host interface facilitates conversation between the two processors (host and SPE) without the need for special devices such as dual port memories. Furthermore, both data and commands may be sent between processors, thus allowing the host to manipulate SPE data structures and pointers.

VLSI Signal Processing

There are four major functional blocks in the SPC. They can communicate across the internal bus (IB). The first block is instruction decode and control. DSP tasks written in the SPE assembly language are stored in program memory. Instructions are fetched from the memory and decoded for execution. Each instruction invokes internal SPC operations such as updating data pointers. New F, C and A fields are also generated for the MARs. Instruction fetch, decode and execution is pipelined for enhanced throughput and better hardware utilization, i.e. while the nth instruction is being executed, the (n+1)th instruction is being decoded while the (n+2)nd is being fetched.

Program sequencing is performed in a separate block which contains the program counter, an eight level stack and hardware looping circuitry. A third block consisting of a 32 word register file (RF), arithmetic unit and special registers (SRF) is dedicated to data structure maintenance. The rather large RF is used for storing a multitude of address pointers that can be modified in a variety of ways to provide unique addressing modes. Finally, a unique host interface provides for a flexible communication strategy with the host processor.

It is impossible to detail all the features and intricacies of the foregoing blocks. We will simply highlight some of the unique addressing modes and program flow constructs and briefly describe the host interface. It will become apparent through these descriptions that the SPC architecture is designed to perform these 'housekeeping' chores with a minimum of overhead thus enhancing SPE throughput without increasing circuit speed.

ADDRESSING MODES AND DATA STRUCTURES

Data structures are resident in external data memory or in the MAR memories, local to where the calculations are performed. As a controller, a major task of the SPC is to provide address information to the MARs for accessing these structures when new data are added, old data are removed or existing data are manipulated. The data pointers are transparently updated.

The SPC provides a good complement of addressing modes, several of which are also common in microprocessors. Memory indirect addressing mode with post modification is of course the primary candidate for data structure manipulation, e.g. stepping through a table. In this addressing mode, one of the registers in the register file is used as an address pointer into the data structure. Accesses to the RF are generally of the read-modify-write variety due to pointer post modification. Different data structures can have different post modification schemes. The address pointer register is 16 bits wide but provides only 13 bits of address. The remaining three bits are programmed to the desired post modification mode for the structure to which the register is dedicated. The uniqueness and power of the SPC in signal processing tasks lies in the flexible post modification schemes. These include modification ± 1, $\pm N$ (where N is programmable) and ± 1 modulo M (where M is programmable).

VLSI Signal Processing

The ± 1 modulo M scheme is particularly useful in maintaining circular buffers which are very common in signal processing and data acquisition applications. A contiguous block of memory is reserved for this structure which operates as follows. Data words are entered into successive locations as acquired until the end of the reserved area is reached whereupon the address pointer is reset and new data words are entered starting at the beginning again. Data are unloaded from the buffer in a similar manner using another pointer. Most general purpose processors do not support such a mode. Therefore, it becomes necessary to test the address pointer value after each access for 'end of buffer', and reset the pointer if required.

However the SPC ± 1 mod M mode updates the address according to:

   new_address = (old_address - base ± 1) mod M + base

which automatically achieves the pointer reset without additional instructions. Up to four M-values may be stored in the four modulo registers of the SRF allowing the realization of buffers of different lengths. One might consider these post modification techniques as variations on the traditional theme of post inc/dec by unity. However, the performance enhancement they provide in time critical signal processing applications is substantial.

Another post modification mode sequences through an array in 'bit-reversed' order for FFT applications.

PROGRAM CONTROL STRUCTURES

We have identified data structure maintenance as a source of overhead in programmable machines and suggested several addressing modes specific to signal processing structures that circumvent this problem. Another major source of overhead is in the execution of program control structures such as loops, subroutine calls and interrupt handlers. These structures are necessary to conserve program memory (the alternative being inline code) and to encourage good programming style. The SPC provides hardware support to realize the benefits of such control constructs without incurring some of the speed penalties.

Many algorithms, not necessarily in signal processing, contain nested loops which occupy the majority of a processor's time. Each iteration of every loop incurs a several instruction cycle penalty [8] to decrement the loop counter, test for completion of the loop and branch back to the start of the loop. The SPC provides a single cycle LOOP instruction which decrements the loop counter and automatically generates a branch to the address stored on the stack. An RF register initialized to the loop count serves as a counter.

The innermost loop of a program is executed most frequently and even the single cycle penalty of the LOOP instruction can be substantial. Therefore, a FASTLOOP facility is provided which has two dedicated registers. One of these is the loop counter which is initialized prior to entering the loop. The second register contains the address of the last instruction in the loop. Whenever this address is reached, the loop counter is automatically decremented and the program counter is appropriately adjusted. The fast loop requires one additional instruction to set up the operation but does not incur the penalty of executing the LOOP instruction every iteration.

SPE-HOST COMMUNICATIONS

The Codex premium end user modem architecture of figure I consists of two processors, the host and SPE, operating in a loosely coupled fashion. Delivering data between a terminal and the telephone line requires much processing which is segmented among the two processors. Therefore, interprocessor communications on a regular basis is necessary and it is desirable to communicate with a minimum of speed and hardware overhead.

Traditional techniques for interprocessor communications include the use of interrupts, semaphore registers or dual port memories. While one processor loads data into a segment of the dual port memory, the other device can unload it. SPE-Host communications is realized without the need for such buffers by making the SPC host interface reside in the address space of the host device. The SPC may be addressed by the host at one of 64 addresses. Each address invokes the execution of a unique SPC instruction by stealing an instruction cycle from the SPE program at the end of the instruction currently in process. Upon completion of the host instruction, the SPE program is resumed.

Generally, host operations require the manipulation of data structures for passing both data and tasks between the two processors. Since the host exercises overall modem control, it maintains a signal processing task queue which is passed to the SPE to perform different algorithms. In effect, signal processing functions appear as high level macros (e.g. FILTER, EQUALIZER etc.) that can be started with a single host access to the SPC, while host instructions executed by the SPC typically update pointers and task queues or cause the SPC program to restart at a specific address.

CONCLUSIONS

We have described two custom chips, the MAR and SPC, which have been designed to facilitate complex signal processing tasks prevalent in high speed telephone line modems. Our approach has been to divide the modem tasks between two loosely-coupled processors: the SPE is devoted to the computationally intensive signal processing algorithms while the host performs control tasks. A flexible host interface provides an efficient communications link between the two processors.

The MAR architecture uses instruction pipelining and a powerful data movement structure to overcome some of the inefficiencies of a single bus architecture. Several special instructions, e.g. delayed multiplication, are key to maintaining a full pipeline during several common signal processing tasks. Throughput is enhanced by providing powerful instructions such as MAX that perform operations traditionally requiring several instructions. Furthermore, two MARs may be conveniently connected in parallel to perform complex arithmetic.

The SPC complements the MAR arithmetic capability with a powerful instruction set that facilitates the implementation of data and program control structures. Together, the SPC and MAR exploit several architectural innovations that improve the throughput of the signal processor. These features are essential in a programmable environment since instruction and control overhead often limit the utility of even the fastest arithmetic units. The SPE averages 2 million delivered operations (e.g. multiplications) per second for modem algorithms which compares favorably with the raw rate of 3.33 million/sec possible with the arithmetic unit. Our conclusion, therefore, is that signal processor architectures benefit from more than simply a fast multiplication facility. The SPE can be substantially enhanced with a single cycle multiplier supported by two operand buses using modern day process technology.

ACKNOWLEDGEMENTS

The new Codex premium end user modem product line, of which these chips are a part, is the result of the vision and efforts of many people both technical and managerial - we are merely the scribes. This paper is dedicated to the memory of our colleague, George Chamberlin.

REFERENCES

[1] Kawakami, Y. et al, "A Single Chip Digital Signal Processor for Voiceband Applications," Proc. of the Intl Solid State Circuits Conf., San Francisco, CA. Feb. 1980, pp 40-41.

[2] Boddie, J.R. et al, "A Digital Signal Processor for Telecommunications Applications," Proc. of the Intl Solid State Circuits Conf., San Francisco, CA. Feb. 1980, pp 44-45.

[3] Magar, S. et al, "A Microcomputer with Digital Signal Processing Capability," Proc. of the Intl Solid State Circuits Conf., San Francisco, CA. Feb. 1982, pp 32-33.

[4] Tsuda, T. et al, "A High Performance LSI Digital Signal Processor for Communication," Proc. of Intl Communication Conference, Boston, MA, June 1983, pp 187-191.

[5] Oppenheim, A.V., Shafer, R.W., Digital Signal Processing, Prentice Hall Inc.,

VLSI Signal Processing

[6] Lucky, R. W., Salz, J., Weldon, E.J., <u>Principles of Data Communications</u> McGraw Hill, 1968.

[7] Booth, A.D., "A Signed Binary Multiplication Technique," Quarterly Journal of Mech. and Appl. Math, Vol. 4, Part 2, 1951, pp. 236-240.

[8] Wakerly, J.F., <u>Microcomputer Architecture and Programming</u>, John Wiley, 1981

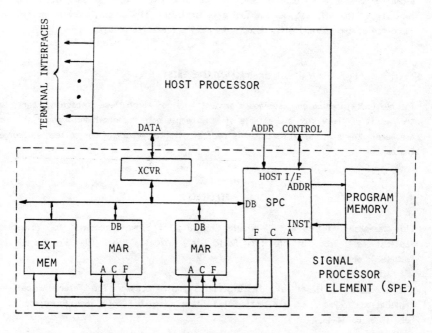

FIGURE 1: CODEX PEU ARCHITECTURE

VLSI Signal Processing

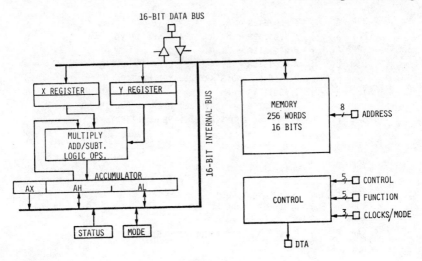

Figure 2: MAR Circuit

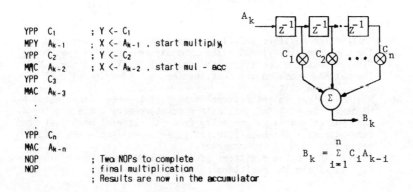

```
YPP  C₁       ; Y <- C₁
MPY  Aₖ₋₁     ; X <- Aₖ₋₁ , start multiply
YPP  C₂       ; Y <- C₂
MAC  Aₖ₋₂     ; X <- Aₖ₋₂ , start mul - acc
YPP  C₃
MAC  Aₖ₋₃
 .
 .
 .
YPP  Cₙ
MAC  Aₖ₋ₙ
NOP           ; Two NOPs to complete
NOP           ; final multiplication
              ; Results are now in the accumulator
```

$$B_k = \sum_{i=1}^{n} C_i A_{k-i}$$

FIGURE 3: TDL FILTER IMPLEMENTATION

VLSI Signal Processing

```
YPP   A₀         ; Y REGISTER ← A₀
MPY   B₀         ; X REGISTER ← B₀ AND START MULTIPLY
YPP   A₁         ; Y REGISTER ← A₁
DMP   B₁         ; X REGISTER ← B₁ AND SET DELAYED MULTIPLY
SAH   P₀         ; P₀ ← A₀B₀ AND START DELAYED MULTIPLY
YPP   A₂         ; Y REGISTER ← A₂
DMP   B₂         ; X REGISTER ← B₂ AND SET DELAYED MULTIPLY
SAH   P₁         ; P₁ ← A₁B₁ AND START DELAYED MULTIPLY
       ⋮
YPP   Aₙ
DMP   Bₙ
SAH   Pₙ₋₁      ; Pₙ₋₁ ← Aₙ₋₁ Bₙ₋₁
NOP              ; WAIT OUT FINAL MULTIPLY
NOP
SAH   Pₙ        ; Pₙ ← AₙBₙ; OPERATION COMPLETE
```

FIGURE 4: CALCULATION OF $\underline{P} = \underline{A} \; O \; \underline{B}$ WITH
DELAYED MULTIPLICATION OPERATIONS

```
YPP.P  MR(aᵣ)      ; YR,YI ← aᵣ
MPY.P  MP(bᵣ,bᵢ)   ; AR ← aᵣbᵣ    AI ← aᵣbᵢ
YPM.P  MI(aᵢ)      ; YR,YI ← aᵢ but YR tagged negative
MAC.R  MI(bᵢ)      ; AR ← aᵣbᵣ - aᵢbᵢ
MAC.I  MR(Bᵣ)      ; AI ← aᵣbᵢ + aᵢbᵣ
NOP                ; Results will be valid
NOP                ; after two cycles
```

<u>Legend</u>

o Instruction modifier .R(.I) denotes instruction is executed in
   the real (imaginary) MAR. The .P modifier signifies that both
   MARs perform the operation concurrently. 'Real (imaginary) MAR'
   is that MAR whose memory contains the real (imaginary) part of
   a complex quantity.
o MR (MI) refers to the internal memory of the real (imaginary) MAR.
   MP means the real MAR gets data from its memory while the imaginary MAR gets data concurrently from its internal memory at the
   same address. The quantities in parentheses denote the values
   being addressed in the memories.
o XR, YR, AR are the registers of the real MAR;
   XI, YI, AI are the registers of the imaginary MAR

Figure 5: Complex Multiplication with Two MARs

VLSI Signal Processing

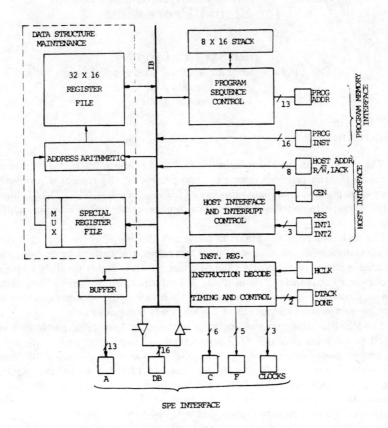

Figure 6: SPC Circuit

MSSP: A Bit-Serial Multiprocessor for Signal Processing

Richard F. Lyon
Fairchild Laboratory for Artificial Intelligence Research
4001 Miranda Ave.
Palo Alto, CA 94304

1 Introduction

The Multi-Serial Signal Processor (MSSP) is a single-instruction-stream multiple-data-stream (SIMD) multiprocessor for signal processing. Its replicated processor and memory element, the Serial Signal Processor (SSP), is a custom NMOS VLSI chip using bit-serial data paths, pipelined bit-serial arithmetic units, and a bit-serial multi-port memory.

The architecture of the MSSP machine and the design of the SSP chip combine novel features to mesh the implementation technology, the application domain, and the compilation software requirements. The resulting system is highly parallel and pipelined and achieves a computation rate of about 250 million 32-bit arithmetic operations per second at a cost several orders of magnitude lower than other approaches. Even better performing next-generation versions of the MSSP are being designed.

The MSSP architecture was originally proposed in April 1982, partly as a vehicle for VLSI tool development and VLSI instruction, and partly because our speech recognition project needed improved computing capabilities in order to speed up experiments with models of hearing [1]. In keeping with modern VLSI techniques and capabilities, the primary design goals for the MSSP architecture were simplicity and efficiency, with preference for simplicity due to a relative shortage of design resources. In particular, we wanted a high-performance system that required only one custom chip design. The resulting machine will provide low-cost numerical super-computing on a reasonable but restricted class of problems, which will allow it to satisfy the needs of many modern research and application areas.

2 MSSP Structural Description

This section describes the components from which the MSSP system is constructed, as an introduction to its architecture. At the top level, the MSSP consists of a controller and an array of processors (each with its own memory). The controller broadcasts an instruction stream and a data stream to the processors, and the processors communicate with a few directly-connected neighbors. The planned configuration, shown in Figure 1, is a one-dimensional nearest-neighbor-connected array of 100 processors, with the end processors also communicating with an I/O section in the controller. The one-board controller communicates with a general-purpose host computer via a MultiBus interface.

The replicated processor is the Serial Signal Processor (SSP), a single 5 mm square NMOS chip with 4-micron features. The SSP consists of a multi-port memory and a small collection of arithmetic function units connected to the memory ports. The current SSP

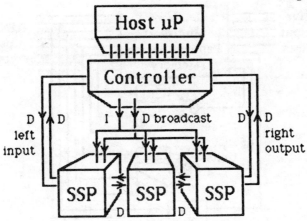

Figure 1: Top-level block diagram of the MSSP machine.

configuration, shown in Figure 2, has a pair of multiply/add/limit units and a general two-input arithmetic-logic unit (ALU). The memory has sixty-four 32-bit words, eight input ports, and eight output ports.

All data interconnections, within and between SSP chips, are made through memory ports, on single-wire paths. The eight memory output ports connect directly to the eight inputs of the function units and also go off-chip for sending data to neighbor chips. Three memory input ports take results from the three function units, and the other five memory input ports take data from chip input pins, which may be connected to the outputs of other chips. With the five available uncommitted input ports, several array interconnection schemes are possible, including grids of one or two dimensions.

The microcode broadcast serially on five wires from the controller to the processors consists mainly of the read and write addresses for all the memory ports, but also includes a few bits for ALU control, etc. Addresses determine which data items are to be sent to each function unit, and where the results from the function units are to be written back into memory. Other than memory addressing, no data path steering is provided.

The MSSP is designed to operate with a 16 MHz clock, and has a major cycle (word time) of 32 clock cycles, or 2 μs. In that time, each SSP chip executes eight memory reads, eight memory writes, two multiplications, up to three adds or subtracts, and some input and output transfers. Output to a specific off-chip port may conflict with the use of a function unit connected to the same memory port.

3 MSSP Architectural Oddities

The name Multi-Serial Signal Processor (MSSP) is meant to convey the primary unusual aspects of the MSSP architecture: it is a bit-serial machine, uses multiple processors, and is specialized for regular numeric applications such as signal processing.

The project goals and prejudices led to several features of the VLSI chip architecture and the system architecture that are atypical for a bit-serial machine, or for a single-instruction-stream/multiple-data-stream (SIMD) multiprocessor, or for a machine

VLSI Signal Processing

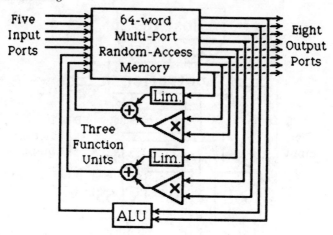

Figure 2: Processor element (SSP chip) block diagram.

specialized for signal processing.

Unlike most bit-serial signal processors, the SSP element is programmable in a fairly conventional way, not hard-wired for a particular class of operations. The MSSP is an experiment on the suitability of the bit-serial approach to programmable numerical computing, with the potential payoff of excellent cost/performance.

In contrast to massively parallel SIMD array machines such as the CLIP4 [2] that use huge two-dimensional arrays of slow one-bit processors, MSSP is designed to use a modest number of relatively powerful processors with a large fixed data word size. Many applications can effectively use a one-dimensional array of such processors, while two-dimensional systems are usually not good for much beyond image processing.

In multiprocessors, it is particularly unusual for the individual processor elements to have multiple concurrently operating function units, like the SSP's two multipliers and three adders. In dedicated bit-serial systems, multiple levels of functional parallelism are more common [3], and can more than make up for the missing parallelism at the bit level.

Unlike some SIMD machines, MSSP provides no shared memory and no switching network (but has only a few hard-wired data paths between elements), and provides only a bare minimum of control logic and no specialized registers in the replicated elements. Thus each processor-memory element, while computationally powerful, is little more than a numeric data-path. This simplicity is not viewed as a defect, but is important in that it leads to good overall system efficiency.

Unlike many architectures specialized for digital signal processing (DSP) applications, MSSP is not a fixed-throughput machine with arbitrary limitations on resources, but rather a modular expandable architecture that can be matched to the application. And rather than embodying stringent restrictions such as coefficient/data segregation as is found in many DSP architectures, MSSP allows flexible programmable use of arbitrary numeric quantities.

Importantly, MSSP has not followed the familiar one-chip solution paradigm common to many other signal processing efforts, and so has not needed to push the chip technology or circuit design art to get a high-performance low-cost system. Better chip technology will of course lead immediately to better MSSP systems.

The most novel structural component of the SSP is the multi-port serial memory, which acts as both the bus and the main memory. As in TRW's unusual *polycyclic* architecture [4], it allows full function unit interconnectivity with programmable delays and no contention, but it does so with lower cost and more generality than the polycyclic interconnect scheme.

These and other unusual aspects of the architecture have a certain harmony with each other and with the speech processing applications that MSSP is intended to serve. It remains to be seen how easy it is to map problems from new application areas efficiently onto this architecture. The biggest constraint on suitable application areas is the single instruction stream—controlling all the processors in lock-step on different data, without specific support for conditional execution.

4 The Programmer's View of MSSP

The MSSP is a vector processing machine, with no support for scalar operations or for control structures other than fixed-count loops. The programmer is expected to design and code parallel algorithms accordingly. But an important goal of this project is to hide low-level architectural features, such as details of the pipelining and the parallel resources and the bit-serial nature of the machine, from the programmer. The processing element, and thus the vector machine, has a simple and familiar programmer's model, while maintaining efficient use of pipelined and functionally parallel resources.

The SSP chip is a very simple processor-memory element that can execute only one type of operation: assignment with a primitive expression. That is, the assignment $A := f(B, C, ...)$ is the only statement type, where A, B, and C are variable names referring to locations in the processor's small memory space, and $f(\cdot)$ is a primitive function. Composing useful functions from the primitives requires that variables be allocated for the intermediate results (there is no accumulator, stack, or other temporary register space). Thus the entire state of any computation is in main memory.

The most often used primitive function is a multiply-add: $A := B * C + D$. Other options, such as negation of either term, are provided, as are a reasonable collection of ALU functions for logical, nonlinear, or conditional operations. The programmer's view is that instructions of this form execute sequentially, in the order written, with each one finishing its assignment before the next one starts. At a higher level, depending on the support software, the programmer can also deal with looping constructs, multiple independent or synchronized tasks, and so on. Since the overall architecture is SIMD, these issues are dealt with by the compiler, the controller, or the host uniprocessor, rather than by individual processor-memory elements.

The only data type handled by the SSP is a 32-bit fixed-point two's-complement number, normally regarded as a fraction with 28 bits to the right of the point. The ALU also allows ways of using numbers as flags and masks.

The initial implementation of the SSP chip has only 64 memory locations, but due to dynamic assignment of locations to variables, more than 64 variables might appear in a program; locally-scoped intermediate result variables with different names will automatically share memory locations when appropriate. So, while the programmer does not need

VLSI Signal Processing

to limit himself to a known total number of variables, he does need to know that he can run out of space.

The programmer will need to be aware, to some extent, of the interconnection topology of the multiple SSP chips. The plan is to have only a linear two-way nearest neighbor interconnection initially, with an overlaid tree interconnection pattern added later to take advantage of the planned memory expansion. Assignment statements like $A := left(B)$ would be written to coordinate I/O transfers between SSP chips; A is assigned the value of B from its left neighbor. The function $left$ is not really a functional operation on the value of variable B, but rather a macro that commands the element to read the value of variable B and send it out toward the right, and take what comes in from the left and assign it to variable A. Thus all SSP chips in the linear array simultaneously transfer data from left to right.

The only way to move data in memory is to do an assignment through a function unit; for example, $A := B$ will compile to move the value of B through a multiplier/adder or the ALU, depending on which is available. There is no way to index memory by data values either, so to implement a delay line in memory it is necessary to explicitly move the data or to unroll a program loop to use different locations in a ring fashion.

A useful primitive function that has been provided in the ALU hardware is $A := select(B, C)$, which selects B if B is non-negative, and C otherwise. This operation can be used for half-wave and full-wave rectification, limiters, break-point nonlinearities, conditional assignment, etc. A value used for B might be the result of a data operation, or it might be a preloaded constant dependent only on the location of the processor in the array, for example to control edge conditions.

5 The SSP Chip Design

The area of the SSP chip is dominated by two large blocks, the memory and the pair of multipliers, as shown in Figure 3. Both of these evolved from previous memory and multiplier blocks that were designed for the *Filters* chip at Xerox PARC, and for a subsequent family of dedicated special-function signal processing chips [5]. Similar multiplier and memory components have been adapted by Denyer *et al.* to a silicon compiler for automatic generation of special-function signal processing chips [6]. In attempting to generalize the functionality of such chips to a wider class of signal processing operations, it was finally realized that the simplest thing to do was to go all the way, and make it completely general by making all connections programmable, through the memory. The SSP chip is the result.

The memory design was a serial-parallel-serial organization of three-transistor RAM cells, which in the Filters chip was addressed strictly sequentially. Multi-porting it involved splitting the data bus into separate read and write busses, and extending each through an array of port registers. The selection logic was replaced by separate read and write address decoders to make it random-access with independent read and write capability. Thus, on each memory cycle, a read and a write can occur concurrently. Eight memory cycles need to fit in thirty-two clock cycles (one major cycle), so each memory cycle can take up to four clock cycles—two to precharge the bus and two to drive and latch data. This 4:1 timing ratio seems to be about the right relationship between the speed of fast serial logic and the speed of long parallel busses without sense amplifiers. The

VLSI Signal Processing

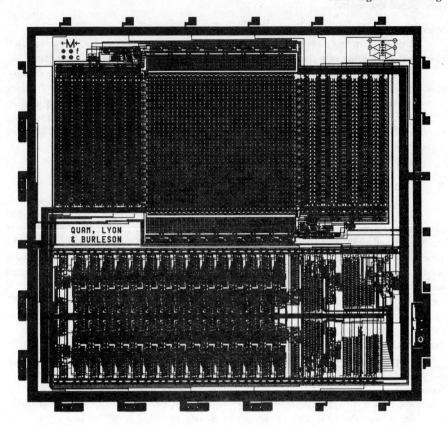

Figure 3: The SSP chip layout, with memory on top, function units below.

resulting memory has a total read/write bandwidth of 16 ports * 16 Mbps = 256 Mbps, with fixed latency and no contention.

The multipliers are based on the 5-level recoded (modified Booth's algorithm) version of the two's complement serial pipelined multiplier [7]. The layout was redone using a pair of small optimized PLA's per stage; one recoder stage, fifteen identical multiplier stages, and a final scaling stage make up the complete multiplier. The resulting structure computes truncated 32-by-32-bit products as fast as operands can be pumped into it, with a pipeline latency of about a word and a half. The scaling is adjusted so that operands have a value range of −8 to +8 (i.e. inputs are regarded as fractions with 4 bits to the left of the point, including the sign). No other scaling mechanisms, such as shifters, are provided.

The multiplier stage cells are laid out on a pitch of only 100λ, with a height of 408λ—about a third smaller than the original [8], and somewhat faster, too. The smaller functional subunits, such as limiters, sign switchers, adders, various ALU blocks, and a control-bit latch, were designed on the same grid of power, ground, and clock distribution as the multipliers. A modest number of random wires (the hard part of the design)

VLSI Signal Processing

interconnect the pieces.

A ring of bonding pads for inputs, outputs, power, and clocks was laid out and routed algorithmically using new parameterized pad designs and other custom DPL functions [9].

The layout was converted to CIF2.0 [10], design rule checked, circuit extracted, and thoroughly simulated before fabrication. Chips were fabricated by VTI's foundry service in the process they call HMOS, which is compatible with Mead-Conway NMOS with $\lambda = 2~\mu$m. Both buried and butting contacts were used, with no problems, using conservative simplified design rules [11]. The chips are just over 5 mm on each side, and have good yield.

6 Controller Design for MSSP

The controller for the MSSP machine must generate an 80 Mbps instruction stream (160 bits per major cycle), and interface three data streams from the host to the processors (to the left and right ends and a broadcast to all processors) and two data streams from the processors to the host (from the end processors). Many levels of capability can be imagined for the design of such a controller. For example, the controller can do runtime resource assignment, subroutine calling, loop counting, code rearrangement, multi-tasking, etc; it could do tests on data from the processors, and alter the control flow accordingly. We chose to start with a very simple design, interfacing to a MultiBus host, and providing no extra control features. This means we need a way to provide at least basic control features at compile time or through the host at run-time.

To accommodate loops, we simply generate code for unwound loops. Multiple looping tasks that run synchronously but at different rates, such as a sample-rate task and a frame-rate task in a speech analyzer, are jointly unrolled into one long program segment whose repetition rate is a common divisor of the repetition rates of the individual tasks.

Since this produces long sequences of mostly repeated code, we use a two-level memory scheme in the controller so that repeated instructions need only be stored once (in the *code memory*), and only indices into the code memory are repeated, in the *program memory*. To accommodate long loops, the program memory has 16K locations. Each location provides a 9-bit index into a table of only 512 distinct instructions, each 160 bits wide.

Each program memory location also contains an enable bit for each of the five I/O streams, and a stop bit. Linear program segment executions are initiated by the host (by loading the program counter and issuing a run command), and are terminated when a stop bit is encountered. The host must execute run-time support code that coordinates segment execution and I/O. The controller provides hardware FIFO's on the I/O streams to reduce the timing demands on the host.

A key feature of the architecture that allows multiple tasks to be jointly unrolled is the lack of special registers, such as accumulators or flag registers, that would otherwise cause conflicts between concurrently executing tasks. A smarter controller would take advantage of this feature by determining at runtime, for multiple asynchronous tasks, which task could run on each major cycle, based on pipeline constraints between program steps. There is no overhead in switching between tasks in this way. An even smarter controller might mix primitive operations from several tasks into a single instruction cycle, with dynamic assignment of actual function units to primitive operations. The sequencer in such a controller would have time for several steps per major cycle, and could implement other features, such as subroutines, scalar data operations, and conditional branches as well.

7 MSSP Support Software System

A Lisp-based compilation system is deing developed to translate high-level code for signal processing algorithms into microcode for the MSSP. Presently, the compiler accepts an intermediate-level of code that is a direct analog of the sequential assignment programming model described above. Generating that level from other source languages is straightforward, as long as the source-level constructs used are restricted to fit the numerical domain closely enough. Application-dependent Lisp-embedded languages are a natural approach.

The compilation scheme is inspired by the MIPS (Microprocessor without Interlocked Pipe Stages) project at Stanford [12], in which it was shown how moving the complexity of a pipelined machine from the hardware to the software could result in significant net performance gains. The compiler uses a *pipeline rearranger* with knowledge of the hardware constraints to convert sequential high-level programming constructs to legal and efficient overlapped machine operations.

Sequential program segments, or *basic blocks*, are independently compiled, then later loaded into the controller's memories. Each step of a sequential program is explicitly represented as a node in a timing dependency graph. Depending on what variables a step reads or writes, and depending on the pipeline latency of the primitive function used, one-sided timing constraints are calculated and entered as edges between the nodes. Each edge specifies how soon a step may be started relative to when a previous step was started. Based on this information, steps can be assigned to function units and code can be packed in time into an instruction stream segment, accommodating whatever parallelism is found.

A critical-path finding algorithm and first-fit packing strategy yield nearly optimal code; more clever rearranging algorithms could yield a slight improvement in some cases. Typically, if only one task has been unrolled into a segment, many potential uses of function units will remain as effective no-ops, due to the long pipeline latencies. When multiple nearly-independent tasks are jointly unrolled, most no-ops can be converted to useful parallel operations.

The definition of the algorithm at the high or intermediate level can have a big impact on how much of the machine resources can be used concurrently. For example, a straightforward FIR filter implementation that uses a memory location as an accumulator will have to wait several cycles between multiply-adds, but one that uses a tree of adds will do as many parallel and overlapped multiply-adds as the hardware allows (with some cost in terms of more memory usage).

8 A Sample MSSP Program

To implement a filterbank model of wave propagation in the cochlea, we can use a structure as simple as a cascade of second-order canonic filter sections [1]. Assuming we build a machine with at least as many processors as model channels, each SSP chip will run one second-order filter.

VLSI Signal Processing

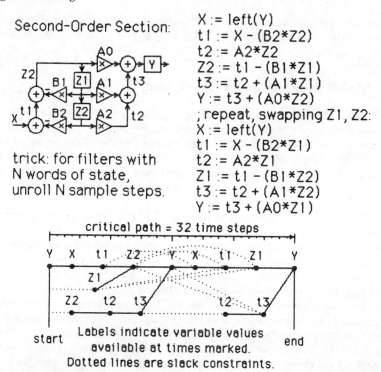

Figure 4: Signal flow graph, sequential assignment program, and instruction graph for the second-order filter section example.

Each stage takes input from its left neighbor and sends the filter output to its right neighbor (each output would also be used within the SSP, as input to further stages of the hearing models). The left-most SSP gets its input from the controller's *left-in* data stream, and the right-most SSP sends output to the controller's *right-out* data port, which ignores it. Before the filters can be run, coefficients must be loaded into the memory, following the same left-to-right path. This section presents the code for the filtering basic block only. The basic block is written to compute two consecutive samples, rather than just one, in order to avoid moving data in the filter's state memory.

Figure 4 shows the signal flow graph, the sequential assignment program, and the instruction constraint graph for the second-order filter section example. The graph is drawn with nodes placed on a time line in the locations where the compiler assigned them. The forward-backward critical path finding and assignment algorithm used puts every instruction as close to the end as possible; the ones with slack could equally well have been put as close to the beginning as possible, or anywhere in between.

The local variables $t1$ and $t3$ can both refer to the same memory location, as their regions of definition do not overlap in time. Similarly, Y, X, and $t2$ can share a location. Thus this program uses 5 locations for coefficients, and 4 locations for data.

This program has been compiled and simulated on our software emulator, yielding exactly the correct numerical output. It will soon be run on the actual MSSP hardware. Initially, we will run a reduced MSSP configuration, so we will rewrite the code slightly to execute several channels per processor. We may also enhance the compiler to automatically map multiple channels of the task to each processor; for example, in the case of two stages per processor, occurrences of $X := left(Y)$ would be converted to $X1 := left(Y2)$ and $X2 := Y1$, and the rest of the code would simply be duplicated with copies 1 and 2 of the variables. There would be no time penalty for running two (or possibly up to six) channels per processor, since the stages would be completely overlapped.

9 MSSP Performance

The above example program executes one second-order section step (5 multiplies and 4 adds) per processor in a loop that repeats every 16 major cycles (32 μs); this is faster than real-time for signal sample rates below 31 kHz. The resulting computation rate, in terms of arithmetic operations on *fractions*, is $9/32 = 0.28$ Mega-frops per processor. The maximum rate possible on one processor is 5 arithmetic operations in 2 μs, or 2.5 Mega-frops (counting multiplies and adds as operations, but not counting sign changes or hard limiting, which can occur concurrently on the operands). Thus, this single task runs the machine at an efficiency of only $0.28/2.5 = 11\%$.

The efficiency of multiplier usage is $[(5 \text{ mults})/(32 \text{ } \mu s)]/(1.0 \text{ Mega-frops}) = 16\%$; in no case are both multipliers in use at once. The other 84% of the capability of the multipliers, and other unused resources, is still available for use by other stages of processing that are compiled to run concurrently with the filtering. For example, if five filter sections are run per processor, the repetition cycle is not lengthened, and the efficiency increases to $25/32 = 78\%$ of multiplier usage.

In our 100-processor machine, designed to match the number of channels of cochlear models that we would like to implement, the total machine performance can be up to 250 Mega-frops. Even at 11% efficiency, the performance beats more expensive array processors by a reasonable factor (28 Mega-frops, vs. less than 12 Mega-flops for an FPS AP-120B); when used efficiently, the MSSP is another order of magnitude better. As usual with array processors, the performance bottleneck will be I/O, and the small memory of the MSSP will exacerbate this problem.

10 MSSP Extensions

To extend the usefulness of the MSSP machine, two stages of extensions are being considered. The first uses the existing processor chips, and augments each with an external 64K-by-1 fast static RAM. The RAM data-in and data-out pins can connect to serial input and output ports, and broadcast address and timing control from a modified controller can then allow either a single memory read or memory write on each major cycle. Although the bandwidth to this 2K words of secondary memory is less than the bandwidth to on-chip memory by a factor of sixteen, the extra space provides a great increase in capabilities.

The first extension plan also includes adding another pattern of interconnect between processors, to allow quick accumulation of values from all processors through a tree of adds. This makes the machine much more useful for spatial convolution, transforms,

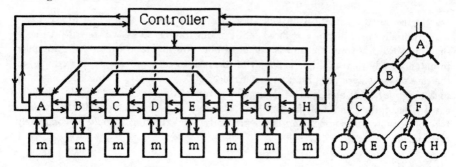

Figure 5: Enhanced MSSP machine with added memory and tree interconnect.

and pattern matching. The planned technique maps processors from left to right onto a tree by pre-order traversal (so that each node is adjacent to its left son if it has one), and requires only one new wire into each node to bring the input from the (non-local) right son node; input from the left-son uses the existing wire from the right neighbor. The resulting configuration is illustrated in Figure 5.

The addition of one wire from external memory and one tree connection wire brings the total number of inputs to a processor up to five (counting left, right, and broadcast inputs), using up all available input ports.

The second stage of extension involves the design of a next-generation processor, which may be a high-performance CMOS chip connected on a hybrid substrate to four 8K-by-8 fast static RAMs. If the RAMs could be cycled in 30 ns, this processor could have 32 memory ports into an array of 8K 32-bit words, with about 8 multipliers and 12 adders, at a bit rate of 32 Mbps.

An array of 100 of these would provide up to 2 Gigafrops, and still fit on a few small boards. With the large memory, the tremendous parallelism of arithmetic resources will be easy to use efficiently in applications such as pattern matching, matrix multiplication, image processing, Fourier transforms, and channel-organized hearing models of the sort for which the architecture was originally designed.

11 Conclusions

The MSSP architecture promises to provide a family of high-performance numerical processing machines, for applications that can utilize the SIMD mode efficiently. We must continue the experiment to see if this promise is actually realized in the case of speech analysis, and then to see how it extends to other applications. If the machine turns out to be as useful and efficient as predicted, we will have demonstrated that the efficiency advantages of bit-serial arithmetic in VLSI extend beyond special-purpose machines to programmable machines. We are highly motivated to finish this experimental development, including the next-generation machine, as the result will be several orders of magnitude improvement in our ability to analyze speech signals through our hearing-model algorithms.

12 Acknowledgments

The MSSP project has benefitted from a number of eager migrant workers who have been willing to go along with the author's odd ideas and add good ideas of their own. Lynn Quam (now at SRI) and Wayne Burleson (MIT co-op student, now at VTI) did the detailed chip design, layout, checking, and simulation. Ian Robinson (Fairchild AI Architecture group) did the controller design, construction, and checkout. Niels Lauritzen (MIT co-op student) developed the support software system on the Lisp machine. Bruce Horn (consultant) wrote diagnostic and support code to run on the host Sun workstation. Steve Rubin (Fairchild AI Architecture group) supported a variety of VLSI tools software locally, and helped us get chips fabricated.

13 References

[1] Richard F. Lyon, "A Computational Model of Filtering, Detection, and Compression in the Cochlea", *Proc. IEEE Intl. Conf. on Acoustics, Speech, and Signal Processing*, Paris, May 1982.

[2] Michael J. B. Duff, "Review of the CLIP4 Image Processing System", *Proc. AFIPS Natl. Comput. Conf.* **47**, 1978.

[3] Noble R. Powell, "Functional Parallelism in VLSI Systems and Computations", *VLSI Systems and Computations* (CMU Conf. Proc., ed. by H. T. Kung and Guy Steele), Computer Science Press, 1981.

[4] B. Ramakrishna Rau, Christopher D. Glaeser, and Raymond L. Picard, "Efficient Code Generation for Horizontal Architectures: Compiler Techniques and Architectural Support", *Proc. 9th Annual IEEE Symp. on Computer Architecture*, April 1982.

[5] Richard F. Lyon, "A Bit-Serial VLSI Architectural Methodology for Signal Processing", *VLSI 81 Very Large Scale Integration*, (Conf. Proc., Edinburgh, Scotland, John P. Gray, editor), Academic Press, August 1981.

[6] Peter B. Denyer and David Renshaw, "Case Studies in VLSI Signal Processing using a Silicon Compiler", *Proc. IEEE Intl. Conf. on Acoustics, Speech, and Signal Processing*, Boston, April 1983.

[7] Richard F. Lyon, "Two's Complement Pipeline Multipliers", *IEEE Trans. on Communications*, April 1976.

[8] John Newkirk and Robert Mathews, *The VLSI Designer's Library*, Addison Wesley, 1983.

[9] John Batali, Neil Mayle, Howard Shrobe, Gerald Sussman, and Daniel Weise, "The DPL/Daedalus Design Environment", *VLSI 81 Very Large Scale Integration*, (Conf. Proc., Edinburgh, Scotland, John P. Gray, editor), Academic Press, August 1981.

[10] R. F. Sproull and R. F. Lyon, "The Caltech Intermediate Form for LSI Layout Description", in *Introduction to VLSI Systems*, by Mead and Conway, Addison Wesley, 1980.

[11] Richard F. Lyon, "Simplified Design Rules for VLSI Layouts", *Lambda Magazine* (now *VLSI Design Magazine*), First quarter 1981.

[12] John Hennessy, Norman Jouppi, Forest Baskett, and John Gill, "MIPS: A VLSI Processor Architecture", *VLSI Systems and Computations* (CMU Conf. Proc., ed. by H. T. Kung and Guy Steele), Computer Science Press, 1981.

VLSI Signal Processing

11

A 2 MICRON CMOS, 10 MHz, MICROPROGRAMMABLE VECTOR PROCESSING UNIT WITH ON-CHIP 3-PORT REGISTERFILE FOR INCORPORATION IN SINGLE CHIP GENERAL PURPOSE DIGITAL SIGNAL PROCESSORS

Frank P. Welten[1], Antoine Delaruelle[1], Frans J. van Wijk[1],
Jef L. van Meerbergen[1], Josef Schmid[2], Klaus Rinner[2]

[1]Philips Research Laboratories,
P.O.Box 80000,
5600 JA Eindhoven
The Netherlands

[2]TEKADE Fernmeldeanlagen,
Postfach 4943,
8500 Nürnberg 1
West Germany

Abstract

In this paper a 2 micron CMOS, Microprogrammable Signal Processor Core (SPC) is described, intended as the number crunching unit in general purpose digital signal processors. This core contains a 16x16 bit parallel multiplier, a 40 bit multiprecision accumulator, a 40-to-32 bit extractor, an overflow detection unit, a format-adjuster, and a 3-port registerfile for local storage of 15 operands. Its 100 nsec. throughput rates makes it highly suitable for signal processing systems with sample rates up to 50 kHz (Speech, Telecom and HiFi Audio).

The architecture of this unit is discussed in detail. The design approach, using full-custom cells, bit-sliced functional blocks, a complete bottom-up logical verification of mask data, will also be described.

The stripped Signal Processor Core contains 18.600 transistors on a 15.5 mm^2. area. This compares with a packing density of 1200 transistors/mm^2.

1. INTRODUCTION

Despite the vast amount of research in, and the growing importance of, digital signal processing (DSP), the application of general purpose DSP's has not fully come of age. They have reached the same point in their evolution, that the microprocessors had attained around 1975, when the first 8-bit general purpose uP's were developed and put on the market, but the more advanced types had yet to appear. The appearance of these devices has finally changed the cost/performance ratio radically, in such a way that they are now economically feasible in markets like speech recognition, mobile automatic telephony (MAT) [5], modems, and HiFi Audio. These markets were up to now mostly served by either analog systems or by customized digital components.

Developing custom signal processors [1] for all off the large number in-house applications within PHILIPS requires and enormous development effort. Moreover general purpose signal processors, supported by a full set of software and hardware development tools are now competitive for applications with sample rates up to 50 kHz.

In this paper the Signal Processor Core (SPC), to be used as the number cruncher in these DSP's - in particular a 16 bit version - will be discussed in detail. The architecture of the SPC was developped after intensive analysis of various algorithms within the speech-, telecom- and HiFi audio application areas. As an example we mention spectral analysis (FFT, DFT), windowing, minimum distance computation, digital filtering (2nd order, multiprecision IIR and FIR filter types), matrix computation (Levinson recursion), and certain adaptive algorithms.

Section 2 gives a detailed description of the SPC core and its arithmetic features. In section 3 we discuss the use of microprogramming techniques to adapt this core to a specific environment. A design-rule independent silicon compiler for the microcode decoding PLA's is used in this context. The most important issue in VLSI design is to deal with complexity. Moreover the design of digital signal processing devices raises stringent requirements for high data throughput. If on top of that also the smallest chip area is desired (high-volume applications), together with low-power consumption (CMOS for markets like mobile automatic telephony), then the design- and verification tools and methods should be chosen with care. These will also be discussed in this section. This will be followed by section 4 on technology issues such as design rules, and choice of gate-lenghts and threshold implants. Since the SPC is currently being processed, we will give the projected performance figures here.

2. PROCESSOR CORE ARCHITECTURE

The architecture of the Signal Processor Core is shown in Fig. 1. It consists of three major functional blocks i.e. a 16x16 bit Booth multiplier and 40-bit accumulator (1), a 40-to-32 bit extractor and formatadjuster (2), and a 3-port, 15-word registerfile (3). The SPC is connected to other functional blocks within a particular signal processor by way of one or two data buses (X-bus and Y-bus), a control bus (C-bus), and a set of flags, located in the SPC status register (PST).

Both databusses are 16 bit wide. The C-bus consists of a 6-bit field, which controls the multiplier-accumulator operation and which is fed to the function select PLA, two source fields (SX and SY), and a destination field (DX).

The source fields SX and SY control which register may write on the X-bus and/or Y-bus (R1 - R15, MSP, LSP, PST). The DX field determines which register (PST, BSR, R1 - R15) may read data from the X-bus. The multiplier-accumulator is always performing a particular function, determined by the previously mentioned 6-bit control field, unless it is set in an idle state by two opcodebits C0 and C1. The control fields have been chosen in such a way that they ar easily incorporated into several types of horizontally microcoded instruction words, which are typical for Harvard-type signal processors.

The operands for the multiplier-accumulator either come from outside the core by way of the X-bus and/or Y-bus, or are stored locally in the 3-port register file. Addressing two operands in the register file, reading those onto the X-bus and Y-bus, multiplying both and storing the result in the product register (PR) takes one basic processor cycle. In the next cycle this product PR is accumulated with a previous intermediate result, which

VLSI Signal Processing

is stored in the accumulator register ACR. From the 40-bit result ACC contiguous 32-bit word is extracted, and stored, after format adjusting, in the MSP and LSP registers. Hence there is only one pipeline dealy from input to output of the SPC. We feel this is nice optimal solution for the dilemma, which is always raised when specifying high-throughput general purpose signal processors, where a lot of algorithms consist of recursive operations, but also a lot of non-recursive operations.

Generality is the keyword here, and it results from the use of loadable microcontrol fields and advanced hardware implementation of various, often used arithmetic subfunctions. These issues will now be discussed in more detail in the following paragraphs.

2.1 Multiplier-Accumulator

This unit (Fig. 2) consists of a 16x16 bit multiplier (M), using a modified Booth algorithm and carry-look-ahead techniques, a 40-bit accumulator (A) and 3 logic units at the multiplier inputs Xin and Yin and at the accumulator register output ACR. These units implement in hardware various subfunctions, which are used extensively while executing certain classes of vector product operations.

Let us now consider these functions in detail. At the multiplier inputs there are two registers MXL and MYL, which are either fully transparant or keeping their previous values. At the inputs of these registers are input selectors ILX and ILY. Under control of the 4 signals the following input operands can be selected:

```
Xin: X-bus, -1, MXL
Yin: Y-bus, -(Y-bus), MYL
```

Note that the selection of -(Y-bus) at the Yin input does not require two's complementing of the data on the Y-bus with an extra ripple-adder. This follows from the decoding schemes within the Booth recoding logic.

The accumulator also contains logic circuitry to facilitate negative and positive accumulations, either single or multiple precision. In this context multiprecision words are represented by 16+15n bits, where n= 1,2,3,..... The accumulator unit performs one of the following subfunctions:

```
ACC = PR
ACC = PR + ACR              Single precision
ACC = PR - ACR              accumulation
ACC = PR + ACR.2^-15        multiple precision
ACC = PR - ACR.2^-15        accumulation
```

where the product PR can be one of the following eight:

```
+(X-bus.Y-bus)     +(MXL.MYL)
-(X-bus.Y-bus)     + Y-bus
+(MXL.Y-bus)       -(Y-bus)
-(MXL.Y-bus)       -(MYL)
+(X-bus.MYL)
```

Indeed, this is a multiplieraccumulator which supports multiprecision operations in hardware. The multiplier itself is a rather straightforward implementation of the modified-Booth algorithm [2], although the recoding logic at the Yin input and the bit-select multiplexers (see Fig. 4) differ from those of the known multipliers: instead of recoding 3 successive bits Yn-1, Yn and Yn+1 into 3 controlsignals, we use 4 control signals (Cx, Cxn, C2x, and C2xn), denoting the selection of Xi, -(Xi), Xi-1 or -(Xi-1). When none of the control signals is active the output of the multiplexer row is zero. This implementation choice proved to be advantageous, when optimizing the critical delay path, since both the uppermost Booth recoder and the top row of multiplexers are in this path.

To increase the multiplier speed even more a carry-save full adder scheme is used in the matrix, together with a single level of carry look ahead over 3 bits in the bottom-row adder, as shown schematically in Fig. 3. This carry look ahead scheme was also used in the 40-bit accumulator adder.

Since the multiplier delivers a 32-bit product, there are 8 overflow bits in the accumulator. Hence at least 256 product accumulations of single precision data are required to generate overflow. Hardware flags are set in the Processor Status Register (PST), denoting sign of the accumulator and status of the accumulator overflow detection hardware (see Fig. 1).

2.2 Extractor and Format-Adjuster

To facilitate the post scaling and selection of product accumulation results, without having to use datadependent single-bit shifts, a special logic unit was conceived (Fig. 4). It consists of a barrelshifter selection matrix (EXTR), a format-adjuster (FORM), and an Extractor out-of-range detection unit (OOR). Furthermore there is a writable control register (BSR), which determines the setting of both the extractor and format adjuster. The extractor selects a 32-bit word (E31-E0) out of the 40-bit accumulator output ACC39-ACC0, by using one of the 16 selection schemes shown in Fig. 5. Note that here 9 extra bits to the left of the MSB of the selected 32 bit word are also extracted from the accumulator output ACC. These 9 bits, together with bit E31 as LSB in a 10-bit word, are fed to the extractor out-of-range detection unit. This unit also sets a flag in the processor status register (PST). (All 3 flags in the PST may be fed to a microprogrammed branch PLA, outside of this SPC).

Next the extracted 32-bit word E31-E0 is routed to the format adjuster, which can rearrange the bits in the following manner:

```
FM15 ...... FM0    FL15 ...... FL0
E31  ...... E16    E15  ...... E0
E31  ...... E16    0 E15 ..... E1
E31  ...... E16    E0   ...... E15
```

VLSI Signal Processing

Both 16-bit words FM and FL are then sampled by the MSP and LSP results registers, each of which can feed their contents to the X-bus and/or Y-bus, under control of the SX and SY fields at the beginning of the next processor cycle.

2.3 Three-port registerfile

Careful analysis of numerous algorithms has learned us that a fairly high amount of the basic operations need some form of arithmetic logic unit. These operations are generally of the register-to-register type. Hence, to improve the overall performance of these "spaghetti" operations, we have incorporated on the SPC a 3-port registerfile, which has the capability of reading two operands and writing one operand in the same cycle.

The registerfile, which currently has 15 words is controlled by 3 addresses, embedded in the SX, SY and DX control fields. As can be seen from Fig. 1 the data-input bus and output buses are connected to the X-bus and Y-bus.

Instead of using a conventional two-port RAM structure, we have implemented a new register cell (Fig. 6a). The new cell uses a latch, controlled by the two non-overlapping main clocks $\emptyset 1$ and $\emptyset 2$, and 3 wordlines (WLWj, WLRAj, WLRBj) as shown in Fig. 6b. This bi-phase latch makes overlapped read/write access in the same register within one cycle possible. By ensuring that none of the address lines is active during $\emptyset 1$ -between address changes in the source and destination fields- proper operations of this memory is guaranteed.

3. DESIGN AND VERIFICATION APPROACH

In the preceding section the architectural highlights of the SPC were discussed. Many useful arithmetic features have been included in this core, which can be selected under microprogram control. We feel that using distributed PLA's is probably the best solution to the problem of attaining highly integrated control structures in VLSI signalprocessors. Moreover they ease the implementation of MICROPROGRAMMING techniques considerably.

To generate these PLA's a silicon compiler is used, as described in [3]. Most PLA generators are just what their name suggests, using predesigned cell libraries with an inherent short life cycle. This one *, however is quite special (see Fig. 7). It uses a minimizer to perform the logical compaction of the binary specification, decreasing the number of product terms. Next topological folding is carried out, taking into account such boundary restrictions as location and order of inputs and outputs, and sets of terminals that should stay together. A procedural cell generator creates a cell library as function of user defined design rules and parameters, such as W/L ratios, width of internal and external supply lines,

---

* This compiler was originally developed by L. Reynders and M. Bartholomeus et al of the ESAT laboratory, K.U. Leuven, Heverlee, Belgium. With some modifications it has been installed by them on the Philips CAD system.

and pitches of inputs, product terms and outputs. A placement routine then combines the generated cells and the symbolically compacted PLA into a mask-data base.

Logic, circuit and layout methodologies and tools have been stretched to their limits, while designing VLSI chips. In addition, VLSI signalprocessors experience special problems, like optimizing the architectural design for a target performance and cost. The challenge, which faces the designers, is to design AND verify these signalprocessor chips in the shortest possible design time. Our approach is the following.

Given the basic algorithms to be performed, the SPC was divided into several functional blocks, such as the Booth multiplier, extractor and 3-port memory. A gate-level simulation of these blocks, but also of the complete SPC was performed, to debug the logic specification. Next the major functional blocks were designed as bitslices, while each cell in these bitslices was designed by hand to attain the optimal cost performance ratio. The design loop for these cells involves a circuit extractor (transistors and capacitances) and a circuitsimulator. Particular attention was given to the possibility of varying the wordlengths of this SPC, without major redesigns. This feature required careful consideration on the internal structure of the functional blocks.

Finally the layout must be checked for logic integrity (see fig. 8). Our approach is the following. A topological domain description with terminals (t) is made of the most often used cells. An example is shown in Fig. 6c. These domains are again input to the proprietary layout verification program, that reconstructs from the circuit topology, inside the unique domains of these cells a transistor level macro description. In the layout data base the basic cells are replaced by their domains, and the extractor now generates the complete network description of the SPC. (Using these domains considerably reduced the computation times of the layout extractor: down to 2:45 CPU hours on an IBM3083 for the complete SPC.) Both the transistor descriptions of the domains and the network (defined as everything outside domains) are then expanded into one transistor level data base and submitted to a logic gate extractor. This programm reduces the database by recognizing classical CMOS gates. The final logical verfication is performed by a switch level logic simulator LOGMOS [4]. this powerful simulator supports the bilaterial charge flow through MOS switches, whereby the logic state depends on the capacitive weight of logic nodes.

The fact that 14 errors were found, when the final bottum-up logic verification was performed (from bonding pad to bonding pad), shows the merits of this verification approach. Design of the SPC took 2.5 man-years, of which roughly 3 man-months (1 calender month) was spent on the logic verification. Following the design approach described above, has resulted in a high performance Signal Processor Core using 18.600 transistors on chip area of only 15.5 mm^2, when stripped of all bondingpads, etc. (Fig. 9). This compares with an overall packing density of 1200 transistors/mm^2. The expected power dissipation (when operating at 10 MHz clock frequency) is 150 mW.

VLSI Signal Processing

4. TECHNOLOGY

CMOS is now widly accepted as the technology of choice for VLSI. The reason is that power supply voltages have not decreased, as chip complexity has increased, and this has exacerbated the power dissipation problems with NMOS chips. This is particularly the case for high-performance signalprocessors.

Another major design consideration for this SPC was that its product life cycle is likely much longer than the interval between design rule changes. Improvements in speed and cost of signalprocessor, in which the SPC is embedded can be obtained by implementing those design into silicon with even more advanced design rules. Philips has chosen a scalable CMOS technology with 2 micron effective channel lengths to implement its next generation designs. Its main technological features are summarized in Fig. 10. The savings in chiparea by going from Perkin Elmer projection aligners to wafersteppers is shown in Fig. 6d.

The Signal Processor Core is currently undergoing processing in the pilot line. The measured cycle time of the Signal Processor Core is 100 nsec. with effective gate lengths of 2 micron for the N-channel transistors and 2.5 micron for the P-channel transistors. This figure was also obtained via careful "worst case" circuit simulations of the critical delay paths in the SPC. Circuit data for these simulations was extracted from the actual mask data base.

5. CONCLUSIONS

In this paper a 10 MHz Signal Processor Core was described. It forms the arithmetic heart of Harvard-type single chip signal processors. Its arithmetic capabilities, selectable under microcontrol, together with its projected 100 nsec. cycle time and updatable CMOS design rules guarantee a long product life cycle.

6. ACKNOWLEDGEMENTS

We wish to acknowledge the valuable contributions made by J. Wittek (Valvo, Philips GmbH, Hamburg, West Germany), G. Schrooten, J. Joosten, L. Reynders and M. Bartholomeus (both with the ESAT lab., K. U. Leuven, Heverlee, Belgium), H. Veendrick, and the members of the maskshop under direction of N. Dekkers, and F. Smolders and his collegues in the waferfab line. The authors wish to thank P. Vary and K. Hellwig (TeKaDe Nurnberg) and R. Sluyter for stimulating discussions about the applicability of this Signal Processor Core in DPS chips.

7. REFERENCES

1 Meerbergen, J.L. van, F.J. van Wyk, "A 2 micron NMOS 256-point Discrete Fourier Transform Processor", IEEE International Solid State Circuits Conference, Digest of Technical Papers, New York, February 1983, pp. 124-125
2 Rubbenfield, L.P., "A proof of the modified Booth's algorithm for multiplication", IEEE Transactions on Computers, Vol. C-24, October 1975, pp. 1014-1015
3 De Man, H., L. Reynders, M. Bartholomeus, J. Cornelissen, "PLASCO: A Silicon Compiler for NMOS and CMOS PLA's", Proceedings IFIP Conference: VLSI'83, Trondheim, Norway, August 1983, pp. 171-181
4 De Man, H., D. Dumlugol, P. Stevens, G. Schrooten, I. Bolsens, "LOGMOS: A transistor oriented logic simulator with assignable delays", Proceeding IEEE International Conference on Circuits and Computers, New York, August 1982, pp. 42-45
5 K. Hellwig, P. Vary, private communication

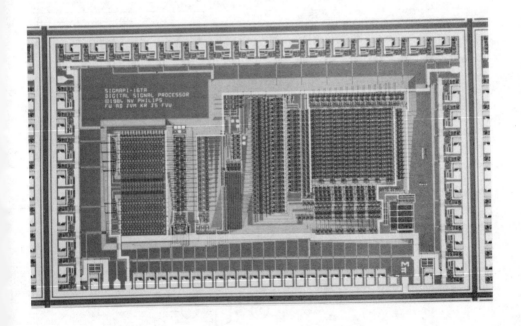

Figure 9. Layout photomicrograph, showing the Booth multiplier (1), carry look ahead adder (2), multiprecision accumulator (3), function select PLA (4), extractor matrix (5), format adjuster (6), 3-port registerfile (7), SPC status register PST (8), and barrelshifter register (9).

VLSI Signal Processing

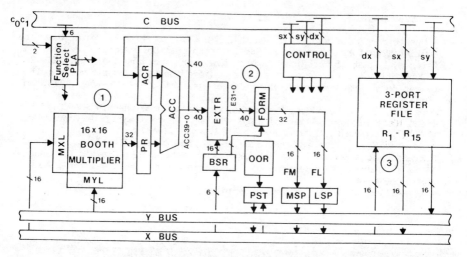

Figure 1. Blockdiagram of Signal Processor Core (SPC)

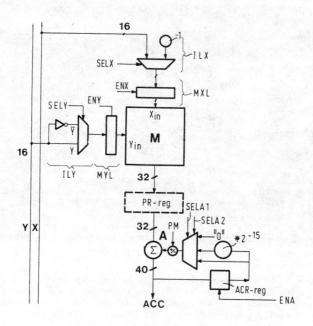

Figure 2. Blockdiagram of Multiplier (M) and Accumulator (A)

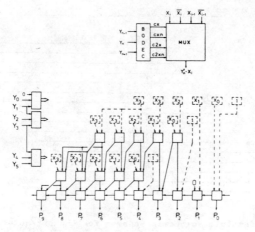

Figure 3. Diagram showing the implementation of the Booth algorithm in a 6x4 bit multiplier. Xi:multiplexers, BOD: Booth encoders, FA: full adders, CLA: single level carry look ahead over 3 bits.

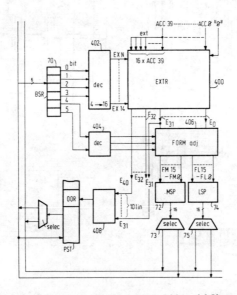

Figure 4. Overview of the arithmetic shift unit, showing a barrelshifter selection matrix (EXTR), a format adjuster (FORM), an extractor-out-of-range detector (OOR), the control register (BSR), the status register (PST), and the result registers (MSP, LSP).

VLSI Signal Processing

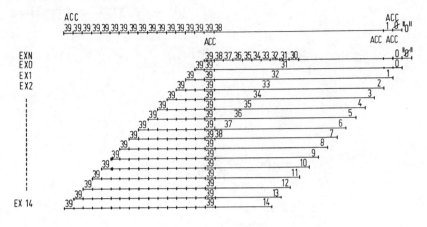

Figure 5. Possible selection schemes for the extractor unit

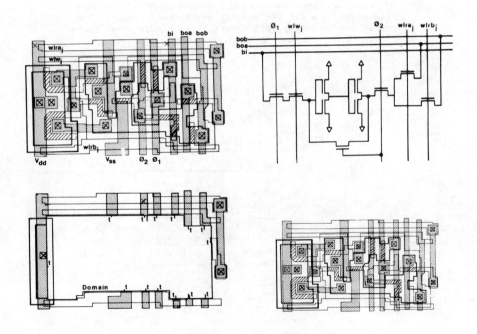

Figure 6. Three-port memory cell, used in SPC registerfile, showing masklayout (Fig. 6a), circuitdiagram (Fig. 6b), domain description with terminals t (Fig. 6c), and waferstepper layout (Fig. 6d).

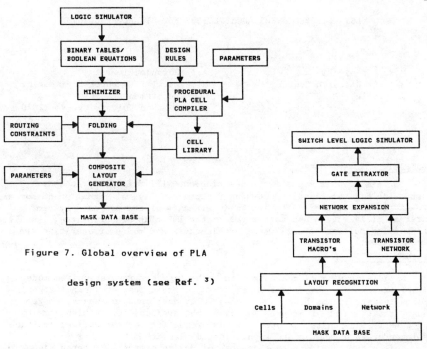

Figure 7. Global overview of PLA design system (see Ref. [3])

Figure 8. Overview of logic verification tools used during the design of the Signal Processor Core.

• Technology	2 micron N-well bulk CMOS
• Number of masks	9
• LOCOS oxidation	
• R sheet polysilicon	30 Ohm/square
• Threshold voltages	
Vtp	-1.0/-0.75 V
Vtn	0.75 V
• Design rule pitches	Perkin Elmer
(including contacts)	PE341 autoalign
metal	7.5 μm
polysilicon	8.75 μm
diffusion	10 μm
• Gate oxide	500 A
• Gate length	
P-channel	2.5 μm
N-channel	2.0 μm

Figure 10. Technological highlights of the CMOS process

VLSI MULTIPROCESSOR ARCHITECTURE FOR SIGNAL PROCESSING

M.R. Baraniecki
M/A-COM Research Center
Rockville, MD 20850

A.Z. Baraniecki
Electrical & Computer Engineering
Department
George Mason University, VA 22030

ABSTRACT

The aim of this paper is to describe the VLSI multiple processors system for digital signal processing functions. Digital signal processor and two interconnection circuits are basic elements of the proposed system. Internal architecture and parameters of the DSP chip are presented. Parallel and modular multiprocessor system architecture and modes of operation are demonstrated.

I. INTRODUCTION

Rapidly growing microelectronics technology has revolutionized modern signal processing. Over last dozen years, a great deal of digital signal processing algorithms have been successfully developed in speech, image, sonar, radar and other areas. Nowadays, the availability of high density and fast VLSI circuits has opened a new avenue for cost effective, realtime implementation of complex and sophisticated algorithms.

This paper considers some aspects of design of the VLSI based signal processor system to perform some computationally involved digital signal processing functions requiring 10-25 million operations per second (MIPS). Initially conceived as a versatile programmable signal processor for communications applications including speech processing and data compression, it has developed into general purpose modularly expandable system with multiple processors activated by a control processor.

Some of the potential applications considered in the design of our DSP processor are: linear prediction analysis and synthesis, medium and low bit rate speech coders, digital speech interpolation, DTMF & MF tone generators and receivers, modems, echo canceller, digital testers, FDM/TDM transmultiplexer, encryption, speech & speaker recognition. Some of these functions are now or will be required per channel basis in digital switching systems, modern voice/data PABXs or satellite interface units. It is not economically justified to assign a single DSP processor e.g. TMS-320 or several of them to execute particular function per channel in such a multitask and multichannel environment.

An alternative approach can be the system of multiple DSP processors interconnected within a suitable communication network to achieve enhanced parallelism of computational tasks.

The multiprocessor system with dynamic assignment of requested functions by a host processor to temporarily idle DSP processors can substantially reduce the hardware and increase efficiency of the entire system.

High throughput of the multiprocessor system is achieved mainly by high degree of concurrency and parallelism in the architecture of the DSP processor and the overall system architecture. Major features of such a system should be: programmable, with instructions set optimized for digital signal processing, VLSI form, high throughput of 15-25 million operations (multiply and add) per second, simple processor control, efficient interprocessor communication, modular reconfigurability and expandability, minimum of associated circuitry.

The two's complement is the basic arithmetic representation used in the processor. However, for some applications requiring high precision computation, increased dynamic range can be achieved from processors operating in parallel with residue arithmetic instead of slower double precision or multi-byte arithmetic. Residue number system as discussed in last section, is ideally suited for parallel systems.

An advantage of our approach is that only 3 different chips are to be designed in order to build even complex multiprocessor system. First of them is the digital signal processor element. Two others are the interconnection chips.

II. VLSI PROCESSOR ARCHITECTURE

Our proposed VLSI DSP processor is a programmable device. The programmable processor has obvious advantages over the fixed-program processor. The ease of reprogramming and associated flexibility of applications is the major advantage.

Signal processing algorithms used in most of applications mentioned in Introduction have some identified attributes:
- arithmetic intensive, but not logic intensive
- mostly well defined and functionally repetitive
- frequent access to the coefficient memory

An analysis of these signal processing algorithms has led to definition of a set of macro instructions (MACROS).

They are essentially subroutines that consist of several (up to 20) lines of micro code. Macros are divided into six categories: signal processing, arithmetic, test and measurement, control, interrupts, I/O and others. The present set may be progressively extended for future applications. Partial list of the MACROS is presented in Table I. These MACROS are frequently called over the programs, therefore they ought to reside in the memory close to the arithmetic unit of the DSP processor. The DSP chip is a basic processor element (PE) in the proposed multiprocessor architecture. It was decided to store all MACROS in the internal macro program of each processor element in the system. The primary advantage of using MACROS is that the rate of access of the external program memory by each processor is considerably reduced.

The DSP processor architecture is basically a double-accumulator Harvard type architecture and is shown in Fig. 1. The DSP has a set of instructions optimized for signal processing. All of them are single cycle instructions with cycle time of 150 nsec. High performance is achieved by the 16x16-bit parallel multiplier with the 32-bit product, the 32-bit ALU, the 0-15 bit barrel shifter with sign extension and two independent the 36-bit accumulators.

The memory hierarchy is another key factor of the chip architecture. There is the 410x16-bit ROM program memory where all MACROS reside. The Coefficient Memory is divided into two sections: the 128x16-bit ROM, where fixed coefficients are located e.g. look-up tables, and the 128x16-bit RAM, where varying parameters reside, e.g. adaptive filter coefficients in echo cancellers. The Data Memory is organized as the 200x16-bit random access memory. There are two stacks and two independent Address Generators: the first one for addressing the coefficient ROM and external program memory (up to 64K); the second for addressing the Data Memory and external Date Base (up to 64K addressing space).

The Instruction Decode and Control block contains pipe line registers, decoders, microprogram counter, macroprogram counter and hardwired control section which controls the processor and the access to the external program memory. There are four 16-bit buses: Program Address Bus, Data Address Bus and Data Bus. The processor operates in three modes. In the Macro Mode the DSP works in the most efficient way. It executes high level instructions from external memory. In this mode most of the fetches are to the internal memory and the access to the external memory is relatively not frequent. Statistically, for most considered signal processing algorithms three accesses for every four are to internal memory. The average single high level macro instruction is equivalent to 10 lines of micro code. Therefore, another measure would be that one out of every eleven instructions executed are fetched from outside.

In the Micro Mode entire code residing in the external memory consists of low level instructions. Therefore, each instruction executed requires corresponding memory access. As a result, the speed of the bus increases and faster memory is required. The Micro Mode is not recommended for a multiprocessor cluster, except special cases because the probability of blocking in the interconnection circuit is higher and overall system throughput decreases.

In the Internal Program Mode the processor executes only microinstructions stored in its internal program memory (ROM). This configuration is optimal for e.g. some per line applications in telecommunications not requiring long programs. The maximum throughput of the single DSP chip can be achieved.

The DSP processor has programmable 16-bit parallel I/O interface and 2 serial ports for flexible communications with e.g. PCM codecs. In order to speed-up the transfer rate of large amount of data from and to the data base memory there is built-in DMA control. Interprocessor communication control block performs control functions for communications with the interconnect chip. It is supported by the 2 level code sensitive interrupt directly software polled. There is also general purpose 4 level prioritized, vectored interrupt. The N, V, C, Z flags indicate the status of the resulting ALU operations.

III. MULTIPROCESSOR SYSTEM

The multiprocessor architecture is shown in Fig. 2. The system comprises of N-clusters, each consisting of 1 to M processor elements. The processors within one cluster share two separate memories: common main program memory where all application programs reside and common data base memory. The accesses to both memories are controlled by two interconnection chips. The entire system is controlled by a single host processor. The communications between clusters and the external world are via the control bus and the TDM bus. The global level of communcations, where the processor element "talks" to the host or other processor elements outside of its own cluster is kept to a minimum and at high level. The local level is the interprocessor communications within the same cluster.

Dynamic assignment of the various tasks to the various processors is handled by a host processor and depends on particular system applications, for example: on traffic in the telecommunications network. This also defines number of clusters, number of processors in the cluster and the ratio of processors running in Macro and Micro Mode. The system is modularly expandable at both processor element and cluster level. It provides built-in redundancy and desired graceful degradation. The targetted efficiency of the processor is estimated to be of 70-80%.

Interconnection Elements

One of the major problems to overcome in multiple processors systems is an efficient and cost effective interconnection network. Among the most known are: crossbar, omega, shuffle exchange, single bus and multiple bus (2)-(4). The full crossbar switch of the size 8x8 or larger would probably cost more than all the system elements combined, considering present relatively low costs of microprocessors and memories. A time-shared bus has a very limited transfer rate, which may be inadequate for even small number of processors.

However, the memory hierarchy and instruction hierarchy are key features of our system architecture. A set of high level instructions (macroinstructions) will be stored in the internal memory of each PDSP chip. Low level instructions (microinstructions) are to be stored in the external main program memory, common for all processors within a cluster. Executions of microinstructions have been statistically estimated to be several times (8-10) less frequent than executions of the internal macros. Therefore, there will not be severe bus contention problems and single bus configurations will be adequate. Two interconnection networks are basically bus arbitrators but their modes of operation are different. The program memory interconnection chip acts as a classic bus controller. Requests form the quene and they are serviced one by one. The memory is enabled every cycle only if there are requests. The required switching speed is 40 ns or less to service 8 processors. The data base memory interconnection chip acts as a protocol starter machine.

Modes of Operations

The processor in the system can be configured in four modes: cluster mode (Fig. 3), processor with external program memory, processor with external data memory and stand-alone mode.

A. Cluster Mode

This is basically a multiprocessor single bus configuration. It can support up to 8 processors sharing a common program memory and data base memory. The program bus and data bus arbitration is controlled by two interconnection chips. The up to 64K either ROM or RAM program memory contains mainly application programs low level instructions but it can contain MACROS as well. The memory is allocated to the requesting processor for one entire cycle. The up to 64K of the relatively slow (150-200 ns) RAM contains the data base memory. This memory can be allocated in two ways: one part serving as common data base and the other either divided for certain applications as voice storage systems or reference parameters library for speech and speaker recognition.

B. External Program Memory Mode

This mode is suitable for applications not requiring interprocessor data transfers as well as for long programs written in micro code and run from external memory. Multiple tasks can be run in both MACRO and MICRO mode on up to 8 processors.

C. External Data Memory Mode

This mode involves only one processor and it is convenient for applications on per line basis requiring long look-up tables and large data base.

D. Stand-alone Mode

In this configuration the PDSP works off its own internal program memory and uses only its own internal data and coefficient memories. The communications with the host processor is kept to a minimum and is limited practically to controlling and initialization functions. Envisioned applications are ones on a digital line card performing continually single task, for example: 32 kbps ADPCM codec. This is the minimum hardware

VLSI Signal Processing

configuration and with minimum control overhead can perform some tasks requiring around 6.5 MIPS. Moreover, for some applications the serial data input can be multiplexed and processor can be programmed to operate as a single instruction multiple data stream (SIMD) processor.

IV. RESIDUE NUMBER SYSTEM IMPLEMENTATION

Recently, numerous authors have demonstrated the potential of residue arithmetic for realizing high speed high accuracy digital processing systems (3). High precision computations can be obtained from microprocessors operating in parallel with residue number systems (RNS) arithmetic instead of slower double precision or multibyte arithmetic. It has been recognized that the special properties of the RNS have many advantages when applied to common digital signal processing requirements such as nonrecursive filters, NTT and FFT processors. The unique advantage of this RNS and the main reason for its attractiveness is that residue coding decomposes the arithmetic operations into a set of parallel suboperations that are well suited for high speed implementation using arrays of ROMs or microprocessors. In the RNS each sample can be represented by a set of integer residue digits, $x = \{x_1, x_2, \ldots, x_M\}$ where the residues modulo m_i are formally written $x_i = |X|_{m_i}$ and the dynamic range is $P = \prod_{i=1}^{M} m_i$, provided all moduli are relatively prime.

The binary operations of addition, subtraction or multiplication can be performed by independent operation on respective residues, i.e. $Z = X * Y$ implies $z_i = x_i * y_i$ where $* \in (+,-,.)$. There is no interdigit carries or borrows, no parital products. This allows for high speed parallel array computation. There are drawbacks associated with RNS when general division is to be performed; also operations such as magnitude comparison and sign detection present major problems. The advantages of the RNS, however, outweigh the disadvantages for many signal processing applications. Most operations stored in arrays of ROMs are used to perform modular operations. The current advances in VLSI techniques allow us to construct both look-up tables and algorithms in minimum chip microprocessor systems. One of the clusters, consisting of M PDSPs (where M is a number of moduli) could be assigned to perform RNS computations. Since operations in RNS can be carried out independently, a separate microprocessor can be used to compute results in each modulus, as shown in Fig. 4.

Three 16 bit processors provide up to 48 bit full precision and high speed operation. Modulo m_i addition in a microprocessor can be defined as Macro and accomplished in 2 steps:
i) compute $y = |a+b|_{2^k}$; k-number of bits, e.g. 16
ii) compute $|y|_{m_i}$

    The second step can be implemented by either adding $2^k - m_i$ (generalized end around carry) in a 2^k bit 2's complement adder to the result of addition if overfow is detected ($y > m_i$) or looking up $|y|_{m_i}$ by using y as an address for a 2^k element look-up table.

RNS multiplication can be implemented by adding indices (3), and a special Macro can be created to accomplish this. For example, for modulus $m_i = 64507 < 2^{16}$ with prime decomposition 257 x 251, indices can be added for two submoduli, 257 and 251. Alternatively, RNS multiplication can be implemented using another Macro called custom-made VLSI residue multiplier. RNS to binary conversion can be implemented using mixed radix decoding (3) or Chinese Remainder Theorm.

V. CONCLUSIONS

Multiprocessor system for signal processing functions in multitask and multichannel environment has been presented. High performance is achieved by a high degree of concurrency and parallelism in the architecture of the processor. The three VLSI chips set can be used to implement modular, expandable signal processing system.

References

(1) F. Mintzer, A. Peled " A Microprocessor for Signal Processing, the RSP", IBM J. Res., July 1982

(2) D. Patterson, C. Sequin "Design Considerations for Single-chip Computers of the Future" IEEE Journal of Solid-State Circuits, Feb. 1980

(3) A. Baraniecki, G. A. Jullien "Residue Number System Implementation of Number Theoretic Transforms in Complex Residue Rings" IEEE Trans. on ASSP, June 1980

(4) J. Patel "Performance of Processor Memory Interconnections for Multi-processors" IEEE Trans. on Computers, Oct. 1981

(5) M.R. Baraniecki, M. Shridhar "A Speaker Verification Algorithm for Speech Utterance Corrupted by Noise with Unknown Statistics", IEEE Int. Conf. ASSP, Apr. 1980

Signal Processing	Arithmetic
. IIR FILTER	. Division
. FIR FILTER	. Square Root
. CORRELATOR	. Modulo
. Least Mean Square	. X**Y
. Window	. ARCTAN
. Step Adaptation	. ARCCOS
. Quantizer	. LOG
. Dequantizer	. Pseudo Random Sequence
. FFT Butterfly	
. Durbin-Levinson Recursion	
. Bit Reversal	
. Cosine Generators	
. POWER	
. AVERAGE	Input/Output
. PEAK PICK	. µ-law/Linear
. Zero Cross Count	. Linear/µ-law
. Goertzel Algorithm	. A-law/Linear
	. Linear/A-law

Table 1 MACROS

VLSI Signal Processing

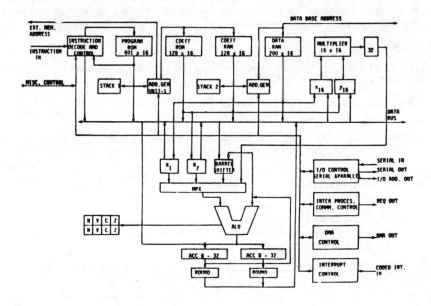

Figure 1. PDSP Internal Architecture

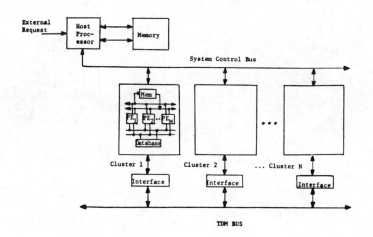

Figure 2. Multiple Cluster PDSP Configuration

VLSI Signal Processing

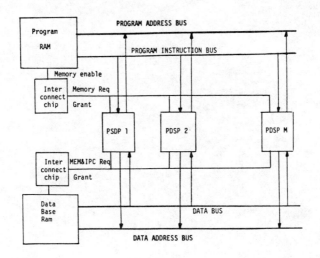

Figure 3. Multiprocessor Cluster

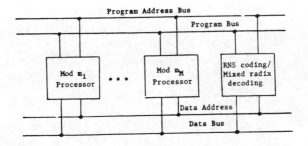

Figure 4. RNS Cluster

Part IV
TELECOMMUNICATIONS

Edited by:
Dr. Gwyn Edwards
Racal-Vadic

VLSI Signal Processing

AN INTEGRATED COHERENT DEMODULATION TECHNIQUE

Kerry Hanson
Texas Instruments

THE COMPLETE TEXT OF THIS MANUSCRIPT WILL BE FOUND BEGINNING ON PAGE 427.

Design Of A Single-Chip Bell 212A Compatible Modem Employing Digital Signal Processing

David Taylor, Stevan Eidson
Advanced Micro Devices
901 Thompson Place
Sunnyvale, CA 94088

Abstract

Modems operating at 1200/300 bps full-duplex have to this point been implemented using a combination of microprocessors, analog hybrid filters, and analog demodulators. The goal of this project is to design a Bell 212A compatible modem with a Bell 103 fallback option in a single-chip CMOS VLSI component. Digital signal processing is employed for all major filtering, modulation, and demodulation tasks required of the modem. All of the DSP algorithms were designed to perform optimally with a shift-and-add type of architecture. This paper presents the algorithmic and system options that have been explored through the process of arriving upon the final design.

Introduction

The second generation single-chip modem is a 1200-bps full-duplex modem compatible with the Bell 212A specification. The modulation technique employed by Bell 212A compatible modems is known as differential-quarternary phase-shift keying (DQPSK). Included within the 212A specification is communication compatibility with the existing base of Bell 103 series modems operating at 300 bps. The modem on the answering end of a data call automatically senses the transmission speed (300 or 1200 bps) of the originating modem and adapts to that speed.

Key goals established for the 212A modem project were: all signal processing must be performed on chip, analog-to-digital and digital-to-analog conversion must be contained on chip, no external components should be required, and performance should be acceptable to customers who traditionally have relied on box-level 212A solutions.

In order to meet these goals, extensive thought was given to the telephone network environment, silicon technology to be used, architecture, signal processing compatibility of both the DQPSK and FSK modems, and algorithm development for efficient implementation.

Digital Signal Processing In VLSI Modems

All modem features (modulation, demodulation, filtering, interface control, etc.) must be integrated onto a single chip, so care must be taken in the design of the modem so as not to waste silicon. There are unique constraints placed on developing

VLSI Signal Processing

VLSI silicon solutions relative to discrete implementations. The amount of analog circuitry on the chip must be limited because it is more sensitive to process variations, power supply, and noise considerations than digital circuitry.

Because of the feature set to be implemented, a digital-signal-processing architecture was chosen instead of using a switched-capacitor technology. A DSP implementation provides some significant advantages over switched capacitor technologies, especially for the 212A compatible modem. For instance, additional stages of filtering require relatively little additional silicon area (ROM and RAM only) in a DSP architecture. However, any additional stages of filtering in a switched-capacitor technology require modifications to the major portions of of the component layout. Taking this analogy further, once the ALU overhead in a DSP architecture is paid for, any additional processing costs very little. Future modifications and improvements are also easy to implement as they normally require simply changing the instruction microcode or filter coefficients stored in a ROM.

Most importantly, implementation of both the DQPSK and FSK modulation techniques required by the Bell 212A specification are easily implemented within a DSP architecture. Virtually identical microcode may be used for both modem types when thought is given to the algorithm development. The FSK algorithms may be designed as simply a subset of the DQPSK algorithms. The finite impulse response (FIR) and second-order infinite impulse response (IIR) filter structures are the same for both modem types, so coefficients from a coefficient ROM are the only major differences between the FSK and DQPSK modems.

The 212A-compatible modem using digital signal processing is a multiply-intensive processor. One multiplicand is often a fixed filter coefficient, while the other operand is the input data sample. Only a few times within the signal processing chain are multiplies required between data values. Since building a multiplier on chip is an expensive silicon investment, fixed-coefficient multiplies may be implemented within the ALU in a simple manner. Using a shift-and-add representation for these filter coefficients, the multiplication can be performed in two to four microcode instruction cycles. The techniques used for coefficient representation are described further on in the paper.

DQPSK and FSK Modulation

The two modulation techniques used within the 212A compatible modem (DQPSK and FSK) are incompatible with one another. FSK modulation is a digital form of frequency modulation where the analog carrier generated is a function of the baseband input signal. If the input is a binary "one", a particular frequency is generated. If the input is a binary "zero", a second frequency is generated. Figure 1(a) shows a binary sequence which will be used to modulate both an FSK and a DQPSK waveform. Figure 1(b) shows the time-domain FSK waveform as a function of a digital input sequence. The frequency spectrum generated by a random FSK modulation is an infinite spectrum. (Refs. 1,2)

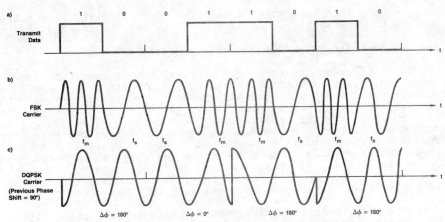

Figure 1: Modulating the digital data pattern in (a) is performed in two manners for the Bell 212A modem. (b) shows an FSK modulated waveform where f_s is the space frequency and f_m is the mark frequency. (c) shows an unfiltered DQPSK modulated waveform.

DQPSK modulation differs from FSK modulation in that DQPSK is a modulation of the phase of a constant carrier frequency. DQPSK modulation encodes two bits per modulation period (baud interval); hence, one of four phases transmitted during a modulation period represent the two bits encoded. Additionally, the phase shift performed on the carrier is differential in nature. In other words, the phase shift performed during the current baud interval is relative to the phase which was transmitted during the previous baud interval. Figure 1(c) shows a time-domain waveform which represents the unfiltered DQPSK signal. The time-domain expression for the DQPSK carrier in the current baud interval may be expressed as:

$$\sin(w_c + phi + n*pi/2), \qquad (1)$$

where w_c is the radian frequency of the carrier,
phi is the phase of the carrier transmitted during the previous baud interval
and $n*pi/2$ represents the phase shift performed during the current baud interval, where n = 0,1,2,3.

Similar in nature to the randomly modulated FSK signal, the randomly modulated DQPSK signal also contains an infinite spectrum. (Ref. 3)

Normally, additional shaping is provided for the transmitted DQPSK signal. This shaping is provided to help limit the amount of inter-symbol interference (ISI) in the transmitted signal. The shaping normally obeys a cosine rolloff function which decays the tails of the baseband pulse to zero more rapidly than would normally occur with arbitrary band-limiting in the band-pass filter. The Bell 212A specification requires a 22% rolloff characteristic. The shaping may be implemented either in the baseband of the pulse or in the passband of the modulated signal.

VLSI Signal Processing

Shaping is not provided for the transmitted FSK signal because of the asynchronous nature of the data. For pulse shaping to be effective, the data sequence must be synchronous.

The Public Switched Telephone Network (PSTN)

The telephone network is a harsh environment for modem operation. Problems such as lightning, induction from power line cables, switching noise caused by central offices, FDM and TDM conversion inaccuracies, plus a myriad of other noise-inducing problems may tax the ability of the modem to provide accurate data demodulation. The human ear tends to act as an effective low-pass filter and interpolator to ignore impairments of this type. However, if a dropout occurs during data transmission, the modem is likely to begin making decision errors. Hence, the design of a data modem such as the 212A must take into account possible harsh operating conditions such as those cited.

Typically, modem performance if characterized as a plot of bit error rate versus signal-to-noise ratio. Bit error rate is simply a measure of the number of errors received in a bit stream versus the total number of bits sent. Signal-to-noise ratio is a measure of the in-band signal power strength versus the in-band noise power strength. Figure 2 shows plots of the theoretical bit error rate performances of both the FSK and DQPSK modems. (Refs. 2,4) The assumption is that the noise is additive gaussian white noise (AGWN). While in practice it is normally not realistic to assume that modem performance will ideally match the theoretical, it is the goal of modem design to come as close to the theoretical plots as possible.

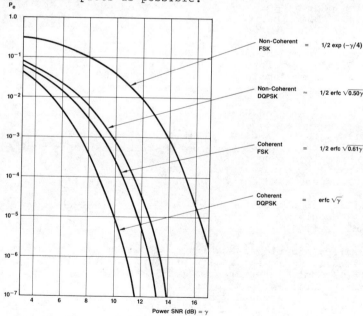

Figure 2: The theoretical minimum BER plots for FSK and DQPSK modems show that coherent demodulators yield superior performance to their non-coherent counterparts.

VLSI Signal Processing

The Single-Chip Modem Transmitter

There are three basic components to both the FSK and DQPSK transmitter. First, a sine wave synthesizer creates a digital representation of the modulated signal. Because the synthesis is done in a sampled manner and the analog signal is modulated by a digital input, high-level harmonic tones are generated that exceed the Federal Communications Commission (FCC) requirements for transmission on the telephone network. The second block of the modem transmitter is a digital band-pass filter for out-of-band energy. The final portion of the modem transmitter is an on-chip digital-to-analog converter (DAC) that puts a modulated analog signal out to the network protection circuitry.

A problem exists with this method of modulation as it has been stated since the output of the DAC creates harmonic frequency components that interfere with the out-of-band energy specifications mentioned before. To eliminate the harmonics, a steep analog filter could be attached to the output of the transmitter, but these filters are difficult to integrate into silicon. A more attractive solution is to increase the sample rate from the basic value (near 8 kHz) to a far higher value. The sampling rate can be increased in stages, each stage performing a filtering function known as interpolation. Then a single pole analog filter which can be integrated into silicon is attached to the output of the DAC.

The modulated signal is synthesized using a phase-to-amplitude technique which is completely digital. A block diagram of the sine synthesizer is shown in Figure 3. The basic concept is to accumulate phase at the frequency to be generated and then to convert the phase to an amplitude using a ROM look-up.(Ref. 5) This method is quite simple to implement because the non-linear amplitude is created from the linearly accumulated phase. Constants stored in a second small ROM determine the increment of phase added to the phase accumulator.

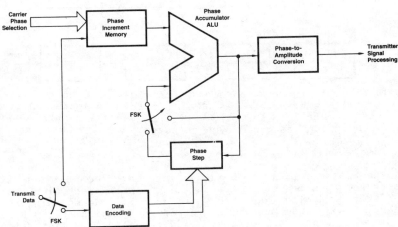

Figure 3: A block diagram of the sine synthesis portion of the transmitter. A phase increment is selected from memory and accumulated to give the absolute phase. The absolute phase addresses a phase-to-amplitude memory.

VLSI Signal Processing

To avoid a large phase accumulator and sine ROM, and to accurately determine changes in the digital input data, the sine synthesis is performed at a high sampling rate. The sample rate is then decimated down to the lower basic value in several stages. Each decimator stage filters out the energy created by the sampled system at frequencies near the reduced sample rate. Both the decimators and interpolators can be implemented as simple linear-phase FIR filters. The decimators and interpolators for the DQPSK modem can be exactly the same as their FSK counterparts.

Like the decimators and interpolators, the band-pass filters employed on the 1200 bps and the 300 bps modem types have an identical structure. Only the filter coefficients are altered for the two different types of modems. These band-pass filters are IIR filters that are group delay equalized to eliminate phase distortion.

A complete transmitter block diagram is shown in Figure 4. To summarize, the modem transmitter consists of a high sample rate phase-to-amplitude sine synthesis followed by decimation to the basic sample rate. At the lower sample rate, the digitally generated sine wave is band-pass filtered to reduce out-of-band energy. The modulated and filtered signal is interpolated up to a high sample rate again before being presented to the transmitter DAC. The analog waveform has a simple one-pole low-pass filter applied to it to reduce harmonics created by the sampled system.

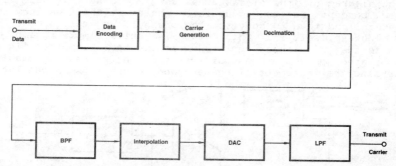

Figure 4: Both the FSK and DQPSK modems are able to use the same transmitter structure as shown in this diagram. Different coefficients are selected for the band-pass filter section depending on the modem type selected.

The Single-Chip Modem Receiver

The modem receiver demands far more complicated signal processing than the modem transmitter. In the reception process, a distorted analog waveform with an amplitude spanning several orders of magnitude must be converted to a digital pattern having a minmal number of errors. Additionally, the distortion and amplitude of the analog signal can vary from telephone connection to telephone connection. Compare this to a simple frequency synthesis and filtering function for the transmitter and it is completely clear why the modem receiver is more complicated.

VLSI Signal Processing

The modem receiver consists of a band-pass filter, an automatic gain control (AGC) stage, a demodulator, and in the case of the DQPSK modem a timing recovery stage. The modem receiver system is shown in Figure 5. The analog-to-digital converter (ADC) employed on the single-chip modem uses a high sample rate to alleviate multi-order external analog filters. A single-pole, on-chip analog filter is sufficient to attenuate any high frequencies that might alias back through the ADC into the sampled system. The over-sampled digital waveform produced by the ADC must be decimated down to the basic sample rate of the system. This decimation is done in stages similar to the transmitter decimation, with the decimating filters being linear-phase FIR filters.

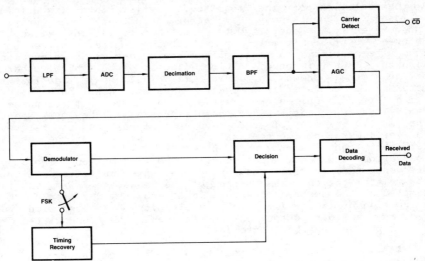

Figure 5: Although the FSK and DQPSK demodulators are not the same, it is possible to execute the same microcode in both cases by carefully choosing coefficients. As in the transmitter, the coefficients differ between the FSK and DQPSK modem types.

Because the public switched telephone network is a two-wire system, and the modem is operating in full-duplex, there is crosstalk from the local modem transmitter into the local modem receiver. The crosstalk frequency spectrum must be band-pass filtered from the receiver input before demodulation can occur. The band-pass filters employed on the single-chip modem are IIR filters that are group delay equalized to reduce phase distortion.

Both the FSK and DQPSK demodulators are quite sensitive to input signal amplitude, thus an automatic gain control (AGC) is used to hold the steady-state amplitude constant. Most AGC algorithms use a feedback technique to derive a fixed-level multiplier from the band-pass filter output. Because the modem is built using DSP, it is possible for the AGC algorithm to be a finite impulse machine without any feedback. The finite impulse algorithm is of great interest because it is physically impossible in an analog system and it offers greater stability than the infinite impulse (feedback) system.

The first step in the AGC algorithm is to develop a power signal from the band-pass filter output. This power signal can be generated in a variety of ways, the most common being a squaring operation or a rectification. The power is next low-pass filtered to eliminate signal variations due to the modulating frequencies. The result of these operations is a slowly varying DC signal that can be used to select a multiplier constant in either a feedback or feed forward loop. The constant is multiplied by the band-pass filter output signal to yield a fixed-level AGC output.

Demodulation of the modulated carrier into its original baseband components may be performed either coherently or non-coherently. Generally, coherent demodulation extracts a reference carrier from the received signal, removes the noise from the carrier, and multiplies the band-pass filtered signal with the clean reference. A block diagram of a typical coherent demodulator is shown in Figure 6(a). (Ref. 6) Notice that the coherent demodulator is a feedback system. Also note that the demodulation requires two carrier signals 90° out of phase to demodulate the quadrature components present in the DQPSK signal. The multiplication of the received signal and the reference carrier generates two spectral components: one at the baseband rate which contains the data information, and one at twice the modulation frequency which contains no information. The double frequency term is filtered out by the low-pass filter following the multiplication. The output of the low-pass filter is then sampled at the desired location in time by the decision block as a function of the timing recovery circuitry. The decision block provides a dibit output which represents the demodulated data during the current baud interval.

Non-coherent demodulation normally does not extract a clean reference carrier in the same sense as does coherent demodulation. A block diagram of a typical non-coherent DQPSK demodulator is shown in Figure 6(b). (Ref. 3) Notice in this case that the band-pass filtered received signal is delayed by a baud interval, then is divided into two paths which are 90° out of phase. The signal processing for the coherent and non-coherent schemes is quite similar except for the establishment of the demodulating signal. Coherent demodulation provides 2.5 dB greater signal-to-noise ratio performance than non-coherent demodulation. This is explained rather simply by recalling that the coherent demodulator provides a noise-free reference signal for demodulation. The non-coherent demodulation signal still contains in-band noise which has not been filtered out. Thus, in the non-coherent scheme, the carrier is being demodulated with a noisy reference.

Similar to DQPSK demodulation, FSK demodulation may also be performed either coherently or non-coherently. Because FSK modulation is encoding only a single bit per baud interval, the FSK demodulator requires only a single demodulation path. An example of a coherent FSK demodulator using a phase-locked loop is shown in Figure 7(a). Note the similarity of this demodulation scheme to that discussed previously for DQPSK demodulation. The differences are that the loop parameters are

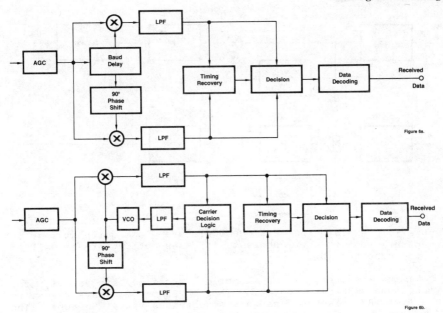

Figure 6: While the non-coherent DQPSK demodulator shown in (a) looks similar to the coherent DQPSK demodulator shown in (b), the carrier extraction in (b) yields superior performance. The coherent demodulator is also known as a Costas loop.

different for the two schemes, and certain blocks are absent in the FSK demodulator, but it is really a "subset" of the coherent DQPSK demodulator. In the coherent DQPSK demodulator, the VCO output is required to generate a stable carrier reference which does not vary with the data pattern and the noise of the received signal. In the coherent FSK demodulator, the VCO frequency should track the received frequency, and not lock to the center frequency between the two frequency shifts. The indication of the data is the output of the loop filter which is then low-pass filtered. The decision block makes its decision on the output of the loop filter as to the content of the data during that baud interval.

Non-coherent FSK demodulation may be performed in a number of different ways. A scheme which maintains compatibility with that required for non-coherent DQPSK demodulation is shown in Figure 7(b). This demodulation scheme is known as a product demodulator. The phase shifter provides a 90° phase shift at the mean frequency between the two frequency shifts. This signal then is used to demodulate the received signal. The key to this demodulation scheme is the phase shift at the FSK mean frequency. If the instantaneous received frequency is above or below the mean frequency, the phase shift will not be 90°, but above or below 90°. After low-pass filtering, the output will provide the cosine of an angle which is either above or below 90°. If above 90°, the cosine is negative; if less than 90°, the cosine is positive. Hence, the decision logic simply tests the sign of the signal from the loop filter for an indication of the bit received during the current baud interval.

VLSI Signal Processing

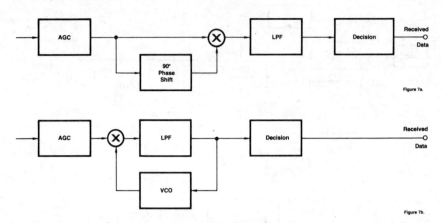

Figure 7a.

Figure 7b.

Figure 7: Again, the FSK demodulators look quite similar in nature, but the non-coherent (a) does not perform as well as the coherent (b).

Timing recovery is a function required exclusively for synchronous (DQPSK) modems. The DQPSK modem always transmits data over the telephone network synchronously at 1200 bps. The user may provide data to the modem at character asynchronous intervals (and bit asynchronous, within limits), but this asynchronous data is converted from its asynchronous format and transmitted synchronously over the line. This contrasts with FSK modulation where the user may transmit data through the modem at asynchronous intervals. The FSK carrier will change as a function of the digital data change provided by the user.

Because of the synchronous transmission nature of the DQPSK modem, and the cosine shaping provided in the modem transmitter, there exists an ideal sampling instant where inter-symbol interference (ISI) is minimum (ideally zero). The timing recovery clock extracts the 600-baud clock from the received data. The decision block samples the output of the low-pass filters at synchronous zero-ISI intervals determined by the timing recovery clock. It turns out that the spectrum of the received data provides a spectral null at the 600Hz component required by the timing recovery circuitry. The trick then is to perform a non-linear transformation on the data pattern which will create a spectral component at 600Hz which may then be extracted and used as the timing recovery clock. The system shown in Figure 8 provides this function. One channel of the demodulated baseband data is delayed and multiplied by itself.

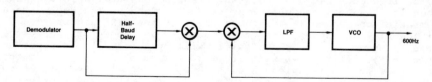

Figure 8: The timing recovery for a DQPSK modem involves creating a double frequency carrier from the demodulator. A phase-locked loop then provides a precise timing signal.

VLSI Signal Processing

This creates a non-linearity which generates a discrete spectral line at 600Hz. This component may then be locked on by a phase-locked loop which provides a clean version of the timing clock.

Algorithm Implementation

The IIR filter, FIR filter, and multiplication algorithms implemented on this single-chip modem have been tailored to work in the shift-and-add architecture of the silicon. Some hardware modifications to the modem ALU are necessary to perform particularly difficult portions of the multpication, AGC and VCO algorithms. The coding of the IIR and FIR filters will first be detailed, followed by some discussion of the multiplication and VCO coding with the requisite hardware changes.

In a standard shift-and-add architecture, an eight-bit filter coefficient requires eight ALU operations per coefficient. This is not only terribly wasteful, but would make the modem implementation virtually impossible in silicon. Fortunately, the architecture allows both add and subtract operations. By grouping the coefficient bits together, it is possible use the addition and subtraction to represent each coefficient in a canonic sign-digit format. An example of a canonic sign-digit coefficient is given below.

$-0.11110011111_2 = -0.9521484_{10}$ binary representation
eight operatons in a add only architecture

$-1.000\bar{1}01000\bar{1}$ = $-2^0 (1-2^{-4} (1-2^{-2} (1-2^{-4})))$
canonic sign-digit representation
four operations in an add/subtract arch.

($\bar{1}$ means subtraction of this bit instead of addition)

With this enhancement to the standard add-and-shift ALU, it is possible to represent coefficients in both IIR and FIR filters using only three to four bits while incurring a less than 0.1 % coefficient rounding error. Coefficients greater than 1.0 in magnitude use a four-bit representation, while those less than 1.0 are three-bit representations. A standard second-order sectons can be implemented in 17 operations.

IIR filters are designed using a proprietary algorithm developed explicitly for this type of ALU architecture. After the initial IIR filter is designed and the coefficients for the filter are rounded, the filter is conditioned to reduce the possibilty of ALU overflow. The poles and zeros are grouped such that they are within close proximity of one another. This minimizes the amplitude peaks in every second-order section. Next, the filter sections are ordered from lowest to highest amplitude peak (Ref. 7) to provide the least possibility of overflow.

Most of the FIR filters are the simple-coefficient decimators that are quite simple to design and generally use only one or two bits per coefficient. There are a few filters in the modem which are high-order FIR sections. The McClellan-Parks algorithm (Ref. 8) is used to design these filters. Coefficients for the FIR

VLSI Signal Processing

filters are normally rounded to three bits in the canonic sign-digit format.

Multiplications of two data values are required in the AGC, demodulator, and timing recovery blocks of the modem receiver. A modified Booth algorithm was employed to do the multiplication in the shift-and-add architecture. Special hardware was developed in the ALU to take the magnitude of both multiplicands, and to readjust the sign of the product at the completion of the multiplication. After the magnitude of both operands is taken, another section of ALU hardware takes control of the adder/shifter and performs the actual Booth multiplication. When the complete product has been formed, the sign is corrected by another piece of hardware.

The voltage controlled oscillator (VCO) also requires that the ALU hardware be slightly modified for implementation. The sine synthesis portion of the VCO operates on the same principle as the transmitter synthesizer. Two values, the current phase and the current phase increment are stored in the ALU's data RAM. A simple low-pass filter in the phase-locked loop feedback path determines whether to increase or decrease the current phase increment. Increasing or decreasing the phase increment alters the frequency that is being generated by the VCO. The updated phase increment is added to the current phase value before the phase-to-amplitude conversion occurs in the sine ROM.

Conclusion

The single-chip modem described in this paper could be upgraded to work at higher bit rates. An on-chip multiplier is warranted in the silicon implementation of higher speed modems. The multiplier can also be used for the adaptive filtering algorithms necessary to place these modems on the telephone network. Adaptive algorithms would also necessitate some coefficient RAM which was not provided on this Bell 212A modem. The key point to implementing higher speed modems in silicon is the simplicity which DSP provides to upgrade a component. The addition of a multplier and coefficient RAM does not imply a complete redesign of the ALU structure. In fact, the ALU may be used "as is" if the number of microcode instructions has not increased significantly. This concept of upward compatibility cannot be applied to the integrated analog versions of new modem types.

The upward compatibilty of architectures will continue to be applied to build faster modems. The modem described in this paper borrowed its architecture from a previous modem project; as future VLSI modems will borrow from the Bell 212A integrated modem. By using a shift-and-add architecture with some modifications to implement Booth's multplication algorithm, a complete Bell 212A compatible modem has been designed. The canonic sign-digit representation of coefficients and proprietary filter design algorithms help reduce microcode requirements for this VLSI component. Trade-offs between several system block implementations have been considered while trying to maintain microcode compatibility between the FSK and DQPSK modem types. By placing both the FSK and DQPSK modems on a single-chip, this component is a boon for the user who desires full Bell 212A compatibility.

References

1. R.W. Lucky, J. Salz, E.J. Weldon Jr., *Principles of Data Communication*, McGraw-Hill, New York, 1968, Ch 8.

2. J. G. Proakis, *Digital Communications*, McGraw-Hill, New York, 1983, Ch 3-4.

3. K. Feher, *Digital Communications, Satellite/Earth Station Engineering*, Prentice-Hall, Englewood Cliffs, 1981, Ch 4.

4. K. S. Shanmugam, *Digital and Analog Communication Systems*, John Wiley and Sons, New York, 1979, Ch 8.

5. H. W. Cooper, "Why Complicate Frequency Synthesis?," *Electronic Design*, pp.80-84, July 19, 1974

6. F. M. Gardner, *Phase-Lock Techniques*, John Wiley and Sons, New York, 1979, Ch 10.

7. L.B. Jackson, "Roundoff-Noise Analysis for Fixed-Point Digital Filters Realized in Cascade or Parallel Form," *I.E.E.E. Transactions on Audio and Electroacoustics*, vol. AU-18, pp.107-122, June 1970

8. J.H. McClennan, T.W. Parks, and L.R. Rabiner, "A Computer Program For Designing Optimum FIR Linear Phase Digital Filters," *I.E.E.E. Transactions on Audio and Electroacoustics*, vol. AU-21, pp.506-526, December 1973

Stevan Eidson was born in Inglewood, California on 4 May 1959. He received his BSCS and BSEE from the University of California, Irvine in 1980 and 1981, respectively. He received an MSEE from UCLA in 1982. From 1980 to 1982, Steve was a member of the technical staff at Hughes Aircraft Company's Ground Systems Group in Fullerton, California. Since 1982, he has been a senior system design engineer responsible for algorithm development in Advanced Micro Devices' Product Planning Division in Sunnyvale, California. Mr. Eidson is a member of Eta Kappa Nu and the IEEE Acoustics, Speech, and Signal Processing Society.

David M. Taylor was born in Sacramento, California on 19 February 1958. He received his BSEE degree from California Polytechnic State University (Cal Poly) in San Luis Obispo in June 1981. He is currently pursuing his masters degree at the University of Santa Clara. Dave was employed by Lockheed Missiles and Space Company in Sunnyvale, California from 1979 to 1981.

Mr. Taylor has been employed by Advanced Micro Devices in Sunnyvale, California since 1981. His current title is Section Manager in AMD's Product Planning Division. His duties include defining new data communication products and performing the systems design on these products. He has had a number of technical articles published since 1980.

VLSI Signal Processing   15

THE INTEGRATION OF A 144 Kb/s LINE TRANSMISSION SYSTEM

A.M. Brown
T.M.C. LTD.
Wiltshire
England

Introduction

As a precursor to an Integrated Services Digital Network I.S.D.N. a Pilot Integrated Digital Access I.D.A. will be launched in the U.K. during the last quarter of 84. The basic access to this network will initially be based on a 80Kb/s transmission scheme between local telephone exchange and subscribers premises. Two transmission techniques are available to support the access, a so called 'burst mode' system and echo cancellation. Echo cancellation offers a superior performance, particularly in terms of transmission range, at the expense of a more complex implementation. The viability, in terms of cost, physical size and power consumption, of implementing such D.S.P. techniques in telecommunication equipment has already been demonstrated (ref. 1).

The basic access rate for the U.K. is expected to be increased to 144 Kb/s in 86/87, thus being in line with C.C.I.T.T. recommendations. Due to the different characteristics of National local networks no attempt has been made to standardize the transmission line interfaces (i.e. line code, transmission levels etc.).

The author is concerned with the development of a 144 Kb/s Line Transmission System based on echo cancellation, specifically designed for the U.K. network. This does not of course preclude its use on other similar networks.

Design Constraints

The subscriber end of the transmission system will form the major part of the so called Network Terminator (N.T.) function (ref. 2).
The N.T. is eventually seen as the digital equivalent of the existing analogue 'telephone socket'.

Line powering of the N.T. plus provision for passing up to 400mW over the subscriber interface is a definite requirement. Given that the maximum available power at the subscriber end of the line is approximately 1W, then assuming an 80% power extraction efficiency, a budget of 400mW is available to power the whole of the N.T. function. A target power consumption for the line transmission element is 250-300mW.

Thus the design of a low cost, low power and small sized unit is demanded.

At the exchange end of the system several (4 to 8) transmission units must be accommodated on a single line card; so size is again at a premium. Although the powering requirements are no longer so restrictive the cost constraint still exists, since the transmission terminations are required on a per subscriber basis.

A transmission range of 40dB at 100KHz is the design objective, this equates to 5.3Km on a 0.5m.m. gauge cable giving a potential coverage of approximately 98%. The subscriber and exchange units must be easily configured for use in a single repeatered connexion, thus giving 100% potential coverage of the network.

Design Philosophy

A major objective of the design has been to employ D.S.P. techniques for as much of the system as is practicable and thus benefit from the attendant advantages of accurate computer modelling and relative ease of subsequent integration, repeatability of performance etc. The main restriction in applying D.S.P. is imposed by the A/D conversion requirements.

Ideally the the A/D converter would be placed as close to the line interface as possible, perhaps directly following a simple hybrid transformer arrangement.

To achieve the required performance a 12 bit converter with a conversion time of some 3 microseconds, good linearity and low power consumption ($<$ 60mW) is called for.

Several approaches to achieving such a converter are being actively pursued. These include the use of oversampling Delta-Modulation coupled with digital filtering, Flash conversion and Algorithmic Pipelined conversion (ref. 3)

The Delta Modulation technique requires a high sampling rate of some 20MHz plus a degree of signal processing comparable in complexity to the echo canceller itself.

Direct Flash techniques are known to easily achieve the speed and power requirements, but are unlikely to meet the 12 bit accuracy and would require an extremely large integrated circuit.

Algorithmic pipelined techniques could be applied to this system because the time delay associated with this approach would not present a problem. Power consumption targets should be easily achieved. Unfortunately the 12 bit accuracy would again present a severe problem.

VLSI Signal Processing

At present a 9 bit converter, utilising a half-flash technique, meeting the speed and power requirements is achievable.

Because of the outstanding problems still seen to exist in this area the system has been designed to take full advantage of a 12 bit converter when available but will still meet the required specification using a 9 bit A/D with additional front end analogue components.

The design has proceeded in three distinct phases. Extensive computer modelling and simulation, followed by the construction of a breadboard model and system testing, then finally the custom integrated circuit design, simulation and processing.

Removal of the breadboarding phase, which can of course be unwieldly and time consuming, is not seen as practical for such a complex system. Indeed the information gained in submitting the model for system testing has been invaluable, moving with confidence to the integration phase.

System Description

The line code is A.M.I. chosen for spectral efficiency, ease of clock recovery and good cross-talk performance. An aggregate line rate of 160 Kb/s is used.

The composite line signal is digitised following a hybrid line transformer arrangement and an analogue receive filter. The filter function could easily be performed digitally and the hybrid circuit simplified if a suitable 12 bit A/D converter becomes available. At present the resultant 9 bit word is input to the canceller for echo replica subtraction.

The echo canceller is realised as an adaptive digital interpolating transversal filter, using a stochastic update algorithm. An interpolation factor of two means the incoming line signal is sampled at 3.12 microsecond intervals. The canceller coefficient update is controlled such as to occur only during a zero in the received A.M.I. coded signal, thus an effective zero reference configuration is achieved. On activation of the system the canceller updates on all samples for a fixed number of iterations, sufficient to guarantee reliable receive signal decisions.

The maximum cancellable echo delay is 16 data bit periods requiring a 32 coefficient filter. Sixteen bit coefficient accuracy is required.

VLSI Signal Processing

Following the cancellation process the resultant receive signal is equalised to remove intersymbol interference. Only post-cursor interference presents a problem with the A.M.I. code and local network transmission lines. Thus an adaptive decision feed-back equaliser (A.D.F.E.) can be utilised.

The impulse response of the digital equaliser extends up to 7 data bit periods and thus requires 14 coefficients.

All subsequent processing following the equaliser is performed digitally, this includes the clock extraction and phase locked loop in the subscribers unit (i.e. subscriber unit is synchronised to the exchange unit as master).

Because of the synchronous nature of the system and the use of a sampled implementation, the ideal sampling instance for receive signal decisions at the exchange, will not be coincident with the corresponding instance of cancellation. The latter being defined by the fixed exchange master clock. A technique for adaptively adjusting the cancellation instance to be concurrent with the ideal decision point (i.e. at maximum eye opening) is implemented in the system.

System Integration

The system comprises two custom integrated circuits plus a small number of analogue components. Both the echo canceller and A.D.F.E. are combined on a single integrated circuit. The second circuit provides control functions for activation/de-activation, framing extraction/insertion and programmable exchange/subscriber interfaces.

The echo canceller/A.D.F.E. cct. is designed in four phase dynamic logic, a technique that lends itself to the pipelined architecture adopted for the design. Parallel pipelined techniques are used throughout the design to minimise the system clock rate and hence power consumption. As might be predicted the high operational delay incurred by using a pipeline technique becomes critical in the A.D.F.E. where the nth output value is required for the $(n + 1)^{th}$ processing cycle. Indeed the maximum tolerable time delay for the A.D.F.E. operation defines the minimum system clock frequency allowable, which for the described system is 2.5MHz.

The ccts. will be realised in a 3 micron silicon gate NMOS process, an estimated size for the larger echo canceller/A.D.F.E. cct. is 240mil x 240mil comprising 3,000 four phase gates.

VLSI Signal Processing

Conclusions

The design of an integrated 144 Kb/s Line Transmission system has been described. Following extensive computer simulation a breadboard model was constructed and submitted for testing to British Telecomm. Results of the tests show the transmission performance objectives have been met. (ref. 4)

An integrated version of the system, comprising two custom integrated circuits plus a small number of analogue components, has been designed. The integrated ccts. are at present in the layout phase and samples are expected the first quarter of 85.

References

[1] Vogel E.C. and Taylor C.G.: 'Digital moves into the local loop', The International Symposium on Subscriber'Loops and Services, Toronto (1982).

[2] I Series C.C.I.T.T. Recommendations

[3] A CMOS Pipeline Algorithmic A/D Converter: Masuada, Kitamura, Ohya and Kituchi.

[4] A New Transmission System for ISDN Access at 144 Kb/s: M. Vry
I.S.S.L.S., Nice, (1984).

16
AN ECHO CANCELLER

YUSUF A. HAQUE
AMERICAN MICROSYSTEMS, INC.
A SUBSIDIARY OF GOULD, INC.
3800 HOMESTEAD ROAD
SANTA CLARA, CA 95051
(408) 246-0330

An echo canceller suitable for integration in MOS technology will be described. The device is intended to perform adaptive balance of analog repeaters and hybrids [1]. Results reported here have been obtained on a prototype breadboard built around an adaptive transversal filter.

Fig. 1 shows an application diagram for the device. The far end speaker is the analog input to the transversal filter. The echo of the far end speaker together with the near end speaker signal is substracted with the output of the transversal filter to obtain a reduced echo version of the near end speaker at node X.

The device operates on least mean square (LMS) gradient minimization approach [2]; the algorithm used is a modified LMS scheme [3]. Fig. 2 shows a block diagram of the device. The transversal filter has 16 tapped delay lines and is implemented in the analog domain with digitally controlled tap weight coefficients. Each coefficient is digitized to 9 bits. The output of the transversal filter is fed into the error amplifier whose other input is the reference signal to which the transversal filter adapts to in its attempts to minimize the power of the error signal. The output of the error amplifier is digitized to 5 bits using an analog to digital converter. A delayed analog input is also digitized to 3 bits using the same converter. These two words are multiplied and added to the old tap weight vector to obtain its new estimate. These operations are performed by a parallel pipelined Booth's algorithm multiplier and an accumulator. The computation is performed in 12 bits (tap vectors are stored in memory in 12 bits; all arithmetic is done in 12 bits, and is truncated to 9 bits for use in the filter) to obtain better stability [4].

Fig. 3 (a) shows the response of the system under sinusoidal excitation for a purely resistive line. Fig.3(b) shows a schematic of the set-up. Fig. 4 shows the response under identical conditions to pseudorandom input signals. Fig. 5 shows the response of the device under pseudorandom noise input conditions, where the reference signal (echo) is passed through a lowpass filter. The transversal filter emulated this transfer function and the output of the echo canceller had a 30-35db residual error. It was experimentally determined that the rejection was correlated with the noise floor of the experimental setup (i.e. after system has adapted, the adaption process is frozen and the analog and reference input are grounded, the residual error is about the same as obtained earlier.)

The transversal filter was implemented discretely and used some integrated circuits whose signal swing limited the input signal for the transversal filter to 1.4V peak to peak. This was a major factor that limited the dynamic range of the device. A 14db improvement can be expected for a chip with a 7 volt peak to peak output. Further, the dynamic range of the transversal filter was limited by fixed pattern noise [5] due to mismatch of discretely implemented components. It is expected for an integrated device, telephone quality dynamic range along with 40-45 db of residual error would be obtainable.

LIST OF REFERENCES

[1] D.G. Messerschmitt, "Adaptive Balancing of Hybrids in Telephony," International Symposium on Circuits & System, 1981, p 716.

[2] B. Widrow, J.R. Glover, I.M. McCool, J. Kaunitz, C.S. Williams, R. H. Hearn, J.R. Zeidler, E. Dong, Jr., R.C. Goodlin, "Adaptive Noise Cancelling: Principles and Applications," Proceeding of the IEEE, Vol. 63, No. 12, Dec. 1975, p 1692.

[3] B.K. Ahuja, M.A. Copeland, C.H. Chan, "A New Adaptive Algorithm and its Implementation in MOSLSI." IEEE Journal of Solid State Circuits, Vol. SC-14, No. 4, Aug. 1979.

[4] V. Feldmus, private communication.

[5] Y. Haque, M.A. Copeland, "Distortion in Rotating Tap Weight Transversal Filters," IEEE Journal of Solid State Circuits, SC-14, No. 3, June 1979, p. 627.

VLSI Signal Processing

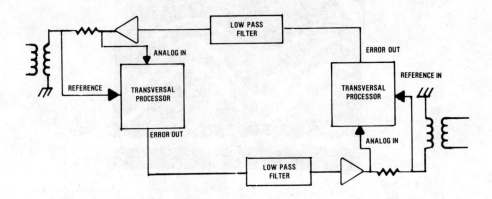

Figure 1. Block diagram of echo cancellor performing 2-4 wire adaptive balance functions.

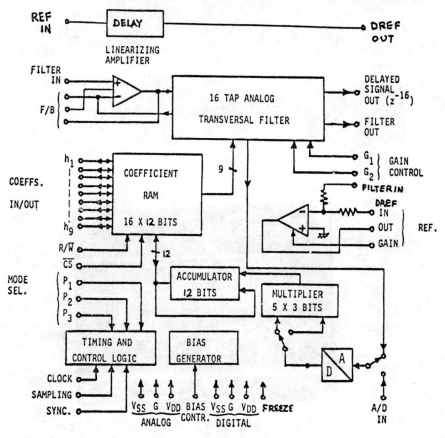

Figure 2. Block diagram of device.

119

VLSI Signal Processing

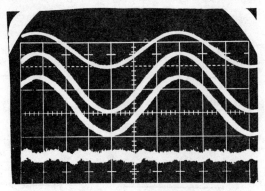

Figure 3a. The response of the system under sinusoidal excitation is shown. Top trace is the input, the next two traces are the transversal filter and reference outputs (0.5U/div) and the lower trace is the error output (10MU/div).

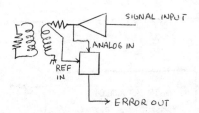

Figure 3b. Schematic of setup.

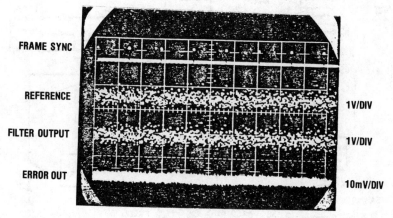

Figure 4. Response under pseudo random noise input conditions.

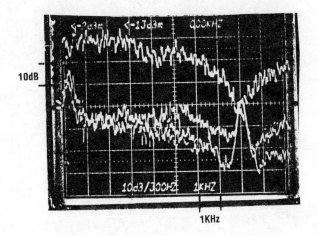

Figure 5. An arbitrary low pass transfer was emulated by the adaptive transversal filter. Top waveform shows the emulated low pass function, the next waveform shows the error out (about 30dB rejection). The lowest trace is an estimate of the noise of the setup.

Part V
VLSI ARCHITECTURES II

Edited by:
Dr. Robert Owen

NEW MODULAR CMOS CORRELATOR CHIP

Dr. John A. Eldon
TRW LSI Products
San Diego, California

ABSTRACT

TRW LSI Products is building a new low-power CMOS correlator chip. The major problem in defining a new correlator is the diversity of the correlator/convolver market. Customer requirements span a wide range of resolutions and numbers of taps, requiring various internal architectural dimensions. The new chip, the TMC2220, is ideally suited to applications involving single or multibit data and a single bit reference, such as commonly arise in robotics and communications with interference or other noise in the data channel. The block diagram shows the four identical internal modules, each of which is a self-contained 32-tap, single bit correlator. These four modules may be tied together in series or in parallel to build up a 4x1x32 correlator, a 2x1x64 correlator, or a 1x1x128 correlator, as desired. Thus, a single chip can address these diverse applications. Furthermore, the chip can be partitioned into two halves, for quadrature applications requiring separate correlation of I and Q data channels.

BACKGROUND

Correlation is a time domain operation widely used to detect weak signals buried in noise, to decode encrypted signals, and to measure time delays of known signals. It continually finds new uses in communications, radar, sonar, robotics, and biomedical imaging systems.

Although all correlation applications are based ultimately on the same fundamental equation,

$$C(x,y,\eta) = \int_{-\infty}^{\infty} x(t) \cdot y(t-\eta) dt \qquad (1)$$

their implementations differ radically, depending on the nature of the signals being examined. Note that correlation always involves two incoming time (t) domain signals (x and y) and a phase offset (η) between them. In a digital system, each signal may be a series of single bit or multibit samples, and the correlation integral becomes a finite sum. Where single bit signals are employed, the two bit values 1 and 0 are generally interpreted as positive and negative. In multibit applications, the bits of each sample are accorded the standard unsigned magnitude (offset binary) or two's complement relative weightings of 1, 2, 4, ...

In applications involving single bit reference and single bit data streams, multiplication is actually implemented with the exclusive-NOR (NXOR) function, which yields positive results (1) if the two bits (polarities) coincide and negative results (0) if they differ (Figure 1). If these 1's and 0's are then summed (i.e., if the 1's are tallied), then the result is a correlation score ranging from 0 (for perfect anticorrelation) to N (for perfect correlation, where N is the number of taps in the system).

Applications involving a multibit data path and a single bit reference are a simple extension of the single bit x single bit case. Here, M parallel single bit channels are employed, where M is the number of bits of data resolution. Within each channel, we are still performing a bit match, this time between a positive/negative value and a 0 or 1. The simplest way to handle this is to continue to use the exclusive NOR logic between data and reference within each channel, then to sum these results over the various channels, according to the respective weightings of their data bits (Figure 2). The result will be a binary number which is near maximum if the data matches the expected reference and near 0 if the data mismatches the reference.

Finally, many correlation applications require a full mathematical convolution of multibit data against a multibit reference, as in a FIR Filter. These cases require a true binary multiplier at each tap, i.e., an MxM array of AND gates and adders. Figure 3 is a Block Diagram of TRW's TDC1028 chip, which is dedicated to convolutions of multibit references with multibit data.

Faced with this wide range of market requirements, TRW has built the TDC1028 and a family of multipliers to address the multibit-multibit market, plus three dedicated single channel correlator chips (TDC1004, TDC1023, and TMC2221) for single bit applications. The new TMC2220 handles single bit-multibit problems adeptly, expanding the market coverage by convolutional IC's.

CHIP ARCHITECTURE

Figure 4 illustrates the interconnections of the four 1x32 correlator modules comprising the TMC2220. The TMC2220 is a general purpose, high pin-out part, from which various specialized versions can be derived. The same silicon and lower interconnections are used for all chips in the family; only the last level metal interconnects and the bonding pattern change. The primary justification for the specialized derivative chips is reduced interconnection count and package size, for greater reliability and board density and lower cost.

The Correlator Modules

Each of the TMC2220's four modules performs the basic function:

$$\text{CORRELATION} = \text{SUM}(\text{AND}(\text{NXOR}(D_i, R_i), M_i)), \quad (2)$$

where the D_i are the current contents of the data register, the R_i are the corresponding values in the reference holding latch, and the M_i are the corresponding latched masking values (Figure 5). Each correlator module also features a reference preload register, to permit the user to load a new reference pattern (requiring up to 32 clock cycles) while completing a correlation with the old reference, already stored by the latch.

The mask function, implemented with one latched AND gate per cell, tells the chip to include only a specified subset of its taps in the final correlation score. The user programs this function by filling the data register with "1's" and the reference register with the desired masking pattern ("1"

VLSI Signal Processing

for each enabled tap and "0" for each omitted tap). Bring the LM (latch mask) control from high (track) to low (hold) locks in the mask pattern.

Each correlator module feeds a "half-scale corrector," which passes or subtracts 16 from the current output. The correction mode maps the range of outputs from -16 (anticorrelation) to +16 (correlation), just as the mathematical correlation function runs from -1 to +1.

The individual modules are laid out as illustrated in Figure 6. Note that the reference latch is integrated with the NXOR function into a single cell, as the masking latch is integrated with the AND gate into a cell. Thus, the central core of the correlator module is simply a 4x32 array of three types of standard cells, facilitating a high layout packing density. The separation of the chip into four independent single channel modules ensures maximum versatility.

Weighting Multipliers

Most applications require the sums from each pair of modules to be combined into a linear sum. The general purpose TMC2220 permits the user to select the coefficients for this sum, as shown in Table 1. (The superfluous combinations come "free" with the necessary ones.)

Typically, a specialized version of the chip will require only a single preprogrammed set of weighting factors. For example, the weighting factors for the TMC2221 single bit correlator are 1:1:1:1, since this is the desired equal treatment of all taps of the full correlator. Each weighting multiplier comprises two adders and two multiplexers.

TABLE 1. WEIGHTING LOGIC SUMMARY

W_2	W_1	W_0	FUNCTION PERFORMED
0	0	0	$Q_L + Q_M = Q$ and $I_L + I_M = I$
0	0	1	$Q_L + 3Q_M = Q$ and $I_L + 3I_M = I$
0	1	0	$Q_L + 4Q_M = Q$ and $I_L + 4I_M = I$
0	1	1	$Q_L = Q$ and $I_L = I$
1	0	0	$Q_M = Q$ and $I_M = I$
1	0	1	$2Q_L + 3Q_M = Q$ and $2I_L + 3I_M = I$
1	1	0	$2Q_L + 4Q_M = Q$ and $2I_L + 4I_M = I$
1	1	1	$2Q_L + 5Q_M = Q$ and $2I_L + 5I_M = I$

Multiplication by 1, 2, or 4 is accomplished by an appropriate shift left. Multiplication by 3 is the sum of a multiply by 1 and a multiply by 2. Similarly, a multiply by 5 is the sum of a multiply by 1 and a multiply by 4.

Recombining Matrix

The outputs of the two weighting multipliers are fed to the recombining matrix, whose four operating modes are: 1) PASS Q; 2) I+Q; 3) I + Q/2;

and 4) MAG. The output of this section drives the chip's main output port, while the output of the I section multiplier drives the auxiliary output port.

In the PASS Q mode, the chip outputs two independent correlation scores, one for each pair of correlator modules. These modules in turn could be wired either in series (as a single 1x64 channel), or in parallel (as a 2x32 channel) by the user. In this mode, the chip functions as a pair of separate correlator chips with common master clock and power supply connections.

In the Q+I mode, the outputs of the two weighting multipliers are added directly, with equal weighting. This mode is particularly useful where the chip is to be used as a single channel 1x128 correlator, with all taps receiving equal weighting. In the Q/2 + I mode, the sum from one module pair is added to half that of the other, to produce the final output. This is particularly useful when the chip performs a 4x32 correlation, with normal binary weighting of 8:4:2:1 among the 4 channels. Each pair of channels is combined with a 4:1 relative weight; the pairs are combined with a 2:1 weight, yielding 8:2:4:1 overall weighting.

In the MAG mode, the chip computes the absolute values of the outputs of the two weighting multipliers (the TC mode should be used) and adds the larger to half of the smaller. The logic flow for this function is as follows:

1) Compute absolute values of the two inputs:

   IF Q<0, Q = NOT (Q) + 1
   IF I<0, I = NOT (I) + 1

2) Compare the absolute values and combine:

   IF Q-I<0, OUT = Q/2 + I
   IF Q-I>0, OUT = I/2 + Q

Figure 7, The Recombining Matrix Block Diagram, can be visually checked against this operation sequence.

CELLS USED IN THE CORRELATOR

The internal circuitry of the correlator is built from the following types of standard and semistandard cells:

1) LATCHED NXOR GATE
2) LATCHED AND GATE
3) SHIFT REGISTER CELL
4) 4-BIT ADDER
5) 2-INPUT MULTIPLEXER

Figures 8 and 9 are Logic and Schematic Block Diagrams of the the latched NXOR and AND gate cells. By integrating the separate functions of latch and logic gate into a single cell, we were able to minimize the design.

OPTIONAL VERSIONS OF THE CHIP

Currently, only the general purpose version (TMC2220) and the single bit single channel version (TMC2221) are committed new products. However, many other chips are possible, (Table 2) including:

VLSI Signal Processing

2x64 dual correlator -- This would be essentially the equivalent of a pair of TRW TDC1023 bipolar correlator chips, although the CMOS version would lack the threshold logic and offers much lower power dissipation. The modules would be connected serially into two pairs and the (unweighted) sum from each pair would emerge through a dedicated port.

4x32 single channel correlator -- This device, dedicated to 4 bit x 1 bit applications, would provide 8:4:2:1 binary weighting among the four modules and present only this total as an output. The data and reference I/O ports for the four modules would remain independent, but the various loading and enabling clocks would be tied together to save pins. Figure 2 shows how such a correlator would be fashioned from the basic TMC2220.

TABLE 2. SOME POTENTIAL VERSIONS OF THE TMC2220

FUNCTIONS	I/O PINS	INSTR PINS	PACKAGE PINS
TMC2220--general purpose programmable version 1x128, 2x64, 4x32	12 shift in 8 shift out 18 corr out	15 enables 7 instrs	64
TMC2221--single bit, single channel version 1x128	3 shift in 2 shift out 8 corr out	5 enables 2 instrs	28
Uncommitted potential product--single channel, 4-bit binary version 4x32	12 shift in 8 shift out 10 corr out	5 enables 3 instrs	48

CONCLUSIONS AND SUMMARY

The TMC2220 is a versatile chip capable of correlating a stream of single bit reference words against a single or multibit data stream. With its user-programmable weighting functions, the chip supports various special applications, including: binary-weighted multibit correlation; extended length single bit correlation; dual (separate) channel one- or two-bit correlation; dual channel quadrature correlation with a single (vector) magnitude output. Using a hierarchical set of standard cells and modules, TRW has designed a chip which can be reconfigured to meet various special user requirements, yet which still offers the high operating speed and high layout density one expects of a full custom chip. Special purpose derivative chips can be housed in smaller packages if superfluous I/O connections are eliminated (Table 2).

VLSI Signal Processing

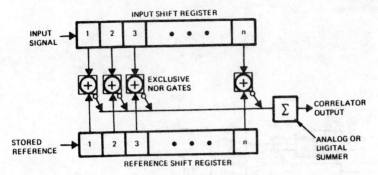

Figure 1. Basic Correlator Block Diagram

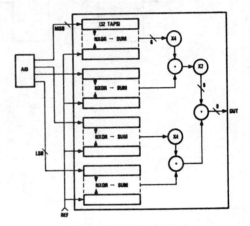

Figure 2. TMC2220 As A 4X1X32 Correlator

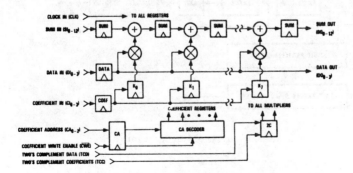

Figure 3. TDC1028 Block Diagram

VLSI Signal Processing

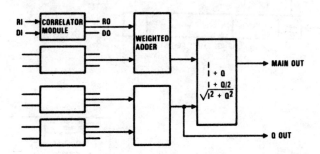

Figure 4. TMC2220 Block Diagram

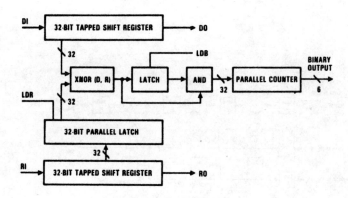

Figure 5. One-Bit Digital Correlator Module

VLSI Signal Processing

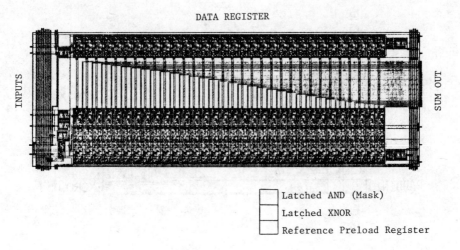

Figure 6. Metal Connectors for the Correlator Modules

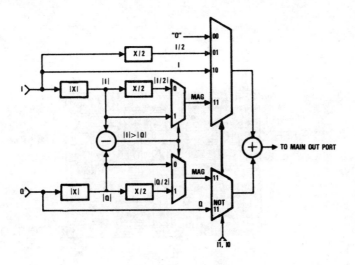

Figure 7. Recombining Matrix - TMC2220

VLSI Signal Processing

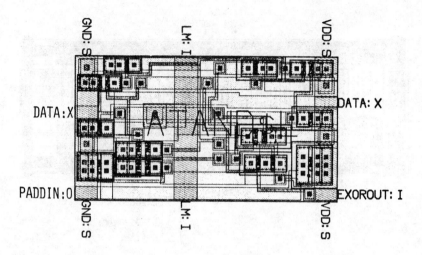

Figure 8. Latched AND Cell

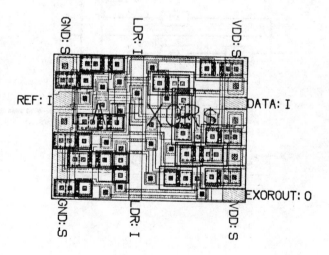

Figure 9. Latched EXOR Cell

A PIPELINED DATA FLOW ARCHITECTURE FOR DIGITAL SIGNAL PROCESSING THE NEC uPD7281

MITCHELL CHASE

NEC Electronics, Inc.
Natick Technology Center
One Natick Executive Park
Natick, Massachusetts 01760

ABSTRACT

The advent of VLSI technology has led to an array of integrated circuits designed for digital signal processing applications. Many of these devices are only variations of general purpose processors and therefore retain some fundamental limitations. A new device by NEC Electronics overcomes many of these limitations. The NEC uPD7281, using data flow architecture on a circular pipeline, specifically addresses the needs of digital image and signal processing.

INTRODUCTION

Microprocessors have traditionally been based on the von Neumann architecture. This means that the microprocessor spends a lot of time fetching instructions and data from memory. For operations best handled in parallel, this processor-memory bottleneck is particularly troublesome. The result is inefficient processing of many algorithms, regardless of how fast the microprocessor operates.

Various other computer architectures have been investigated to overcome this bottleneck. One method of interest for increasing the parallelism of execution is pipelining. Operations are performed in an assembly line fashion with pipelining. However, some functional versatility can be lost depending on how the pipeline is implemented.

The concept of attaching a 'tag' to the data to instruct the processor on what to do with it is the basis for the data flow technique. The computer word, called a token, consists of both data and a tag, in which the data is identified and the operations to be performed are encoded. Combining data flow techniques with pipelining can produce a powerful machine. NEC Electronics has taken this combination and developed it into a single VLSI peripheral device with ample capabilities for digital signal processing.

THE NEC uPD7281

The uPD7281, called the Image Pipelined Processor (ImPP) because it was initially conceived for tackling the demands of image processing, is well suited as a general digital signal processor. It consists of ten functional blocks and is designed to be used with a host processor. The device features a powerful instruction set, simple input/output handling, easy multiple processor configurations, a high speed 16x16 multiplier and rewritable control store in a 40 pin package fabricated with 1.75 micron N-channel MOS. The uPD7281 operates on a single five volt supply.

The uPD7281 is the first VLSI implementation of a processor employing data flow techniques and a circular pipeline. It successfully overcomes the problems of looping and branching that were thought to be inherent in data flow architectures. Although it requires a different method for programming, the uPD7281 is simple to use and to design with. This is clear when one looks at an example, but first the parts of the device must be known.

ImPP FUNCTIONAL DESCRIPTION

The uPD7281 is divided into ten functional blocks, five of which make up the circular pipeline (Figure 1). The blocks can be briefly described as follows:

- Input Controller (IC)

A 32-bit token is entered into the ImPP in two 16-bit increments using a simple request/acknowledge handshake. Once the token has entered the Input Controller, the module number (MN) is checked. The MN (contained in the four most significant bits of the token) identifies the intended ImPP in a multi-ImPP system. If the MN does not match that of the ImPP, the token is routed to the Output Controller.

- Link Table (LT)

The Link Table is a 128 x 16 bit dynamic RAM. The ID field of an incoming token is used as an address to the table. The contents of the addressed table are then used to form a new token (Figure 2). The LT adds to the token a six-bit Function Table Address (FTA), a new seven-bit ID field, a one-bit Function Table Right-field Control (FTRC) and a two-bit Select (SEL) field. The FTA is used to address the Function Table, FTRC for control information and SEL to identify the type(s) of instruction.

- Function Table (FT)

The Function Table is a 64 x 40 bit DRAM. It too takes the incoming token, and using the contents of the DRAM, forms a new token. The FT is composed of three fields: Function Table Left (FTL) for specifying Processing Unit instructions, Function Table Right (FTR) for specifying Address Generator and Flow Controller instructions and Function Table Temporary (FTT) for temporary control storage of counters and other control information.

- Address Generator and Flow Controller (AG & FC)

The Address Generator and Flow Controller generates the addresses and controls the reading and writing of the Data Memory. The AG & FC passes one-operand instructions to the Queue, or, for two operand instructions, pairs the operands together.

Figure 1

uPD7281 Block Diagram

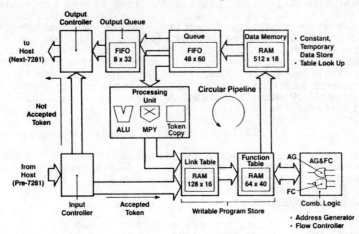

- **Data Memory (DM)**

   The Data Memory is a 512 x 18 bit DRAM used for temporary storage of data or constants. While waiting for the second operand of a two-operand instruction, the first operand will be placed in the DM. The DM can also be used for storage of look-up tables, numeric constants needed for calculations or as a buffer for I/O data.

- **Queue (Q)**

   The Queue is a first in - first out (FIFO) memory made of two parts: the Data Queue (DQ), a 32 x 60 bit DRAM, for temporary storage of tokens bound for the Processing Unit or Output Queue, and a Generator Queue (GQ), a 16 x 60 bit DRAM, for the Generate instruction.

- **Processing Unit (PU)**

   The Processing Unit performs two different functions. It does all the customary arithmetic and logic functions, including multiply, bit manipulations, shifts, compares and much more. Additionally, it can also generate new tokens or make multiple or block copies of existing tokens under program control.

- **Output Queue (OQ)**

   The Output Queue is a first in - first out memory configured with an 8 x 32 bit DRAM. The OQ is used to temporarily store tokens targeted to be output from the ImPP.

- **Output Controller (OC)**

   The Output Controller accepts tokens from either the IC or OQ, and outputs them in two 16-bit segments using a request/acknowledge handshake as on the IC.

VLSI Signal Processing

- **Refresh Controller (RC)**

The Refresh Controller injects special tokens into the circular pipeline to refresh all function blocks containing DRAM (LT, FT, DM, Q, and OQ). The RC can be set during program download from a host processor to provide refresh tokens every n-th clock cycle thus relieving the user of the burden.

Tokens flow in from the IC to the circular pipeline and out through the OC. The circular pipeline consists of the Link Table, Function Table, Data Memory, Queue and Processing Unit. The contents and width of the token changes as it passes through each block. Figure 2 shows how tokens are modified as they travel through the ImPP.

INSTRUCTION SET OF uPD7281

There are four classes of instructions for the uPD7281: Address Generator and Flow Controller, Processing Unit, Output and Generate (GE). The instructions are listed in Figure 3.

- **AG & FC Instruction**

The AG & FC instructions can be divided into those that generate addresses and those that control execution flow. There is also a third type that combines these two functions.

The addresses generated are for read/write operations on the Data Memory. The flow control instructions modify the data stream in a way that corresponds to the looping and related instructions of a conventional processor. The QUEUE instruction, used for queuing two operand instructions, combines the two functions.

- **PU Instructions**

These instructions use the Processing Unit to perform operations on the data field of arriving tokens. Among the operations that can be performed are logic (OR, AND, XOR, etc.), arithmetic (ADD, SUB, MUL, etc.), bit manipulations, barrel shift, compare, etc. All operations are performed in one pipeline clock cycle.

- **GE Instructions**

These instructions use the Processing Unit to copy data by making multiple copies of tokens.

- **OUT Instructions**

These instructions direct tokens to the Output Controller.

THE uPD7281 IN ACTION - AN EXAMPLE

To illustrate how data flow works, the following equation was chosen:

$$R(i) = [W(i) - X(i)] \times [Y(i) + Z(i)] \tag{1}$$

VLSI Signal Processing

Figure 2
Token Formats and Transitions

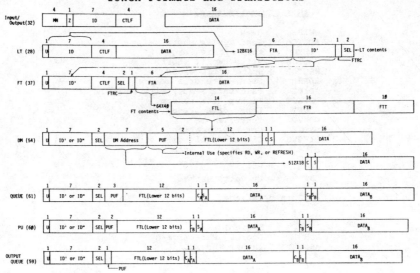

It is assumed that each input variable (W, X, Y and Z) can arrive in any order (i.e. Z, W, X, Y or Y, W, X, Z, etc.) but that, for each variable, the order is sequential (i.e. X(1) before X(2)). Note that in a conventional processor, the order of inputs must be known and maintained. Data flow allows the inputs to arrive in any order (this particular program is written so that for each variable within an array, the data must be in order).

With a conventional processor, for each R(i) calculated, the following sequence would have to be executed:

```
Load       W      ;put W into accumulator
Subtract   X      ;subtract X from W
Store      T      ;store W - X temporarily in T
Load       Y      ;put Y into accumulator
Add        Z      ;add Z to Y
Multiply   T      ;multiply (Y+Z) by T (W-X)
Out        R      ;output result
```

The order of each variable is prescribed and unalterable once the program is loaded. To construct a data flow program, the first step is to diagram the algorithm (Figure 4). This type of diagram is called a flow graph. The functions are represented by nodes. Links or arcs connect the nodes. The links will corresponds to entries in the Link Table, nodes with entries in the Function Table.

W and X are queued and then subtracted in node F1. The QUEUE instruction holds whichever variable (W or X) arrives first. When the second variable arrives, it is grouped with the waiting variable and sent to the Processing Unit. The result, L1, is then paired, or queued if necessary, with L2. Y and Z following a similar path. Notice that multiple Ws, for example,

Figure 3

uPD7281 Instruction Set

- AG & PC Instructions

QUEUE	Queue
RDCYCS	Read cyclic short
RDCYCL	Read cyclic long
WRCYCS	Write cyclic short
WRCYCL	Write cyclic long
RDWR	Read/write data memory
RDIDX	Read data memory with index
PICKUP	Pickup data stream
COUNT	Count data stream
CONVO	Convolve
CNTGE	Count generation
DIVCYC	Divide cyclic data stream
DIV	Divide data stream
DIST	Distribute data stream
SAVE	Save ID
CUT	Cut data stream

- PU Instructions

OR	Logical OR
AND	Logical AND
XOR	Logical Exclusive-OR
ANDNOT	Invert and AND
NOT	Invert
ADD	Add
SUB	Subtract
MUL	Multiply
NOP	No operation
ADDSC	Add, shift and count
SUBSC	Subtract, shift and count
MULSC	Multiply, shift and count
INC	Increment
DEC	Decrement
SHR	Shift right
SHL	Shift left
SHRBRV	Shift right with bit reversal
SHLBRV	Shift left with bit reversal
CMP	Compare
CMPNOM	Compare and normalize
CMPXCH	Compare and exchange
GET1	Get one bit
SET1	Set one bit
CLR1	Clear one bit
ANDMASK	Mask with logical AND
ORMASK	Mask with logical OR
CVT2AB	Convert 2's complement to sign-magnitude
CVTAB2	Convert sign-magnitude to 2's complement
ADJL	Adjust long (for double precision numbers)
ACC	Accumulate
COPYC	Copy control bit

- GE Instructions

COPYBK	Copy block
COPYM	Copy multiple
SETCTL	Set control field

- OUT Instruction

OUT1	Output 1 token
OUT2	Output 2 tokens

Figure 4

Data Flow Graph for Equation (1)

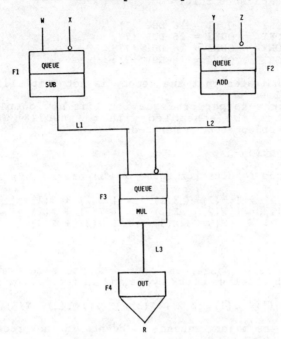

can arrive before the first X. The only limitation on the number of Ws, in this case, is the size of the queue. Queue size is set by the programmer using the functional assembler.

To translate the flow graph into program code, first label the inputs as shown in the data flow graph as follows, using the uPD7281 Assembly Language:

INPUT   W,X,Y,Z;

and the output:

OUTPUT   R;

Next, write all the links connecting nodes or a node and output:

```
LINK     L1 = F1(W,X);
LINK     L2 = F2(Y,Z);
LINK     L3 = F3(L1,L2);
LINK     R  = F4(L3);
```

then the functions at each node:

```
FUNCTION   F1 = SUB,QUEUE(QUE1,16);
FUNCTION   F2 = ADD,QUEUE(QUE2,16);
FUNCTION   F3 = MUL,QUEUE(QUE3,16);
FUNCTION   F4 = OUT1,QUEUE(QUE4,16);
```

VLSI Signal Processing

Finally, the space in Data Memory desired for each of the QUEUE instructions above is reserved. The length of memory for each queue was arbitrarily chosen to be sixteen words long. A MEMORY directive accomplishes this:

```
MEMORY     QUE1 = 16 DUP (?);
MEMORY     QUE2 = 16 DUP (?);
MEMORY     QUE3 = 16 DUP (?);
MEMORY     QUE4 = 16 DUP (?);
```

The DUP (?) indicates that the memory is not initialized to any particular value.

The program to perform equation 1 is now complete. The instructions can be assembled with the uPD7281 Functional Assembler and object code generated.

- Program Execution

The desired results for equation (1) are:

$$R(1) = [W(1) - X(1)] \times [Y(1) + Z(1)] \quad (2)$$
$$R(2) = [W(2) - X(2)] \times [Y(2) + Z(2)] \quad (3)$$
$$R(3) = [W(3) - X(3)] \times [Y(3) + Z(3)] \quad (4)$$

For illustrative purposes, an input data value is assumed to be available each pipeline clock cycle and in the following order:

$Y(1)$, $X(1)$, $W(1)$, $Z(1)$, $Z(2)$, $Z(3)$, $X(2)$, $Y(2)$, $Y(3)$, $W(2)$, ..

Conceptually, the major sequence of events in the processor is:

1. Y(1) is the first data to arrive. It is given a new identifier and stored in the Data Memory until a Z data arrives.

2. Next, X(1) arrives, given new ID, tagged for needing W and stored in Data Memory.

3. W(1) arrives next, given new ID, paired with X(1) stored in the Data Memory and sent to the Queue and later to the Processing Unit. After processing, it is then given another new ID of L1 and stored in Data Memory to await the first L2 value.

4. Z(1) is the next to arrive, given a new ID, paired with waiting operand Y(1) and sent to the Queue and then the Processing Unit. As it starts the trip through the pipeline again, it is given another new ID, paired with the waiting operand and sent to the Queue and then the Processing Unit to be multiplied. It is then targeted for output by being given the ID of R and exits the processor through the Queue, Output Queue and finally the Output Controller. R(1) has thus exited the uPD7281.

5. Z(2) now arrives, given new ID, and stored in Data Memory until the next Y arrives.

6. Z(3) is similiarly given a new ID and stored in the Data Memory. The Data Memory now contains Z(2) and Z(3).

7. X(2) comes in, is given new ID and sent to the DM to wait.

8. Y(2) finally arrives, is given its new ID, paired with the Z value that has been waiting the longest, Z(2), and sent to the Queue. After the Processing Unit, the data is now tagged with another new ID and sent to the Data Memory to wait for the other operand.

9. Y(3) is similarly identified and paired, leaving only X(2) in Data Memory.

10. W(2) now enters the processor, given a new ID, paired with the waiting X value and sent to the Queue and then Processing Unit. As it circulates through the pipeline again, it is tagged with new ID and paired with the waiting other operand, sent to the Queue and then the PU to be multiplied. On the next pass through the pipeline, it is tagged as R and output by the same route as the last R value. R(2) has now exited the processor.

Figure 5

CYCLE	IC	LT	FT	DM	Q	PU	OQ	OC
1	Y(1)							
2	X(1)	Y(1)						
3	W(1)	X(1)	Y(1)					
4	Z(1)	W(1)	X(1)	Y(1)				
5	Z(2)	Z(1)	W(1)	X(1),Y(1)				
6	Z(3)	Z(2)	Z(1)	Y(1)	W(1)&X(1)			
7	X(2)	Z(3)	Z(2)		Y(1)&Z(2)	L1(1)=W(1)-X(1)		
8	X(2)	L1(1)	Z(3)	Z(2)		L2(1)=Y(1)+Z(1)		
9	X(2)	L2(1)	L1(1)	Z(2),Z(3)				
10	Y(2)	X(2)	L2(1)	Z(2),Z(3) L1(1)				
11	Y(3)	Y(2)	X(2)	Z(2),Z(3)	L1(1)&L2(1)			
12	X(3)	Y(3)	Y(2)	Z(2),Z(3) X(2)		L3(1)=L1(1)xL2(1)		
13	X(3)	L3(1)=R(1)	Y(3)	Z(3) X(2)	Y(2)&Z(2)			
14	W(2)	X(3)	R(1)	X(2)	Y(3) & Z(3)	L2(2)=Y(2)+Z(2)		
15	W(2)	L2(2)	X(3)	X(2)	R(1)			
16		W(2)	L2(2)	X(2),X(3)			R(1)	
17			W(2)	X(2),X(3) L2(2)				R(1)

VLSI Signal Processing

This skims the surface of what is happening. The above explanations neglect the concurrent processing that is, or could be, occuring in the pipeline. Figure 5 illustrates how the data enters and circulates around the pipeline. Several values or operations can be ongoing. The Processing Unit is utilizied only during four of the seventeen cycles shown, resulting in low efficiency in this example. In general, to achieve the highest efficiency, the algorithm should be structured, and data input at a rate that has the Processing Unit executing an instruction every pipeline cycle. To achieve even higher performance, the uPD7281 can be connected in a multiprocessor configuration.

MULTIPROCESSING WITH THE uPD7281

The high speed processing realized with the use of one uPD7281 can be further enhanced by the use of several processors in a multiprocessor configuration. To connect several processors together, simply connect the sixteen output data lines and the two handshaking lines to the corresponding input lines of the next processor (Figure 6). To distinguish the different processors, a module number is set in the Input Controller during the reset period.

The partitioning of a system for multiprocessor operation requires careful thought to acheive optimal effectiveness. Two basic configurations for image memory processing are the cascade and ring. The cascade configuration (Figure 7) requires a processor or circuitry to format and feed the tokens into a chain of uPD7281s. Additional circuitry at the end of the chain must decode the tag and store the processed data.

In a ring configuration (Figure 8), control circuitry multiplexes the image memory between the host processor and the uPD7281 chain. To support this arrangement, NEC is developing the 'MAGIC' (Memory Access and General bus Interface Chip) support integrated circuit. 'MAGIC' will support up to four uPD7281s, provide for object code loading and direct memory acces between the image memory and display processor.

The configurations for digital signal processing are system dependent. For example, an incoming stream of data can have a 'tag' appended to form tokens and then be fed into a chain of uPD7821s. The algorithm might be structured to allow each uPD7281 to operate on one sequential part before passing tokens to the next processor. The desired output would then be available at the end of the processor chain. Many other system configurations are possible.

PERFORMANCE OF THE uPD7281

Typical performances of the uPD7281 are shown in Table 1 and Table 2. The actual processing throughput may differ depending on applications and system architecture. Generally, the overall processing time decreases as the number of processors increase, much more so than with conventional processors. There is a limit to the cost effectiveness of adding additional processors. After an optimal point is reached, additional processors will not result in a significant performance improvement.

CONCLUSIONS

The NEC uPD7281 will provide a powerful, cost effective solution for many signal processing requirements. It is well suited for image processing applications implementing algorithms such as two-dimensional convolution, shrinking, enlarging, rotation, etc. Digital signal processing is also well handled by the uPD7281 in applications requiring real time signal processing due to the large concurrencies in these algorithms. Moreover, the processor is suitable for numerical processing applications such as matrix-matrix or matrix-vector multiplications, floating point arithmetics and real time evaluation of transcendental functions such as cosine, sine, exponentials, logarithms and square roots. The architecture of the uPD7281 may be the model for future signal processors.

Figure 6

Multiprocessor Connection

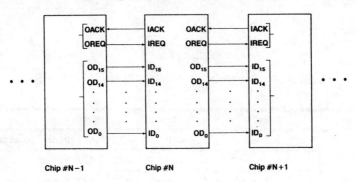

Figure 7

Cascade Configuration

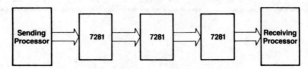

Figure 8

Ring Configuration

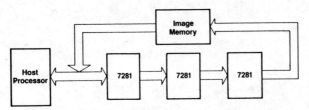

Table 1

Execution Times in a Cascade System

Operation	1 ImPP	3 ImPP's	Note
Multiplication	22 usec	8 usec	48-bit X 48-bit = 96 bits
3 X 3 Matrix Multiplication	24 usec	18 usec	17-bit Fixed Point
64-stage FIR Filter	50 usec	18 usec	17-bit Fixed Point
Floating Point Multiplication	4.6 usec	1.8 usec	17-bit Mantissa 17-bit Exponent
COS (X)	40 usec	15 usec	33-bit Fixed Point

Table 2

Execution Times in a Ring Configuration

Operation	1 ImPP	3 ImPP's	Note
Rotation	1.5 sec	0.6 sec	512 X 512 Binary Image
1/2 Shrinking	80 msec	30 msec	512 X 512 Binary Image
Smoothing	1.1 sec	0.4 sec	512 X 512 Binary Image
3 X 3 Convolution	3.0 sec	1.1 sec	512 X 512 Gray Image
1024 Complex FFT	60 msec	24 msec	17-bit Fixed Point

19 VLSI Signal Processing

New Word-Slice Components Architecturally Optimized for High-Performance DSP.

by

Dr. Paul M. Toldalagi
ANALOG DEVICES, INC.
Danville, CA 94526.

ABSTRACT

A family of four CMOS 16-bit slices was designed to deliver all the functions necessary for high performance digital signal processing. This set consists of a program sequencer chip with the largest address and stack space so far; an address generator that is unique in not being dedicated to a single signal -- or numerical -- processing algorithm; an arithmetic, logic, and shift unit that incorporates a barrel shifter in parallel with an ALU; and a multiplier/accumulator that is enhanced by extra on-chip functions.

This paper discusses some of these devices' architectural features and highlights various design tradeoffs that were encountered.

INTRODUCTION

Every design involves a series of optimization steps with respect to conflicting design objectives or constraints. For example, to optimize the design of high-performance signal processors or CPU's, the hardware engineer must evaluate complex tradeoffs among at least three figures of merit: the system's throughput, the required performance, and the flexibility with which components can be configured to achieve the desired goal. Other considerations, like board-space or power dissipation may become critical in this evaluation.

Similary, starting from macroscopic specifications provided to him by a marketing team, the chip designer must build up his design by evaluating various chip-related objectives and constraints like total die-size, pin count, speed, and reliability; however, his design cannot be considered as complete until tested by above system criteria. Although time-consuming, this iterative process (cf., Fig. 1) is an absolute requirement for achieving good market acceptance for a new component.

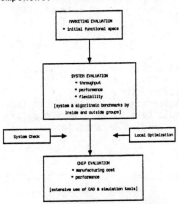

Figure 1. Basic Chip Design Procedure

Analog Devices has recently announced [1, 2, 3, and 4] a family of four IC's geared at delivering high througput (10MHz), low power (two-micron, double layer metal CMOS process) solutions for a large variety of

VLSI Signal Processing

Digital Signal Processing and fast number crunching problems. These devices were all conceived after multiple iterations through the procedure described above. Therefore, this paper constitutes a collection of four case studies highlighting some of the tradeoff's previously outlined.

THE ADSP-1401 MICROPROGRAM SEQUENCER

Initial marketing specifications for this device were to design a very fast (10MHz), 16-bit CMOS microcode sequencer:

* with more than 16 levels of subroutine nesting,

* full interrupt handling capability,

* and an instruction set at least as powerful as that of the best bipolar bit-slice sequencer.

The resulting design (cf., Fig. 2) is that of a 48-pin CMOS part computing successive microcode addresses every 70ns, dissipating less than 125 mW of power, and controlled by a 7-bit instruction word. Not only were all initial specifications respected, but several enhancements were discovered that should make it even more popular than originally anticipated.

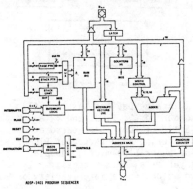

ADSP-1401 PROGRAM SEQUENCER

The main elements of the 1401 are 4:1 multiplexer for selecting the microcode address to be output through the Y-port; a 64 x 16 RAM for stack and miscellaneous storage; four 16-bit up- and down-loadable event counters; the 16-bit input/output D-port for data; 10 registers for storing 8 external and 2 internal interrupt vectors; an internal 16-bit bus linking the RAM, counter, interrupt vectors and other registers (stack limit, stack pointer, masks); and, a 16-bit adder for offsetting the program counter when in relative addressing mode.

Flexible Internal RAM

Since one of the constraints of this design was to provide extensive subroutine nesting capability, as well as internal storage, it was decided that a total of 64 words would be sufficient; moreover, rather than choosing a solution with separate stack and address storage memories, a single-access RAM structure was chosen for its speed, flexibility, and layout compactness.

Lower addresses (Fig. 3) are reserved for the subroutine stack (SS), which is used to provide interrupt and subroutine return addresses and to store counter and status values that the user wants to associate with those routines; upper addresses are reserved for two register stacks (RS) used to save up to four jump addresses per level of local ("nested register stack" with pointer NRSP) or global ("global register stack" with pointer GRSP) subroutines, as well as for indirect addressing storage. The switch from local to global register stack use is determined by a status bit. The GRSP can be saved and restored on the subroutine stack.

VLSI Signal Processing

Microcode branching is possible either in an immediate fashion by using two bits of the instruction word as relative address for the appropriate register stack, or, indirectly, by using 6 bits of the D-port to designate an absolute location in the RAM. Finally, instead of picking fixed stack-sizes, the RAM allocation was made programmable by the introduction of an auxiliary 6-bit stack limit register and the highest priority interrupt was assigned to stack overflows or underflows.

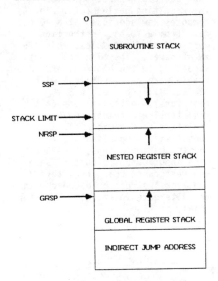

FIG 3 - RAM Organization in the ADSP-1401

Simple Timing

To achieve high execution speeds, internal timing conventions were simplified maximally. Instead of having instructions latched externally in a pipeline register as is customary for bit-slice designs, they are first decoded and then latched (t-latch) during phase 1 of the clock.

This technique not only saves an external register, but it also provides more flexibility for the system designer to position his clock edges, as long as the proper setup times (for decoding) are respected.

In addition, this chip was designed to minimize the output time of addresses on the Y-port (about 35ns for the worst case instruction), while transfers occur on the internal bus. For maximum speed, this bus is always precharged in phase 2 and active in phase 1. For instance, during an interrupt, the interrupt vector is output through the Y-port, while the program counter is saved onto the subroutine stack through the internal bus.

Flexible Event Counting

Four internal 16-bit two's complement counters are provided to monitor various events (flags, interrupts, timeouts, loops) with a complete set of conditional and unconditional decrementing instructions; in addition, many of these instructions can be executed in parallel with regular sequencing instructions (e.g., JUMP, JSR), since they involve distinct portions of hardware. The sign bit of the selected counter can also be used as an internal flag for branching (e.g., "IF NEG THEN JUMP xx"), or if enabled, as an internal interrupt. Finally, because each of these counters is actually wrapping, the user has the possibility of simulating the equivalent of "IF POS THEN JUMP xx" instructions by initializing the counter with a negative value and forcing jumps until clearing of the sign bit. Finally, the combination of event counters, branch addresses in RAM, and program counter provides this device with the capability of handling a wide range of fast, overhead-free branching instructions, including three-way branching.

VLSI Signal Processing

Complete Interrupt Handling Capability

Since only 4 pins were available for incoming interrupts, a time-multiplexing scheme (phase 1/phase 2) was chosen, thereby allowing the device to handle up to 8 external interrupts. In addition, two internal interrupts are available, the lowest priority one corresponding to negative counter values, the highest priority one to stack overflows. Each of these interrupts is automatically prioritized, possibly masked, and has one interrupt vector associated to it. External interrupts can also be either latched (to monitor transient interrupts) or unlatched, in which case interrupt requests have to be asserted externally until acknowledged.

Simplification of System Interface

Since all instructions to the sequencer are already internally latched, the user does not need an external pipeline register, unless he wishes to operate in a double-level pipelining mode to achieve even higher throughputs (40 ns cycle time) at the expense of programming flexibility. Similarly, the external flag and all interrupt signals are latched internally, thereby simplifying interfacing problems, as long as proper setup requirements are verified.

THE ADSP-1410 ADDRESS GENERATOR

Address generators specify read and write locations for data and coefficient memories. Since many algorithms in DSP have intricate addressing schemes and require rapid data transfers to and from high-speed number crunchers, a fast, flexible addressing element was required.

Most single-chip address generators are still dedicated to single algorithms like radix-2 or radix-4 FFT's. Alternatively, general purpose address generators capable of handling 16-bits of address space could only be built by juggling four 48 pin 1-W ALU slices along with external SSI circuitry.

This time, the initial specifications called for the design of a fast (10 MHz), 48-pin, general purpose 16-bit-wide slice CMOS address generation slice. The resulting design (cf., Fig. 4) computes new addresses at a 70 ns rate from a 10-bit instruction word and dissipates less than 125 mW of power. The main components appearing in the block diagram are the 16-word pointer file R with its auxiliary offset (6 B registers), initialization (4 I registers) and comparison registers (4 C registers), a 16-bit wide simplified ALU for pointer manipulations; the 16-bit wide input/output D-port to save and restore on-board registers; and, the instruction decoder with the Alternate Instruction Register (AIR).

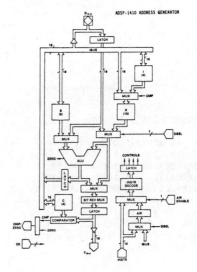

Generic Instructions

Since the most frequent address manipulation consists of scanning data arrays sequentially, the whole architecture was conceived to execute most efficiently an address modification using the ALU, a comparison with a pointer marking the end of the array, and possibly an initialization, all in one cycle:

e.g., Y←R; IF (R≥C) THEN R←I ELSE R←R+B

Of course many other instructions are available, including logical/shift instructions (e.g., Y←R; R←LSL R) and register transfer instructions used to save or restore externally the content of all registers.

Device Timing

As in the microprogram sequencer, instructions are typically decoded before being latched, thus providing addresses as fast as 20 ns after the rising edge of the clock. Internally, simple rules were adopted for transferring data from one register file to another, in a manner similar to the one used in the 1401.

Slow versus fast modes of execution

A multiplexer located on the path of the Y-port gives the possibility of selecting either the R-register file and the D-port ("Fast" mode), or the result of the ALU ("Slow" mode), corresponding to output/ALU/update or ALU/output/update type operations, respectively:

FAST MODE: Y←R: IF (R≥C)
             THEN R←I ELSE
             R←R+B

SLOW MODE: Y←R+B; IF
             (Y≥C) THEN
             R←I ELSE
             R←Y.

The comparison circuitry always monitors the output port Y: therefore, depending on the mode, a comparison occurs either with the R-register (fast mode) or with its next value (slow mode).

The fast mode (approx. 20ns after the rising edge of the clock) is slightly more cumbersome to handle in software because pointer values are the result of previous and not current operations; however, it has big advantages by relaxing memory access time requirements.

The Alternate Instruction Register

The AIR can be considered as a one word cache containing a single instruction. Its use is found mostly in repetitive operations to save or share microcode fields with other devices or for supporting modulo arithmetic operations, like in circular buffers, without any overhead penalty. In the normal mode, the device gets its instructions from the microcode memory and, as in the program sequencer, no external pipeline register is needed; the AIR instruction, however, can be executed either by asserting an external pin or upon the result of a comparison if the appropriate mode is enabled.

The AIR can be loaded for example with the instruction:

Y←R+1; IF (Y≥C) THEN R←I ELSE R←Y;

the device can then be disconnected from the microcode memory as long as a sequential reading of the circular buffer delimited by the addresses contained in registers I and C is needed.

Notice also the absence of overhead for branching and reinitializing the pointer R to the beginning of the buffer. A more complex problem arises, however, when performing modulo type addressing operations like:

output address = base address + [relative address.modulo(b)],

where b, as well as the base address, are arbitrary. Typically, this type of operation would require two comparators for testing the range of relative addresses in both directions as well as at least two ALUs. The die size, it was determined, would have made this chip prohibitively large and costly. Therefore, instead of making the device more complex, the AIR was made even more flexible. Consider for example, the case of a decimation FIR filter with N taps.

The filter input samples are stored in a circular buffer of length M (M>N.b), where b corresponds to the decimation stepsize:

$$Y(n) = \sum_{i=0}^{N-1} a(i).X(n-i.b)$$

Let R0 be the pointer for the input samples X and let B0, B1, and C0 be initialized with b, M and the last address of the array respectively. By loading the AIR with the instruction "Y←R0; R0←R0-B1" and by enabling the "execute AIR on compare" mode, successive samples are properly retrieved from memory using the single line of microcode:

Y←R0; IF (R0≥C0) THEN AIR ELSE R0←R0+B0.

Double Precision Addressing Capability

Two of these parts can be used for double precision addressing, since they are 16-bit slices.

However, because of the limited number of pins (48), the most significant pins of the D-port and the Y-port are actually bidirectional pins for inter-chip communication, thereby reducing the effective addressing range to 30 instead of 32 bits.

Another important enhancement of the original specification is the capability of performing double precision addressing (i.e., up to 1 Gbit) in a single device, using, however, two cycles per address instead of one. Register files are then organized as files of double precision registers, even registers being dedicated to least significant words, odd registers to most significant words and all single precision operations are extended to double precision, including all comparisons and ALU operations: of course, because of the timing, comparison flags are valid only after the second cycle in single-chip, double precision operations.

THE ADSP-1201 ARITHMETIC, LOGIC AND SHIFT UNIT (ALSU)

This part (Fig. 5) was first described in [4]. The original idea was to design a very fast, 16-bit-slice ALU with a complete set of arithmetic and logic functions and several on board registers.

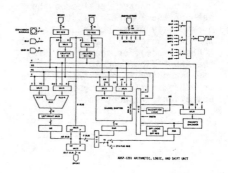

ADSP-1201 ARITHMETIC, LOGIC, AND SHIFT UNIT

The pinout was set at 96 pins (100 PGA), with 48 pins just for I/Os.

After carefully analyzing a large number of DSP algorithms, it was decided that a barrel shifting unit (BSU) would be an important addition to this part, especially in view of the fact that it did not occupy much space on the die. Also, at that point, several internal data paths became desirable to perform ALU operations either before or after shifting operations. However, too many buses could not be run internally without considerably increasing the total die size. Therefore, a single feedback path was introduced which could be accessed either from the ALU or the BSU.

The final design is represented in Fig. 5 -- its main features are the ALU with its single bit output shifter, and the barrel shifter unit (BSU) hanging off the same input structure; the input registers X0, Y0 and the two 8-word register files; the feedback bus R; the output port Z; and, the priority encoder with its separate register (PER).

Input Ports and Register Files

Data provided over the X and Y ports pass through the X0 and Y0 input registers. Normally, inputs are latched into X0 and Y0 on the rising edge of the clock.

However, they can also be made transparent (special mode). Note that the input registers, whether latched or transparent, do not introduce any pipeline delay in the part's operation: in any cycle, the data read from X0 or Y0 to the ALU or BSU is identical to the data that passed through these registers earlier in the cycle (T-latches).

VLSI Signal Processing

The two 8-word register files (X,Y) provide arguments to the ALU, the BSU or the priority encoder; they also provide destinations for incoming data (through X0 and Y0), the ALU result register (AR) or the BSU result register (BSR). Each register is read at the rising edge of the clock and written at the following falling edge. Because of their independent addressing capability, each register can be read and written in the same cycle.

Priority Encoding

The priority encoder detects the most significant bit (MSB) of either unsigned magnitude or two's complement 16-bit numbers as they pass through the Z-port (BFP mode) or are fed to port A of the barrel shifter (non BFP mode). The output of the priority encoder is a 5-bit two's complement number between 0 and -15, that encodes magnitudes to the nearest power of two. This number can be stored in the priority encode register (PER) and/or used to control left shifts in the barrel shifter.

Double Precision

Once again, this device has many more features than originally anticipated; in particular, although a true 16-bit wide slice, this chip can also perform practically all double precision operations on its own (i.e., 32 bits) using multiple cycles.

Special instructions were added to maximize double precision throughput for single devices. Consider for example, the case of the absolute value operation ABS(X), where X is a 32-bit number formed by (MSW,LSW):

> cycle 1: save the sign bit of the MSW as an internal flag (Iflag);

VLSI Signal Processing

cycle 2: execute the "ABS LSW" ALU-instruction that passes the LSW as is if Iflag=0, or complements the LSW and adds a 1 to it if Iflag=1;

cycle 3: execute the "ABS MSW" instruction that passes the MSW as is, if Iflag=0, or complements it and adds the past carry, if Iflag=1.

Of course, using two devices in tandem, the same result could have been achieved in a single cycle, although stretched by 20% for inter-chip communication.

Block Floating Point Operations

Block floating point arithmetic is a very useful concept in DSP, since it has many of the advantages of full floating point without its arithmetic complexity and attendant costs.

The ALSU has three block floating point flags (BFP1, BFP2, BFP3). When the chip is in BFP mode, these flags are decoded from the content of the priority encode register (PER) and can be used to control the ALU shift operations (one bit shift right, if true), or the barrel shifter, to renormalize numbers by as many positions as flags are set.

As a typical example of how internal flags are used, let us consider the case where 7 numbers (A, B, C, D, E, E, G) are added together in two different sequences.

Sequence 1: [(A + B) + (C + D) +[(E + F) + G]]

When adding A and B, BFP1 is used to control the scaling at the output of the ALU. Similarly, BFP1 is used when adding C and D as well as when adding E and F.

G may also need to be scaled to keep its exponent consistent with the other numbers. BFP2 is selected when adding (A + B) to (C + D). BFP2 is also used when adding (E + F) so that the exponent associated with G is the same as that of (E + F).

This of course could be done in the ALU. However, to avoid tying up the ALU and thus speeding up the operation, the barrel shifter is designed to provide this type of scaling while the ALU is doing some of the other additions. To finish the sequence of additions, BFP3 is used when adding [(A + B) + (C + D)] to [(E + F) + G]. As mentioned above, when in block floating point mode, the barrel shifter can normalize words so that they can be properly added to the correct sum.

Sequence 2: ((((((A + B) + C) + D) + E) + F) + G).

BFP1 is used in the ALU while adding A and B. At the same time, BFP1 is being used in the Barrel Shifter (BRL) to scale C so that the block exponent associated with C is the same as that associated with (A + B). BFP2 is used in the ALU while adding (A + B) to C. At the same time, (BFP1 + BFP2) is used in the BRL to scale D so that the next exponent associated with D is the same as that associated with ((A + B) + C). No shift is used in the ALU while adding ((A + B + C) to D. At the same time, (BFP1 + BFP2) is used in the BRL to scale E so that the block exponent associated with E is the same as that associated with ((A + B) + C) + D). BFP3 is used in the ALU while adding (((A + B) + C) + D) to E. A the same time, (BFP1 + BFP2 + BFP3) is used in the BRL to scale F. No shift is used in the ALU while adding ((((A + B) +C) +D) +E) to F. At the same time (BFP1 + BFP2 + BFP3) is used in the BRL to scale G.

Finally, no shift is used in the ALU while adding $(((((A + B) + C) + D) + E) + F)$ to G.

THE ADSP-1101 ENHANCED MULTIPLIER/ACCUMULATOR EMAC

The purpose of this design was to obtain a very fast, smart 16 x 16 CMOS enhanced multiplier/accumulator (EMAC) device, that would also contain many of the external support functions needed by traditional three port MACs.

Once again, a 100-pin grid array package was chosen, thus leaving ample room for instructions (37 bits) and data ports (48 pins). The resulting design (Fig. 6) operates at a rate of 60ns and consumes about 300mW of power.

Multiplier Array with Format Adjust Capability

The EMAC has a 16 x 16 parallel multiplier array that can perform two's complement, unsigned magnitude, or mixed mode operations. Its format adjust control allows the products of two's complement operands to be left shifted by one bit **before** accumulation, allowing the redundant sign bits of two's complement products to be eliminated. Furthermore, because of the presence of a negation circuitry, both the terms (product $\pm$ accumulator) and (accumulator $\pm$ product) can be obtained.

Input Porting and Registers

Instead of using the same porting structure as in the ALSU, a pair of clocked registers is associated with each of the ports of this device. Input operands may still bypass, however, these registers and feed directly the multiplier array, with a 10ns penalty for the additional propagation delay. This flow-through operation does not affect the content of input registers. Notice also that the Ybuffer registers (Y0 and Y1) can be written from any of three sources: the Xport, the Yport, or the MSW of the adder/subtractor result (bits 31:16).

Two Accumulators A and B

Two 40-bit accumulation registers A and B are provided on this chip. For complex operations, for example, one accumulator could be used for real values, the other for imaginary values. Another novelty is the possibility of pre-loading any of these result registers at the beginning of a cycle using the Y-port at a 30ns speed.

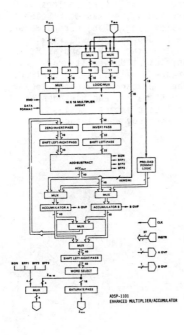

ADSP-1101 ENHANCED MULTIPLIER/ACCUMULATOR

This feature provides the capability of performing operations like " a + b x c" in a single cycle.

Output Control and Saturation

Data in the A or B registers can be read off chip via the 16-bit Zport; either the 16 LSW bits, the 16 MSW bits or the 8 extension bits (EXT) can be output on Z15-Z0.

When outputting the EXT word, the EMAC automatically causes the upper 8 bits of Z either to be sign extended for signed data, or to be set to all zeros for unsigned data. The instruction set also allows a single-cycle output of the MSW and LSW, or the MSW and EXT fields: the LSW is output on the rising edge of the clock, the MSW on the falling edge, and similarly for MSW and EXT, respectively.

Similar to the more modest single-port 16 x 16 MAC (ADSP-1110), the EMAC has an overflow detection and conditional saturation capability on output. However, instead of a single bit left shifter, this device contains a full single-bit up/down shifter, which is quite helpful in data formatting exercises. Of course, this shifter operates on the full 40-bit result (either A or B). Finally, true unbiased rounding of 40 to 24 bit results was implemented in a manner consistent with the output shifter.

CONCLUSION

This paper highlighted and explained some of the novelties incorporated into Analog Devices' new Word Slice family of components, not only with respect to their functionality, but also in terms of their hardware implementation.

REFERENCES

[1] D. Garde, J. Oxaal: "16-bit-slice Family Creates Ultrafast Digital Signal Processors", Electronic Design. May 17, 1984. pp. 136-144.

[2] W. Windsor, J. Wilson: "Arithmetic Duo Excels in Computing Floating-Point Products", Electronic Design. May 17, 1984. pp. 144-151.

[3] T. Dintersmith, J. Nuttall: "Putting the Chips through their Paces in Three Systems", Electronic Design. May 17, 1984. pp. 151-168.

[4] J. Nuttall, J. Oxaal: "A 16-Bit Three Port Arithmetic Logic and Shift Unit", ICASSP Paper.

Paul Toldalagi, is a senior staff engineer at Analog Devices. Previously, he was with Scientific Systems Inc. Paul earned MS and Ph.D. degrees in electrical engineering at MIT and an MS in physics at University Paris XI France. He's a member of IEEE, the American Meteorology Society and Sigma Xi, and his hobbies include classical music, meteorology and finance.

A NEW APPROACH TO FLOATING-POINT DSP

Robert M. Perlman

Advanced Micro Devices

THE COMPLETE TEXT OF THIS MANUSCRIPT WILL BE FOUND BEGINNING ON PAGE 400.

Part VI
ADVANCES IN ALGORITHMS

Edited by:
Dr. Earl E. Swartzlander, Jr.
TRW

VLSI Signal Processing

21

Digital Processing Structures for VLSI Implementation

by

C. T. Mullis
R. A. Roberts
Department of Electrical Engineering
University of Colorado
Boulder, CO 80309

Abstract

The development of VLSI has changed the important parameters in signal processing algorithms and structures. Technology developments are dictating new and different criteria for "efficient" realization. In VLSI the number of computations performed per output sample is not enough to measure the "goodness" of a realization/implementation. This work is an attempt to quantify the important parameters in VLSI implementations for DSP problems and apply the theory to the task of digital filtering.

1. Introduction

In the past we have often used the number of computations per output sample as a measure of algorithmic complexity. A good algorithm reduced the number of computations especially multiplies, e.g., the fast Fourier transform in computing the DFT or the direct form filter in computation of a digital filtering task. In VLSI it is not enough to merely count the number of computations per output samples in assessing the suitability of an algorithm for implementation. And, in fact, this measure is generally of small consequence.

What are the important parameters in designing DSP algorithms for VLSI? To answer this question we first need to define some terms such as **task or problem, realization or algorithm,** and implementation. A task is an input/output description of a digital processing problem, e.g., a digital filter transfer function. A realization or algorithm is a specification of precisely how to compute the output for each new input. A good example of a realization is a software program for the task. Another example, although not quite as specific as a software program is a signal flow graph. An implementation is the physical hardware that embodies the algorithm. In this context the implementation is the VLSI chip specific for the given task.

The point-of-view taken here is that of a dedicated chip designer. We are not constraining the implementation to be a software program on a general purpose DSP chip, for example. The reason for this is that we wish to find the "best" algorithm/implementation for a given task. Certainly given a fixed amount of chip area and a given technology the performance of a dedicated design is at least as good as the software implementation and, most probably, much better.

For each task there are an infinite number of algorithms. And for each algorithm there are an infinite number of implementations. We can use the freedom we have in choosing algorithms/implementations for a given task to optimize various criteria associated with the actual computation. In the present case we can choose to optimize criteria we views as important to VLSI such as **data throughput, hardware complexity, output signal** quality due to finite register effects, and **geometrical properties** associated with the physical layout of the implementation. Certain of these criteria can be optimized only at the expense of others. For example, high data throughput generally implies additional hardware complexity in some form.

The measures we attach to the criteria above will also affect the choice of an algorithm/implementation. Data throughput is most commonly measured in terms of output samples/second. Hardware complexity has been measured in various ways. We shall use the area of the implementation as our measure. Output signal quality is directly related to the computational error performance of an algorithm caused by the use of finite length registers. Output signal quality can be increased by merely increasing the length of the registers in an implementation. However, there are examples where certain algorithms can use very short length registers and perform with an error performance that corresponds to a different algorithm with very long length registers. The use of good finite-length-register effect algorithms is therefore advantageous in reducing the total complexity of an implementation.

If it is not possible to simultaneously improve all four criteria mentioned above, we must know how to intelligently compromise. An approach is needed which allows the designer to trade-off competing criteria in the synthesis and analysis of various designs. The most important aspect of this approach is the development of a mathematical description which allows us to manipulate realizations in order to affect trade-offs in the criteria. We believe this description should possess four major attributes:

(1) One should be able to use the description to directly evaluate finite length register effects. These include such things as scaling (to prevent overflows), roundoff noise power, and limit cycles.

(2) The description should identify the degree of modularity and local connectivity. This involves both spatial and temporal properties of the computation. One should be able to identify the possibilities of concurrent computation, the timing constraints imposed by the order of computing intermediate results, the possibilities of time-multiplexing, and the degree of possible pipelining.

(3) The description must admit a wide class of transformations (under which the input/output characterization of the task is invariant) so that one can manipulate the algorithm to affect trade-offs.

(4) After obtaining an algorithm/implementation which is suitably modular and regular and the modules(s) are characterized, the hardware implementation should be uniquely determined in space and time at the module level.

2. An Approach to the Algorithm/Implementation Problem for VLSI

We have introduced the attributes we would like to develop for a mathematical description for design in VLSI. In this section we shall illustrate the outlines of such a description and apply the description to the task of digital filtering.

The most commonly used descriptions for the "structure" or algorithm used for digital filtering are signal flow graphs (SFG's). The realization step is generally taken to be the process of generating the SFG from a given transfer function. There are several standard ways to do this including factoring of the transfer function into second-order sections which are then realized in some particular ways such as the direct form. The SFG, however, is not a particular useful description in the sense described above. There is nothing in the SFG which inherently describes timing and parallel computation. Moreover, data-flow is ambiguous. Attempts to improve the SFG description by the use of fractional delays ($z^{-n/m}$) and other devices generally result in yet more confusion.

We have developed a new description called the **factored state variable description** (FSVD) that has a level of detail closer to the attributes discussed in section 1. Included in this description are properties such as

(1) the ability to partition an input/output description into subfunctions or modules of varying degrees of complexity;

(2) the ability to vary the degree of parallelism in the realization;

(3) the analytical description needed to completely specify and optimize effects of finite-length registers;

(4) the ability to manipulate the realization for purposes of improving signal quality and/or improving some architectural facet of the implementation such as modularity and local connectivity of the modules.

VLSI Signal Processing
3. The Factored State Variable Description

A FSVD is simply the members or factors of a matrix product that describes the order and substance of the computations which are implicit (but not explicit) in any computable SFG. (The most important property of a computable SFG is that it contains no delay free loops.)

The construction of a FSVD for a given SFG for a digital filter is straightforward. One merely identifies the order of computation of intermediate (node) variables in the SFG. Beginning with the input and variables stored in the unit delays, one then relates the intermediate results by a product of matrices. The filter is then describable in a form

$$\begin{bmatrix} x(k+1) \\ y(k) \end{bmatrix} = Q_n Q_{n-1} \cdots Q_1 \begin{bmatrix} x(k) \\ u(k) \end{bmatrix} \qquad (1)$$

where $x(k)$ is the vector of variables stored in the unit delays. If we multiply $Q_n Q_{n-1} \cdots Q_1$ together we obtain the usual state variable description (SVD) of the digital filter. This multiplication is, therefore, equivalent to eliminating all internal nodes not directly connected to the unit delays.

Using this description we can proceed to investigate the ways one can manipulate a FSVD in order to attain a particular set of goals (high signal quality, high data throughput, modularity and local connectivity, elemental processor simplicity). The first question is what methods are available to transform a given FSVD into a more desirable FSVD. Under various forms of transformations we shall require certain invariants. For example, we shall always require that input/output function remain invariant. We might also require that scaling conditions remain invariant. There are a number of architectural/geometric considerations which one can require of the transformations used.

Suppose, for example, we have a filter G defined by a FSVD of the form (2).

$$G = Q_n Q_{n-1} \cdots Q_1 \qquad (2)$$

If we require that only the I/O characteristic remain invariant, then the following transformations are permissible.

(1) External similarity transformations:

$$Q_n, Q_{n-1}, \ldots, Q_1 \leftarrow \begin{bmatrix} T^{-1} & 0 \\ 0 & I \end{bmatrix} Q_n, Q_{n-1}, \ldots, Q_1 \begin{bmatrix} T & 0 \\ 0 & I \end{bmatrix} \qquad (3)$$

where T is nonsingular and may also be structurally constrained to produce geometric constraints on the new filter G'.

(2) Internal similarity transformations:

$$Q_n, \ldots, Q_1 \leftarrow Q_n, \ldots, Q_k T^{-1}, T Q_{k-1}, \ldots, Q_1 \qquad (4)$$

where T is nonsingular.

(3) Minor cycle merging (combine $Q_k Q_{k-1}$):

$$Q_n, \ldots, Q_1 \leftarrow Q_n, \ldots, Q_k Q_{k-1}, \ldots, Q_1 = Q'_{n-1}, \ldots, Q_1 \qquad (5)$$

(4) Minor cycle splitting (factor Q_k):

$$Q_n, \ldots, Q_1 \leftarrow Q_n, \ldots, Q_{k+1}, F_1, F_2, Q_{k-1}, \ldots, Q_1 = Q'_{n+1}, \ldots, Q_1 \qquad (6)$$

The first two transformations (3) and (4) can be used to scale and optimize roundoff noise in a filter. If we define the scaling matrices K_i and the noise gain matrices W_i, $i=1,2,\ldots,n$, then we can show that

$$K_i = Q_i K_{i-1} Q_i^T, \quad i=1,2,\ldots,n$$

$$W_{k-1} = Q_i^T W_i Q_i, \quad i=1,2,\ldots n \qquad (7)$$

VLSI Signal Processing

with

$$K_0 = \begin{bmatrix} K & 0 \\ 0 & I \end{bmatrix} \text{ and } W_M = \begin{bmatrix} W & 0 \\ 0 & I \end{bmatrix} \tag{8}$$

and

$$K = AKA^T + BB^T$$

$$W = A^T WA + C^T C \tag{9}$$

The filter is l_2 scaled iff all the diagonal elements of each K_i are unity. The variance of the roundoff noise at nodes of accumulation at each stage of the algorithms is proportional to the trace of W_i. This analysis is a simple generalization of previous work [1-2]. Further the elimination of overflow limit cycles can also be guaranteed by requiring that the matrices $[I - Q_i^T Q_i]$, $i=1,2,...,n$ be positive definite.

The transformations (3) and (4) affect the K_i and W_i matrices in a straightforward manner and thus affect scaling and internal roundoff noise. Using the correct transformations we can optimize the finite-length register effects in the filter.

What architectural considerations can be addressed? The FSVD can be used to manipulate timing, memory requirements, data flow, and the number and complexity of the arithmetic units that perform the essential calculations in the filter. By considering the number of factors n in the FSVD, the maximum dimension of the Q_i, $i=1,2,...,n$, the number and width of the nontrivial rows in each Q_i, $i=1,2,...,n$ one can infer the complete structure of filter down to the module level. We have used this description to obtain a filter architecture/implementation which is shown in Figure 1. It has the following attributes:

(1) The filter is an array of simple modules, all identical, which are locally connected. Each module computes an order two inner product of the form $\alpha_1 z_1 + \alpha_2 z_2$ where α_1 and α_2 are real numbers defined by the transfer function of the filter.

(2) The filter is pipelined with a throughput rate of one-half the throughput of a single module. The word rate of the filter is independent of order.

(3) The filter is automatically l_2 scaled.

(4) The finite register effects of the filter are excellent.

4. Summary

We have outlined the beginnings of a theory for the design of VLSI implementations for DSP problems. The particular example chosen is well-suited for the proposed description. Each DSP task will require its own detailed study. However, hopefully certain principles from one task will carry over to other tasks and suggest what are good algorithm/implementations for a particular problem.

Thus far most of the improvement in computing the output for a DSP problem has been a direct result of improving the technology applied to the actual computation. However, just as the FFT algorithm revolutionized the computation of the DFT, there may be new algorithms well-suited for VLSI that can also make a profound improvement in performance for the computation of various DSP tasks. These new algorithms (as suggested here) will be found by considering the VLSI algorithmic/implementation problem from new principles that are imposed on the solution by the introduction of new constraints.

References:

1. C. T. Mullis and R. A. Roberts, "Synthesis of Minimum Roundoff Noise Fixed Point Digital Filters," *IEEE Trans. on Circuits and Systems*, vol. CAS-23, pp. 551-562, 1976.

2. ———————————, "Roundoff Noise in Digital Filters: Frequency Transformations and Invariants," *IEEE Trans. on Acoustics, Speech, and Signal Processing*, vol. ASSP-24, pp. 538-550, 1976.

22
Some VLSI Design Implications for DFT Calculations Based on a WFTA/M68000 Implementation

James T. Rayfield and Harvey F. Silverman[*]

Laboratory for Engineering Man/Machine Systems (LEMS)
Division of Engineering
Brown University, Providence, RI 02912

Abstract

Special-purpose VLSI chips for signal processing and, in particular, for calculating the DFT, are available. In this paper, we describe an approach which marries the Winograd Fourier Transform Algorithm (WFTA) with a state-of-the-art, 16-bit, general-purpose microprocessor for the purpose of DFT calculation. The heart of the approach is the real-input 240-point WFTA which has been carefully optimized for time and space to run in 10.8 ms on a 10 MHz, no wait-state, 68000. The approach is shown to extend easily for the inverse transform and for the complex case. Implications for VLSI design based on this approach are discussed.

1. Introduction

The DFT has become an everyday tool for a broad spectrum of applications, in large part due to advances in hardware technology. In signal analysis[1], speech processing[2] [3], image processing[4], sonar signal processing[5], seismic processing[6], and numerous other areas ranging from mechanical vibration analysis to the analysis of I/O referencing for computer programs[7], DFT techniques are being adopted as the costs go down.

The price paid for using the DFT in an everyday environment, until recently, has been measured in terms of several thousands of dollars. This level has been achieved in several pieces of test equipment such as the Nicolet Oscilloscope[8]. VLSI is currently causing the latest *revolution* in the use of the DFT, in that special-purpose digital signal processing chips are becoming widespread. The most widely used of these is the Texas Instruments TMS-320[9], upon which a fixed-point, real-valued DFT of 256 points can be computed in 4.86 ms[10][**]. The current cost of the TMS-320 is $150, and thus a substantial reduction in systems cost to the end user is easy to predict.

This paper describes an approach to the calculation of the DFT which potentially offers many end-users even lower cost than does current DSP VLSI. The ideas here are not profound; the WFTA and the real/complex symmetry algorithms are well-known to DFT experts; multi-microprocessor structures are well-known to hardware and systems experts. However, the combination of using -- 1) WFTA, 2) a standard 68000 microprocessor, 3) a tightly optimized assembler-language version of the WFTA for a particularly "good" selection of N and for some useful applications, and 4) algorithmic, hardware and software extensions based on using a real WFTA as a building module is rather unique. In fact, there are strong advantages to

---

[*]This work principally supported by NSF Grant ECS-8113494

[**]This figure does not include the double buffered I/O or windowing that is needed in most applications.

VLSI Signal Processing

this combination although, at first, a cursory reading of [11] might lead one to believe that the WFTA is ill-suited to the general-purpose computer environment. The WFTA fits the integer/general-purpose microprocessor environment very well! This tends to imply that if VLSI were to be designed for DFT--rather than *all* DSP applications--then a more standard approach to the architecture might be reasonable.

Although the WFTA in combination with the 68000 has been discussed by Gibson and McCabe[12], the approach taken is to discuss the more traditional complex implementation and a general-N system as suggested in [13], and [14]. In pretty sharp contrast, we are advocating the use of a *single-N, real input,* highly-optimized implementation as a modular building block for a DFT calculation system. This is reasonable for our application -- real-time speech spectrography -- and for several others. Most of all, however, it is a most suitable scheme to allow the highest level of time optimization for the 68000. The WFTA is ideally suited for this general-purpose microprocessor in that 1) reasonable microprogrammed multiplication is an instruction set member, although its calculation time is substantially larger than that for addition, 2) sophisticated modes of addressing are available, 3) the number of general-purpose registers has grown, and 4) register/register operations are quite fast. As a result, for the WFTA which was optimized, the percentage of the time taken by multiplication, addition, memory reads, and memory writes was evenly balanced.

One of the important aspects of this paper is in the comparison of DFT calculation systems. Many real implementations suffer due to the additional processing required for windowing, or due to I/O operations. Thus, we shall compare performance in three ways. First, is the *mathematical* way when only operations for the DFT are considered, i.e., one assumes no windowing is done, and the input data are already in memory; output data is simply written to memory. Second, the cost of using an N-point time window is added to the first measure. Third, the cost of obtaining N points of data and of outputting N transform points (N/2-1 complex plus two real values) in an interrupt-driven fashion is considered. Here, it is assumed that two input and two output buffers are used so that the process can proceed in real time; one buffer is used to accumulate data while the other is used for the transform process. The method for comparison is cited in each case.

In the next section, the 240-point, real WFTA is introduced and its implementation on the 68000 microprocessor discussed. A short description of pertinent properties of the 68000 is given, as is a detailed breakdown of the timing of various parts of the implementation. In the following section, software extensions are discussed, such as complex calculation using the real implementation, and inverse transform. Several tradeoffs are possible. The construction of a module which can perform the basic transform is introduced in section 4, along with some indication as to how to apply the module as a building element for faster operation or larger transform calculation.

2. Implementation

The principal class of applications for which this WFTA routine was programmed is real-time speech processing. This imposed a number of constraints on the implementation. Most speech processing requires a sampling rate on the order of 6-12 kHz and a transform size which corresponds to a duration of 20-40 ms. This length window is short enough to include only a relatively stationary portion of the speech signal but large enough to include several pitch periods. In addition, it may be desirable in some applications to have some overlap between adjacent windows in the time domain. Together, these factors determine the minimum throughput required for a real-time implementation.

Finally, it is often desirable to have sufficient idle time to allow the processor to handle tasks other than the WFTA, such as premultiplication of the input signal by a time-domain window function, post-processing of the transform data, and miscellaneous control functions.

VLSI Signal Processing

2.1. The 240-Point WFTA Algorithm

Window sizes in the range 210-420 points were considered, based on the above constraints on sampling rate and window duration. Of the possible sizes in this range, the 240-point algorithm was found to have the lowest number of multiplications per point, $324/240 = 1.35$, and was chosen for that reason; also, 240 is a three factor algorithm (3, 5, 16) and three factor algorithms are slightly easier to optimize than are four factor algorithms. However, the additional cost per point of several other sizes in this range is not that great. For example, the 336-point algorithm has 1.45 multiplications per point, and the 360-point algorithm has 1.47 multiplications per point.

A flow diagram for the 240-point WFTA is shown in Figure 1. The first step is a reordering of the input data into three 80-point vectors. These vectors are processed by the input additions of the three-point small-N algorithm and are then formed into three groups of 16 five-vectors. Each of these groups is processed by the input additions of the 16-point small-N algorithm and emerges as 18 five-vectors. Each of these 54 five-vectors is processed by the complete five-point small-N algorithm, and each subgroup of 18 five-vectors is recombined into a group of 16 five-vectors via the output additions for the 16-point algorithm. These are recombined into three 80-vectors, processed by the output additions of the three-point algorithm, and then combined into a 240-vector. This vector is reordered into the desired output vector. The time taken for each phase of the coded 240-point algorithm and the size of that portion of the code is given in Table 1.

Table 1 Timing Summary by WFTA Phase		
Phase	Time (ms)	Code size (bytes)
Reordering/3-pt. input additions	1.2076	88
16-pt. input additions	1.3760	378
5-pt. inner loop	4.5212	186
16-pt. output additions	1.8608	516
Reordering/3-pt. output additions	1.8180	170
Subtotal	10.7836	1338
Windowing	2.2572	26
Input/output	2.3040	10
Total	15.3448	1374

2.2. Pertinent Features of the 68000

The 68000 is a fast, general-purpose microprocessor in wide use today[15]. Some basic instruction times for the 68000 are shown in Table 2. The timing information given in this paper is based on a 10 MHz clock rate, although the 68000 is currently available with a 12.5 MHz clock rate, and versions to 20 MHz are planned. Also, some versions with other speed enhancement mechanisms are nearing delivery, such as the 68020[16]. Clearly, when these improvements are available, the algorithm described here would run faster by as much as a factor of two for the 20MHz version.

The 68000 has a 16-bit data path to memory, but all the registers and most instructions can handle 32-bit data. The basic memory cycle time is 400 ns, for the 10 MHz clock rate, with most instructions executing in the time it takes to access the opcode and operands and store the results. The 68000 has two 16-bit multiplication instructions (signed and unsigned) in microcode, which take 17.5 times as long (maximum) to execute as a register-to-register addition. This property makes it a good candidate for the WFTA.

VLSI Signal Processing

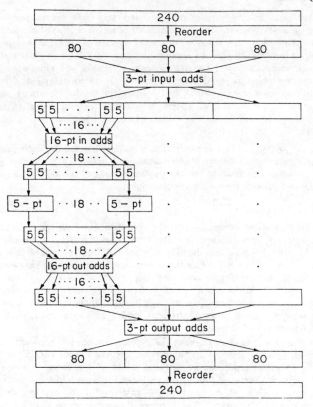

Figure 1
240-Point WFTA Flow Diagram

Table 2 Basic 68000 Timing (16-bit operands, 10 MHz clock)	
Operation	Time (μs)
Memory cycle	0.4
Move reg to reg	0.4
Move mem[base] $\rightarrow$ reg	0.8
Move mem[base+displ] $\rightarrow$ reg	1.2
Move mem[base+displ+index] $\rightarrow$ reg	1.4
Add/subtract reg to reg	0.4
Multiply reg to reg (avg)	5.4
Multiply reg to reg (max)	7.0

VLSI Signal Processing

The 68000 has eight general-purpose data registers and eight general-purpose address registers, counting the stack pointer. This large number of registers allowed almost all data to be kept in registers during loop executions. Thus, most operations are register-to-register, and these are very fast on the 68000. This is a significant factor for WFTA implementations, once again indicating the importance of the match between algorithm and processor. Thus, full vector memory read/writes are required only for reading the input, storing and reading the results of two input-addition stages, storing and reading the results of two output-addition stages, and writing the output data. Two address registers are unused, thus allowing a simple interrupt routine, for taking input, to execute asynchronously without having to save and restore registers.

In addition, the 68000 has a large selection of addressing modes, including indirect, indirect plus displacement, auto-increment/decrement, and indirect plus index register plus displacement. These are a significant aid to the implementation of the WFTA and allowed us to achieve real-time performance for speech spectrography.

2.3. Details of Optimization

A number of optimizations contributed to the overall speed of the implementation. For example, the reordering, windowing, and first input additions were combined into the same loop. This was done by fetching the data in a reordered fashion, rather than having a separate reordering phase, and by reordering the window coefficients so that they corresponded to the reordered data. In addition to reducing loop overhead, this considerably reduces memory references since intermediate results can be kept in registers. The reordering offsets and window coefficients were combined (alternatingly) into a single vector to allow access to them via a single address register in auto-increment mode. This saves many cycles as auto-increment mode is free on the 68000.

Fixed integer scaling was used throughout the implementation, using a word length of 16 bits. The program accepts an array of 12-bit integers, typically from an A/D converter, in the range -2048 to +2047 as input, leaving a factor of 16 of headroom. The intermediate results are scaled down by a factor of four before the 16-point input additions and by another factor of 4 before the five-point algorithm, thus guaranteeing that no overflow can occur by the end of the inner loop multiplications. Furthermore, if overflow occurs during the five-point or 16-point output additions, the final results are guaranteed to be correct. This is because all word-length additions and subtractions are effectively done in modular arithmetic. Thus, the output will be correct even if overflow occurs in intermediate steps as long as the correct results fit in their final destination without truncation[17].

Other routines have been written for a variable scaling. For the 240-point algorithm, after scaling by 4 before the 16-point input additions, one can determine the maximum data value after the 16-point input additions, and shift all the data at this stage up or down by the appropriate number in an allowable shift range (say 3 down to 7 up). Results can be renormalized at the completion of the WFTA. If, for example, only a log magnitude is the only result required, as in spectrography, then all that is needed is to add the appropriate constant to the final result.

All input additions in our implementation are real (non-complex) additions as we have assumed real data. Also, all multiplications are pure; that is, they involve a real number times a real number or an imaginary number times a real number. For the 240-point implementation, each group of six coefficients for the five-point algorithm is always of the form (R, R, R, I, I, I) or (I, I, I, R, R, R), where R = real and I = imaginary. (A WFTA implementation for another window size will result in two different complementary patterns.) The pattern for each group of multiplications is specified by a binary flag stored with each group of six coefficients. This allows two different inner loops to be coded, each of which handles one of the patterns. The code to be executed is chosen by the single test for the six coefficients, which is significantly faster than testing each coefficient separately.

VLSI Signal Processing

Many of the output additions for the five-point algorithm are pseudo-additions; that is, they involve a real number plus an imaginary number. Thus, only data movement need be done, not actual arithmetic operations in these cases. The binary flags control where the five-point algorithm output additions' results are placed; this is, indeed, the only code in the inner loop which is controlled by the flags. After these output additions, all data are treated as complex-valued. The real and imaginary components are handled separately by virtually identical sections of code.

The final output additions are combined in a loop with the output reordering. Only 121 frequency points are calculated, since the real-valued nature of the input data yields a symmetric result; real valued terms for $r = 0$ and $r = 120$, and 119 complex terms for $r = [1,119]$.

2.4. Packaging

The current 240-point WFTA is packed as a subroutine for the C language. It is not set up as a double-buffered module. It accepts three parameters: The address of the input array (240 12-bit integers stored in 16-bit words), the address of the output array (120 complex integers, four bytes each), and the address of a scratch area (1520 bytes). The input and output arrays may both be placed at the beginning of the scratch area to save memory. The size of the object code is 1338 bytes, and 1716 bytes of constant data are needed. If buffered accumulation is needed, as in most applications, an additional 1920 bytes of RAM is required over the 1520 byte scratch area. In this case, data may be gathered and disseminated simultaneously with the signal processing, but for a cost of an additional 2.3 ms.

2.5. Timing

The following tables summarize the execution time required by the WFTA implementation. Table 3 shows the execution time breakdown for the standard WFTA, while Table 4 shows the execution time breakdown for the WFTA preceded by a time-domain windowing function.

The execution times are broken down into transfer instructions, arithmetic instructions, and control instructions. Note that the time spent on addition and multiplication is approximately balanced for the non-windowing version. The windowing significantly increases the time spent on multiplication. Also note that little time is wasted on register-to-register transfers and control instructions.

Table 3 Timing Summary (no windowing)		
Operation	Time (ms)	Percent
Transfer from memory	2.4690	23
Transfer reg-reg	0.5988	5
Transfer to memory	2.1890	20
Add/Subtract	1.5728	15
Multiply	2.2680	21
Shift	1.0932	10
Test/Branch	0.0964	1
Loops	0.4964	5
Total	10.7836	100

VLSI Signal Processing

Table 4
Timing Summary (with windowing)

Operation	Time (ms)	Percent
Transfer from memory	2.5650	20
Transfer reg-reg	0.5988	5
Transfer to memory	2.1890	17
Add/Subtract	1.7648	13
Multiply	3.9480	30
Shift	1.3812	10
Test/Branch	0.0964	1
Loops	0.4976	4
Total	13.0408	100

3. Extensions: Inverse and Complex Transforms
3.1. Inverse Transform

The utility of the approach would be quite limited if the heavily optimized forward algorithm could not be used for an inverse. In this section we show that that the inverse calculation may be achieved with this routine with very small additional computation and program storage.

Consider the real DFT transform pair

$$x(n) \longleftrightarrow X(r) \quad 0 \leq n,r \leq N\text{-}1 \qquad (3.1.1)$$

where

$$X(N\text{-}r) = X^*(r) \quad 0 \leq r \leq \frac{N}{2} \qquad (3.1.2)$$

and * implies the complex conjugate.

We define A and B as *real* sequences such that

$$X(r) \equiv A(r) + jB(r) \quad 0 \leq r \leq N\text{-}1 \qquad (3.1.3)$$

We also define an operator (not a DFT)

$$s(n) \equiv \sum_{r=0}^{\frac{N}{2}-1} S(r) W_N^{nr} \quad 0 \leq n \leq N\text{-}1 \qquad (3.1.4)$$

where S(r) is a *real* sequence. If we use Equation 3.1.4 to define a(n) and b(n) from A(r) and B(r) of 3.1.3, then the following relation may be derived from the definition of the inverse DFT:

$$x(n) = \frac{1}{N} \Big\{ 2Re[a(n)] + 2Im[b(n)] - X(0) - X(N/2) \Big\} \quad 0 \leq n \leq N\text{-}1 \qquad (3.1.5)$$

Suppose we next form the *real* N-point sequence z(i).

$$z(i) = \begin{cases} A(i) + B(i) & 0 \leq i \leq \frac{N}{2}-1 \\ A(N\text{-}i) - B(N\text{-}i) & \frac{N}{2} \leq i \leq N\text{-}1 \end{cases} \qquad (3.1.6)$$

VLSI Signal Processing

We then take the available, forward, real transform of $z(i)$, and attempt to write the resulting data in terms of a and b. Using the DFT definition, substituting Equation (3.1.5), changing the independent variable so that all the summations are in the range $[0, N/2-1]$, and using the definition of the complex conjugate of the operator defined in Equation (3.1.4), we get

$$Re[Z(r)] = 2Re[a(r)] - A(0) - A(N/2) \quad 0 \leq r \leq \frac{N}{2}$$

$$Im[Z(r)] = 2Im[b(r)] \quad 0 \leq r \leq \frac{N}{2} \quad (3.1.7)$$

Thus, we may combine the results of (3.1.5) and (3.1.7) to obtain the pair,

$$x(n) = \frac{1}{N}\Big\{ Re[Z(n)] + Im[Z(n)] \Big\} \quad 0 \leq n \leq \frac{N}{2}-1$$

$$x(n) = \frac{1}{N}\Big\{ Re[Z(N-n)] - Im[Z(N-n)] \Big\} \quad \frac{N}{2} \leq n \leq N-1 \quad (3.1.8)$$

Therefore, the algorithm for the inverse transform using the optimized code is 1) form $z(i)$ from the $N/2+1$ frequency components; 2) take the forward transform (apply the routine) to $z(i)$; and 3) apply 3.1.8 to get the time-domain data $x(n)$. If we neglect the divisions by N, then only 480 extra additions need by performed over the forward transform, 240 to form $z(i)$ and 240 to obtain $x(n)$. Available address registers imply that only 24 cycles per operation need be done, or an additional 1.152 ms is added to the time taken for the forward transform.

3.2. Complex Transform

The complex transform may be calculated from the real transform by the most obvious means. If we define a complex $x(n)$ as

$$x(n) = g(n) + jh(n) \quad (3.2.1)$$

where $g(n)$ and $h(n)$ are the real and imaginary parts of the complex data respectively, then the complex transform may be easily constructed from the real transforms of $g(n)$ and $h(n)$, i.e. $G(r)$ and $H(r)$ by

$$Re[X(r)] = \begin{cases} Re[G(r)] - Im[H(r)] & 0 \leq r \leq \frac{N}{2}-1 \\ Re[G(N-r)] + Im[H(N-r)] & \frac{N}{2} \leq r \leq N-1 \end{cases} \quad (3.2.2)$$

$$Im[X(r)] = \begin{cases} Im[G(r)] + Re[H(r)] & 0 \leq r \leq \frac{N}{2}-1 \\ Re[H(N-r)] - Im[G(N-r)] & \frac{N}{2} \leq r \leq N-1 \end{cases}$$

Thus, two real transforms plus the shuffling of Eq (3.2.2) are required to do the complex transform. The shuffling is **in-place** and 0.24 ms need to be added to the cost of a complex 240-point transform. Using the calculation time from Table 3, the complex transform will take 21.8 ms, with I/O and windowing costs neglected.

It should be noted that there is only a slight increase in cost in using this technique as compared to writing a complex WFTA program. In the real program, full advantage has been taken of the fact that the data is real all the way through until the final two stages of output additions. Therefore, the complex program would give no advantage in any of these code sections. On the other hand, there is duplication in the last two stages of output additions; it may be inferred from Table 1 that 3.66 ms are *wasted* here. Furthermore, this waste could be eliminated if **in-line** code were implemented, but this is a very tedious process which implies the use of extensive program memory and has only a little payoff.

VLSI Signal Processing

4. Discussion: VLSI Implications

DSP VLSI typically centers around a fast hardware array multiplier; a substantial fraction of the chip area is dedicated to this multiplier. Speed is achieved by appropriate care and feeding of the fast multiplier. The selection of an optimzed WFTA algorithm, however, has shown that, for the worst case--windowing included--only 30% of the time is taken by the multiplications when a slow, microcoded multiplier is used. Thus, even if a general-purpose microprocessor could be made with a substantial portion of its silicon dedicated to a fast multplier, then (reference Table 4) the figure for the time for the 240-point DFT would be reduced to about 10 ms. Of course the potential for a better marriage between this new processor and a DFT algorithm must also be considered.

Perhaps worthy of more consideration is using any extra silicon to extend the cache memory (registers) to some much larger number. This implementation requires 1520 bytes of scratch storage. If this amount could be included on-chip as a cache memory, then the data transfers to and from memory, accounting for 37% of the calculation time, could be eliminated.

Finally, there is a new version of the M68000, the M68020[18], which incorporates a 256-byte on-chip instruction cache. Since most of the code in this implementation is inside loops, we estimate that this mechanism would eliminate up to 45% of the calculation time.

5. Summary

The marriage of a conventional, 16-bit microprocessor with a highly optimized WFTA DFT implementation can provide real-time, streamed DFT output for surprisingly high data rates. In fact, the 240-point algorithm runs only about a factor of two slower than does the 256-point FFT on the TMS-320 VLSI signal processing chip. The ability of the 68000 to do many other, perhaps, non-signal processing tasks makes the use of this marriage attractive for many implementations. Highly optimized code can also be used for more extensive purposes, such as the calculation of the inverse 240-point transform, complex transforms, and larger real transforms. We expect that the relationship in performance between this general-purpose microprocessor implementation and that of special-purpose signal processing VLSI will remain about the same in the future; technology appears to be improving throughput at about the same rate. For example, the M68020 is claimed to be four times as fast as the M68000. This would imply that 240-point implementation discussed here would go twice as fast as the current TMS-320. Therefore, we see the current approach will continue to be a viable one for the forseeable future.

6. References

[1] Wavetek Rockland, "Model 5820A Cross Channel Spectrum Analyzer, Technical Information", Wavetek Rockland Inc., Rockleigh, NJ 07647, 1983.

[2] H. F. Silverman, N. R. Dixon, "A Parametrically Controlled Spectrum Analysis System for Speech", *IEEE Transactions on Acoustics, Speech, and Signal Processing*, vol. ASSP-22, pp. 362-381, Oct. 1974.

[3] B. A. Dautrich, L. R. Rabiner, T. B. Martin, "On the Effects of Varying Filter Bank Parameters on Isolated Word Recognition", *IEEE Transactions on Acoustics, Speech, and Signal Processing*, Vol. ASSP-31, No. 4, Aug. 1983, pp. 793-807.

[4] P. E. Dudgeon, R. M. Mersereau, *Multi-Dimensional Digital Signal Processing*, Englewood Cliffs, New Jersey: Prentice-Hall, 1984.

[5] C. H. Knapp, G. C. Carter, "The Generalized Correlation Method for Estimation of Time Delay", *IEEE Transactions on Acoustics, Speech, and Signal Processing*, vol. ASSP-24, pp. 320-327, August, 1976.

[6] A. V. Oppenheim, Editor, *Applications of Digital Signal Processing*, Englewood Cliffs, New Jersey: Prentice-Hall, 1978, Ch. 7.

[7] P. A. W. Lewis, P. C. Yue, "Statistical Analysis of Programs Reference Patterns in a Paging Environment", *IEEE International Computer Society Conference (5th)*, Digest, New York, IEEE, 1971, pp. 133-4.

[8] Nicolet Oscilloscope Division, "Nicolet 2090 Digital Oscilloscope Instruction Manual", Nicolet Instrument Corporation, 1984.

[9] Texas Instruments Corporation, *TMS32010 User's Guide*, 1983.

[10] L. R. Morris, Oral Presentation at the Brown/ASSP Workshop on Fast Algorithms, Brown University, Oct. 1983.

[11] L. R. Morris, "A comparative Study of Time-Efficient FFT and WFTA Programs for General Purpose Computers", *IEEE Transactions on Acoustics, Speech, and Signal Processing*, Vol. ASSP-26 No. 2, April 1978, pp. 141-150.

[12] R. M. Gibson and D. P. McCabe, "Fourier Transform Algorithm Implementations on a General Purpose Microprocessor", *Proceedings of 1981 ICASSP*, Atlanta, Ga. pp. 670-672.

[13] H. F. Silverman, "An Introduction to Programming the Winograd Fourier Transform Algorithm (WFTA)", *IEEE Transactions on Acoustics, Speech, and Signal Processing*, Vol. ASSP-25, No. 2, April 1977, pp. 152-165.

[14] H. F. Silverman, "A Method for Programming the Complex, General-N Winograd Fourier Transform Algorithm", *Proceedings of 1977 ICASSP*, Hartford, Ct. pp. 369-372.

[15] *68000 16/32-Bit Microprocessor, Programmer Reference Manual*, Fourth Edition, Prentice-Hall, Inc, Englewood Cliffs, NJ, 1984.

[16] *MC68020 Product Preview*, Motorola Semiconductor Products Inc., Austin, Texas, 1982.

[17] R. C. Agarwal, private communication.

[18] D. MacGregor, D. Mothersole, and D. Winpigler, "32-bit microprocessor reaps the benefits of a host of enhancements", *Electronic Design*, July 26, 1984, pp. 235-248.

Numerical Stability of Fast Convolution Algorithms for Digital Filtering

L. Auslander
Department of Mathematics, CUNY Graduate Center
33 West 42nd Street, New York, NY 10036

J. W. Cooley,
IBM Thomas J. Watson Research Center
P.O. Box 218, Yorktown Heights, NY 10598

A. J. Silberger
IBM Thomas J. Watson Research Center
P.O. Box 218, Yorktown Heights, NY 10598
and Department of Mathematics,
Cleveland State University, Cleveland, Ohio 44115

Abstract

Automated methods for generating fast convolution algorithms have been developed and used for designing digital filters. The main attention in these designs has been towards reducing the number of arithmetic operations. The present work describes strategies in the algorithm design which help reduce generated arithmetic error and which permit efficient implementation on vector processors or VLSI technology.

0. Introduction

Winograd's computational complexity theory [11,12,13,14] has led to semi-automated methods for the construction of cyclic convolution algorithms [3,5,6,7] which have a number of desirable properties. These algorithms are generally referred to as rectangular transform (RT) algorithms because the matrices of coefficients describing them are rectangular. While emphasis has been on the reduction in computation, the second property which has received somewhat less attention is error generation or numerical stability. It was noted by Cooley [7] that in adapting these algorithms to a fixed point computation, those algorithms whose matrices had large or many non-zero elements produced intermediate quantities which made it necessary to scale down the data, resulting in a loss of accuracy. A floating point calculation would suffer, under the same circumstances, from a build-up of rounding or truncation errors. The present work has the objectives of devising strategies leading to algorithms which are not only computationally efficient but which reduce generated error in either fixed or floating point calculations. The latter property of an algorithm will be referred to as "stability".

A third purpose of the present effort is to achieve the two above objectives and, at the same time, formulate the algorithms so as to permit efficient implementation in a parallel processor such as in VLSI technology. It is also hoped that these algorithms for computing cyclic convolutions will be useful for the design of efficient digital filters in some of the new vector and parallel general-purpose computers.

Most of the algorithms we present are for convolutions on N points, where N is a prime number. We expect to publish more general as well as more extensive results elsewhere. The reader will find here examples of what can be achieved by attempting to balance the usually conflicting goals of constructing stable and multiplicatively minimal convolution algorithms. He will see that we have in some cases erred on the side of stability and included algorithms which are rather far from being multiplicatively minimal. In these cases we have in mind the need to study differing kinds of implementations of our algorithms in serial, vector, and parallel processors. It does not appear immediately clear which algorithms will be most efficient for these various systems. We have included algorithms which are constructed via applications of the Chinese Remainder Theorem (CRT), via iterative procedures, and via a mix of the two methods. Our hope is that the study of these examples will reveal which methods of construction are appropriate to the different applications.

VLSI Signal Processing

We have given here a fairly large number of examples of cyclic convolution on a prime number of points. Our examples are important both in their own right and as building blocks for constructing algorithms for convolution and for the Winograd Fourier transform on large numbers of points via the tensor product construction. We have given one example of such a construction, namely the 30-point example below. It should be emphasized that we intend to complete this list of examples with more prime power examples as well as examples of cyclic convolution on a composite number of points constructed both via the tensor product and other methods.

This paper has been written as material for a workshop. We have prepared some theoretical foundations, developed some tools for designing and automatically producing and testing programs for algorithms and we have done some numerical examples which illustrate our methods and what can be achieved with them. It is hoped that interaction with computer architects and digital signal processing experts will give us a better understanding of criteria for good algorithms and thereby help in the production of efficient methods for implementing these techniques. The paper should be regarded as a glimpse at results which we intend to publish elsewhere in a more detailed and complete form.

Although S. Winograd is not a coauthor, our understanding of the algorithms discussed in this paper developed during discussions with him. The SCRATCHPAD techniques and the program for automatically reducing the numbers of additions developed from manual techniques which he suggested several years ago. Winograd was the first to study stable algorithms. He was aware of the existence of a stable algorithm on nine points like the one we mention here before we were.

1. Some Generalities

For any positive integer N let Q^N denote the set of N-tuples of rational numbers. For any $a = (a_0, \ldots, a_{N-1})$ and $b = (b_0, \ldots, b_{N-1})$ in Q^N we define the *cyclic convolution* $a * b$ of a and b by setting

$$(a * b)_i = \sum_{j=0}^{N-1} a_j b_{i-j} \tag{1.1}$$

for any i, $0 \leq i \leq N - 1$. To make this definition meaningful we set

$$b_k = b_{k'} \text{ for } k \equiv k' \text{ Mod(N)}. \tag{1.2}$$

In this paper we present algorithms for computing the sequence $(a*b)_0, \ldots, (a*b)_{N-1}$ from the input sequences $a_0, \ldots, a_{N-1}$ and $b_0, \ldots, b_{N-1}$. Our main interest will be in constructing algorithms which would be useful in the context where one of the sequences, say b, is constant over time, while the other varies; i.e., where we compute a large number of convolutions involving the same sequence b and many different sequences a. This would be the case, for instance, when b represents the weighting function of a filter and the set of sequences a the set of inputs to the filter.

1.1 Algorithms for Computing Cyclic Convolution on N Points

By an algorithm for computing $(a*b)_0, \ldots, (a*b)_{N-1}$ we understand the following: Let $x = (x_0, \ldots, x_{N-1})$ and $y = (y_0, \ldots, y_{N-1})$ be vectors of "variables". We wish to construct from the set of $2N$ variables $x_0, \ldots, x_{N-1}, y_0, \ldots, y_{N-1}$ the polynomial expressions $(x*y)_0, \ldots, (x*y)_{N-1}$ via a sequence of field operations applied to expressions in the components of x and y constructed in previous steps of the algorithm. In practice, the steps of the algorithm divide into three separate "stages". In the first or "pre-multiplication" stage we form separate linear expressions in the x variables and the y variables; i.e., we perform rational vector space operations on the separate spaces generated by the x variables and the y variables. In each step of the second or "multiplication" stage we multiply a linear expression in the x's by a linear expression in the y's, each expression having been formed in the first stage. The steps of this stage are the multiplications which we count in the algorithm. In the final "post-multiplication" stage of the algorithm we perform rational vector space operations on the set of products formed in the second stage.

One sees from this that we can represent an algorithm by a matrix product of the form

$$x * y = C(Ax \cdot By), \tag{1.3}$$

VLSI Signal Processing

where x and y are the transposes of the vectors $(x_0, \ldots, x_{N-1})$ and $(y_0, \ldots, y_{N-1})$. Letting M be the number of multiplications of the algorithm, the matrices A, and B are of size $M \times N$ and the matrix C is of size $N \times M$ All three have entries which are rational numbers. The matrices A and B represent the pre-multiplication stage of the algorithm; C represents the post multiplication stage. We indicate by "•" the multiplication stage of the algorithm, which consists of the componentwise multiplication of the vectors $A\,x$ and $B\,y$.

1.2 The Stability Criterion

We call an algorithm *iteratively stable* if the A and B matrices above have no entry unequal to 1, 0, or -1. It is clear that the tensor product of matrices of this form is again a matrix of this form, which observation justifies the name. We call an algorithm *relatively stable* if its entries are "almost" of the above required form. Again the attempt is to construct algorithms which will remain relatively stable under iteration, i. e. under the tensor product operation on the corresponding matrices. This goal rather than a precise meaning for the word "almost" is perhaps most desirable. Thus relative stability in essence depends on the use of an algorithm in an iterative scheme.

1.3 The Transpose of an Algorithm and Digital Filtering

In all of what follows, we consider digital filtering applications where, typically, one sequence is a fixed set of "weights", which we will denote by $h = (h_0, \ldots, h_{N-1})$ which is convolved with many x sequences. In this case, one typically precomputes, in perhaps higher precision, the transform $A\,h$ and stores it. Therefore, it is efficient to let the transform of h be the more expensive calculation. In the construction (1.1) above, the C matrix comes from the Chinese Remainder Theorem (CRT) and involves more computing. Therefore we switch its position in the algorithm with the A matrix and let it operate on the input h. With these assumptions in mind we express the convolution in the form

$$y = C(Ah \cdot Bx). \tag{1.4}$$

In the counting of arithmetic operations, we do not include the operations in computing h or consider the errors in its calculation.

We construct algorithms for digital filtering in two steps. First, we construct the A, B, and C matrices of stable algorithms. For simplicity we construct algorithms in which $A = B$. The A and B matrices have integer coefficients which are usually ± 1 or 0. The C matrix has entries which are complicated, namely, consisting of rational numbers. The following notion of the transpose of an algorithm, originally proven by Winograd [14] and derived independently by Agarwal [3], allows us, in effect, to permute the matrices. As we shall explain below, this helps us to design algorithms for digital filters.

In order to make Winograd's idea clear we introduce a slight modification of the usual notion of the transpose of a matrix. Let N be fixed and let A be any $M \times N$ matrix. Number the columns of A from 0 to $N - 1$. Let A' denote the matrix whose i-th row equals the $(N - i)$-th column of A, $0 < i < N$; let the 0-th row of A' equal the 0-th column of A. Then A' is a $N \times M$ matrix. For $N \times M$ matrices we define the $'$ operation to be the inverse of the operation just defined.

Suppose now that we are given an algorithm represented as a matrix product $C(Ah \cdot Bx)$. The transposes of this algorithm are the algorithms (and they do give algorithms for cyclic convolution!) represented by the matrix products $A'(C'h \cdot Bx)$ and $B'(Ah \cdot C'x)$.

We use the stability criterion together with the transpose to construct algorithms for computing $h * x$ when one of the N-tuples, say h, is fixed and the other N-tuple is regarded as variable. Here we are willing to expend the effort to compute a matrix product of the form $C'\,h$ with considerable precision, since it has to be computed essentially only once, whereas, since the multiplications by the matrices A and B have to be repeated, we would want to simplify these computations. If we begin with an iteratively stable algorithm of the form (1.4), we may use the transpose in the form

$$y = A'(C'\,h \cdot B\,x) \tag{1.5}$$

to construct an algorithm in which the operations resulting from the matrix C are built into the filter and the pre-multiplication and post-multiplication steps, resulting from the respective matrices A and B, consist entirely of additions and subtractions. The algorithms we present in this paper are all constructed in this way. A program of one of the authors transposes the stable algorithms whose construction we explain below and, after following procedures designed to minimize the number of additions in the algorithm, presents the algorithm as a sequence of additions, multiplications, and again additions, giving the counts of the operations in the three stages of the algorithm.

2. The Construction of the Algorithms

2.1 General Statement of the Chinese Remainder Theorem (CRT)

The Chinese Remainder Theorem (CRT) is used extensively throughout the derivation of the equations which follow. It will be described for rings of polynomials with integer or rational coefficients. In the description of multi-dimensional techniques, in Sections 3 and 4, the CRT in rings of residue classes of integers is used. Let

$$F(u) = F_1(u)F_2(u) \ldots F_r(u), \tag{2.1}$$

where the $F_i(u)$s are mutually prime. Let $f_i(u)$ satisfy

$$f_i(u)(F(u)/F_i(u)) \equiv 1 \bmod F_i(u) \tag{2.2}$$

In other words, $f_i(u)$ is the inverse of $F(u)/F_i(u)$ in the ring of polynomials Mod $F_i(u)$. In terms of these, we define the idempotents

$$e_i(u) = f_i(u)(F(u)/F_i(u)). \tag{2.3}$$

Thus, from (2.2) we have

$$\begin{aligned} e_i(u) &= 1 \ \text{Mod} \ F_i(u), \\ e_j(u) \ \text{Mod} \ F_i(u) &= 0, \ i \neq j. \end{aligned} \tag{2.4}$$

Given an arbitrary $X(u)$ in the ring of polynomials modulo $F(u)$, the CRT states that

$$X(u) = \sum_{j=1}^{r} e_j(u) X_j(u) \ \text{Mod} \ F(u). \tag{2.5}$$

where

$$X_j(u) = X(u) \ \text{Mod} \ F_j(u). \tag{2.6}$$

In particular, if $X(u) = 1$, then

$$1 = \sum_{j=1}^{r} e_j(u) \ \text{Mod} \ F(u) \tag{2.7}$$

From the definitions, we have

$$e_i(u)e_j(u) = 0 \ \text{Mod} \ F(u), i \neq j. \tag{2.8}$$

Multiplying both sides of (2.7) by $e_i(u)$ gives

$$e_i(u) = \sum_{j=1}^{r} e_i(u)e_j(u) = e(u)_i^2 \ \text{Mod} \ F(u) \tag{2.9}$$

and multiplying both sides of (2.5) by $e_i(u)$ gives

$$e_i(u)X(u) = e_i(u)X_i(u). \ \text{Mod} \ F(u) \tag{2.10}$$

VLSI Signal Processing
2.2 First CRT Reduction

In order to explain the construction of our algorithms for computing cyclic convolution, we reformulate the problem. Let $Q[u]$ denote the algebra of all polynomials in the variable u, let $\phi(u)$ be an arbitrary monic polynomial in u, and let $Q[u]/(\phi(u))$ denote the factor algebra consisting of all polynomials Mod $\phi(u)$. In particular, let $N > 1$ be an integer and let $Q[u]/(u^N - 1)$ denote the factor algebra of $Q[u]$ obtained by introducing the relation $u^N = 1$. Let $X(u) = x_0 + x_1 u \ldots + x_{N-1} u^{N-1}$ and $H(u) = h_0 + h_1 u \ldots + h_{N-1} u^{N-1}$. Then, in the algebra $Q[u]/(u^N - 1)$,

$$X(u)H(u) = \sum_{i=0}^{N-1} (x * h)_i u^i. \tag{2.11}$$

Giving an algorithm for computing $x * h$ is equivalent to giving an algorithm for computing $X(u)H(u)$ in the algebra $Q[u]/(u^N - 1)$. Since $n > 1$, the polynomial $u^N - 1$ has the two relatively prime factors $u - 1$ and $u^{N-1} + \cdots + 1$, to each of which corresponds an idempotent. For the two-factor case, where

$$F_1(u) = u - 1 \quad \text{and} \quad F_2(u) = u^{N-1} + u^{N-2} + \cdots + 1 \tag{2.12}$$

we have $f_1(u) = 1/n$ so that

$$e_1(u) = \frac{1}{n}(1 + u + \cdots + u^{N-1}) \tag{2.13}$$

From the fact that

$$1 = e_1(u) + e_2(u), \tag{2.14}$$

it follows that

$$e_2(u) = 1 - e_1(u) = \frac{1}{n}((N-1) - u - u^2 - \cdots - u^{N-1}). \tag{2.15}$$

Since, in $Q[u]/(u^N - 1)$,

$$(u - 1)e_1(u) = 0 \tag{2.16}$$

and

$$(1 + \cdots + u^{N-1})e_2(u) = 0, \tag{2.17}$$

one sees that $e_1(u)$ and $e_2(u)$ are the idempotents which project $Q[u]/(u^N - 1)$ on $Q[u]/(u - 1)$ and on $Q[u]/(u^{N-1} + \cdots + 1)$, respectively. In particular, $e_i(u)e_j(u) = 0$, when $i \neq j$, and $e_i(u)e_i(u) = e_i(u)$, $i = 1$ or 2. Therefore,

$$\begin{aligned} X(u)H(u) &= (e_1(u) + e_2(u))X(u)H(u) \\ &= e_1(u)(e_1(u)X(u))(e_1(u)H(u)) + e_2(u)(e_2(u)X(u))(e_2(u)H(u)), \\ &= e_1(u)X_1(u)H_1(u) + e_2 X_2(u)H_2(u) \end{aligned} \tag{2.18}$$

where $X_i(u) = e_i(u)X(u)$ and $H_i(u) = e_i(u)H(u)$, $i = 1$ or 2. Since

$$X(u)e_1(u) = (x_0 + \cdots + x_{N-1})e_1(u) \tag{2.19}$$

and, similarly, for $H(u)e_1(u)$, we have

$$e_1(u)X_1(u)H_1(u) = e_1(u)(x_0 + \cdots + x_{N-1})(h_0 + \cdots + h_{N-1}). \tag{2.20}$$

From the relation

$$X(u)e_2(u) = (x_0 - x_{N-1}) + \cdots + (x_{N-2} - x_{N-1})u^{N-2} \tag{2.21}$$

and a similar expression for $H(u)e_2(u)$, we obtain that

$$e_2(u)X_2(u)H_2(u) = e_2(u)\sum_{j=0}^{N-2}(x_j - x_{N-1})u^j \sum_{j=0}^{N-2}(h_j - h_{N-1})u^j. \tag{2.22}$$

For every n we define the multiplication step

$$m_0 = (x_0 + \cdots + x_{N-1})(h_0 + \cdots + h_{N-1}) \qquad (2.23)$$

and observe that

$$X(u)H(u) = e_1(u)m_0 + e_2(u)X_2(u)H_2(u) \quad \text{Mod}(u^N - 1). \qquad (2.24)$$

The coefficients of $X_2(u)H_2(u)$ are linear expressions in the multiplication steps $m_1, \ldots, m_{M-1}$ of the algorithm. By writing the vector of coefficients of the $(N-1)$-th degree polynomial $X(u)H(u)$ as a matrix product one obtains the $N \times M$ matrix C of Section 1.1; to obtain the $M \times N$ matrices A and B it is necessary to express the vector $(Ah) \times (Bx)$ as a componentwise product of matricial expressions in h and x. We give only one algorithm for cyclic convolution on three points and we give all the details concerning the construction of its matrix representation. For one of the algorithms given below for cyclic convolution on seven points we indicate in some detail how one goes about finding its matrix representation. In the other algorithms we discuss only the method of computing the coefficients of the product polynomial $X_2(u)H_2(u)$. Matrices A and C for all algorithms and the PLI program statements for computing the rectangular transforms $b = Bx$ and $y = Cm$ are given in the appendix. The PLI statements were produced automatically by the program mentioned above which sequences the additions so as to minimize their number.

2.3 Cyclic Convolution on p Points, p a Prime Number

When $N=p$, a prime number, the polynomial $1 + u + \cdots + u^{p-1}$ is irreducible. This implies that the algebra $Q[u]/(u^p - 1)$ does not decompose further. Before turning to a case-by-case discussion of the multiplication $X_2(u)H_2(u)$, in which we treat each algorithm separately, let us discuss in general terms how we compute the coefficients of this product. If we refer to an algorithm as a CRT algorithm, we mean that we have constructed the unreduced polynomial $X_2(u)H_2(u)$ by using the following technique: Let $F(u)$ be an arbitrary monic polynomial, which we call the *reducing polynomial* for the product to be computed. We may write

$$X_2(u)H_2(u) = R(u) + Q(u)F(u), \qquad (2.25)$$

where the degree of $R(u)$ is less than the degree of $F(u)$. Of course, $Q(u)$ and $R(u)$ are, respectively, the quotient and the remainder of $X_2(u)H_2(u)$ after division by $F(u)$. If $F(u) = F_1(u) \ldots F_r(u)$, a product of relatively prime factors, then, by the CRT,

$$R(u) = X_{2,1}(u)H_{2,1}(u)f_1(u)\frac{F(u)}{F_1(u)} + \cdots + X_{2,r}(u)H_{2,r}(u)f_r(u)\frac{F(u)}{F_r(u)} \quad \text{Mod } F(u), \qquad (2.26)$$

where $X_{2,i}(u)$ and $H_{2,i}(u)$ are, respectively, the remainders of $X_2(u)$ and $H_2(u)$ after division by $F_i(u)$ and $f_i(u)$ is the inverse of $F(u)/F_i(u)$ in the algebra $Q[u]/(F_i(u))$ for all $i = 1, \ldots, r$. The polynomials $e_i(u) = f_i(u)(F(u)/F_i(u))$ are the idempotents in the algebra $Q[u]/(F(u))$. Each product $X_{2,i}(u)H_{2,i}(u)f_i(u)$ may obviously be computed in the algebra $Q[u]/(F_i(u))$; indeed, if each such product is chosen with degree less than the degree of $F_i(u)$, then we have a polynomial equality (with no congruence condition) in Equation (2.25). It is sometimes convenient to write $R_{2,i}(u)$ for the remainder obtained after dividing the product $X_{2,i}(u)H_{2,i}(u)$ by $F_i(u)$. We shall use this convention in the example of seven point cyclic convolution discussed below. In any case, we have to express each of the products $X_{2,i}(u)H_{2,i}(u)$ as well as $Q(u)$ in terms of the multiplication steps of the algorithm to determine $X_2(u)H_2(u)$ as a polynomial in u with coefficients linear forms in $m_1, \ldots, m_{M-1}$. Using the CRT algorithm for cyclic convolution on seven points as an example, we will illustrate how this may be done. In the CRT algorithms below we indicate the presence of $Q(u) \neq 0$ by adjoining to $F(u)$ the factor $(u - \infty)^s$, where $s = 1 + \text{degree}(Q(u))$.

Let us remark: *A necessary condition in order that a CRT algorithm be iteratively stable is that every root of $F(u)$ be either zero or a root of unity (or ∞).*

VLSI Signal Processing

2.3.1 The case $p = 3$

We give only one algorithm for cyclic convolution on three points, as there is essentially only one which is minimal and iteratively stable. We must express the quadratic

$$X_2(u)H_2(u) = [(x_0 - x_2) + (x_1 - x_2)u][(h_0 - h_2) + (h_1 - h_2)u] \quad (2.27)$$

in terms of the multiplication steps of the algorithm. We use the CRT, taking $F(u) = u(u + 1)(u - \infty)$. Letting

$$m_1 = (x_0 - x_2)(h_0 - h_2), \quad (2.28)$$

$$m_2 = (x_0 - x_1)(h_0 - h_1),$$

and

$$m_3 = (x_1 - x_2)(h_1 - h_2), \quad (2.29)$$

we obtain

$$X_2(u)H_2(u) = m_1(u + 1) - m_2 u + m_3 u(u + 1), \quad (2.30)$$

$$= (m_1 - m_3) + (m_1 - m_2)u \quad \text{Mod}(u^2 + u + 1), \quad (2.31)$$

which implies that, $\text{Mod}(u^3 - 1)$,

$$X(u)H(u) = \frac{1}{3}(1 + u + u^2)m_0 + \frac{1}{3}(2 - u - u^2)[(m_1 - m_3) + (m_1 - m_2)u]. \quad (2.32)$$

Writing this as a matrix product, we have

$$X(u)H(u) = \frac{1}{3}[1 \ u \ u^2] \begin{bmatrix} 1 & 1 & 1 & -2 \\ 1 & 1 & -2 & 1 \\ 1 & -2 & 1 & 1 \end{bmatrix} \begin{bmatrix} m_0 \\ m_1 \\ m_2 \\ m_3 \end{bmatrix}$$

$$= \frac{1}{3}[1 \ u \ u^2] \begin{bmatrix} 1 & 1 & 1 & -2 \\ 1 & 1 & -2 & 1 \\ 1 & -2 & 1 & 1 \end{bmatrix}$$

$$\left(\begin{bmatrix} 1 & 1 & 1 \\ 1 & 0 & -1 \\ 1 & -1 & 0 \\ 0 & 1 & -1 \end{bmatrix} \begin{bmatrix} x_0 \\ x_1 \\ x_2 \end{bmatrix} \cdot \begin{bmatrix} 1 & 1 & 1 \\ 1 & 0 & -1 \\ 1 & -1 & 0 \\ 0 & 1 & -1 \end{bmatrix} \begin{bmatrix} y_0 \\ y_1 \\ y_2 \end{bmatrix} \right) \quad (2.33)$$

This is the matrix representation of the algorithm which we refer to as C3CR4 in Table 1 below and in what follows. The label C3CR4 denotes cyclic convolution of 3 points, derived with the CRT and taking 4 multiplications.

2.3.2 The case $p = 5$

We give two iteratively stable algorithms for this case, each requiring ten rational multiplications. For the first, we use the CRT to represent the coefficients of $X_2(u)H_2(u)$ in terms of the multiplication steps of the algorithm. We take

$$F(u) = u(u + 1)(u^2 + 1)(u^2 + u + 1)(u - \infty). \quad (2.34)$$

Since we are multiplying cubics, we can compute the product in $Q[u]$ modulo the degree seven polynomial $F(u)(u - \infty)$. This computation of $X_2(u)H_2(u)$ requires nine rational multiplications, three for each quadratic factor of $F(u)$ and one for each linear factor and $u - \infty$. Thus this algorithm for cyclic multiplication on five points requires a total of ten rational multiplications. This algorithm is referred to as C5CR10 in Table 1.

For the second algorithm we write

$$X_2(u)H_2(u) = (\alpha_0 + \alpha_1 u^2)(\beta_0 + \beta_1 u^2), \qquad (2.35)$$

where

$$\alpha_i = (x_{2i} - x_4) + (x_{2i+1} - x_4)u \qquad (2.36)$$

and

$$\beta_i = (h_{2i} - h_4) + (h_{2i+1} - h_4)u, \qquad (2.37)$$

$i = 0$ or 1. We then compute the products of the linear polynomials in u^2 by using the CRT as in the case $p=3$. This amounts to computing the products $\alpha_0\beta_0$, $(\alpha_0 - \alpha_1)(\beta_0 - \beta_1)$, and $\alpha_1\beta_1$. Each of these three products of linear polynomials in u is computed by again using the stable algorithm for computing products of linear polynomials, again exactly as in the $p=3$ algorithm. We have three multiplications of linear polynomials, each requiring three rational multiplications, so again we compute $X_2(u)H_2(u)$ in nine multiplications. Thus, both algorithms for cyclic multiplication on five points require ten rational multiplications. Let us remark that neither is a multiplicatively minimal algorithm. This algorithm is referred to as C5I10, the "I" meaning "iterative" in Table 1 below.

2.3.3 The case $p = 7$

We give three iteratively stable algorithms, one based on the CRT and two based on iteration. For the CRT algorithm we use, as reducing polynomial, the polynomial

$$F(u) = u^2(u + 1)(u^2 + 1)(u^2 + u + 1)(u^2 - u + 1)(u - \infty)^2. \qquad (2.38)$$

We obtain an algorithm requiring seventeen rational multiplications. Let us indicate how we construct its matrix representation. Besides the multiplication step m_0, which was constructed above, in Section 2.2, we must describe the remaining sixteen multiplication steps and show how to determine the matrices A, B, and C. We number the factors of $F(u)$ in the order in which they appear, so that

$$F(u) = F_1(u) \ldots F_5(u)(u - \infty)^2. \qquad (2.39)$$

Dividing $X_2(u)$ and $H_2(u)$ by $F_1(u) = u^2$, we obtain the remainders

$$X_{2,1}(u) = (x_0 - x_6) + (x_1 - x_6)u \qquad (2.40)$$

and

$$H_{2,1}(u) = (h_0 - h_6) + (h_1 - h_6)u. \qquad (2.41)$$

After dividing the product $X_{2,1}(u)H_{2,1}(u)$ by $F_1(u)$, we obtain the remainder

$$\begin{aligned}R_{2,1}(u) &= (x_0 - x_6)(h_0 - h_6) + [(x_0 - x_6)(h_1 - h_6) + (x_1 - x_6)(h_0 - h_6)]u \\ &= m_1 + (m_1 - m_2 + m_3)u,\end{aligned} \qquad (2.42)$$

where

$$m_1 = (x_0 - x_6)(h_0 - h_6), \qquad (2.43)$$

$$m_2 = (x_0 - x_1)(h_0 - h_1), \qquad (2.44)$$

and

$$m_3 = (x_1 - x_6)(h_1 - h_6). \qquad (2.45)$$

Dividing $X_2(u)$ by $F_2(u) = u + 1$, we obtain the remainder

$$X_{2,2}(u) = x_0 - x_1 + x_2 - x_3 + x_4 - x_5 \qquad (2.46)$$

with a similar expression for $H_{2,2}(u)$. We set

$$m_4 = R_{2,2}(u) = (x_0 - x_1 + x_2 - x_3 + x_4 - x_5)(h_0 - h_1 + h_2 - h_3 + h_4 - h_5). \qquad (2.47)$$

VLSI Signal Processing

In like fashion we divide successively $X_2(u)$ and $H_2(u)$ by the factors $F_3(u) = u^2 + 1$, $F_4(u) = u^2 + u + 1$, and $F_5(u) = u^2 - u + 1$. We obtain

$$X_{2,3}(u) = (x_0 - x_2 + x_4 - x_6) + (x_1 - x_3 + x_5 - x_6)u \qquad (2.48)$$

with a similar expression for $H_{2,3}(u)$, which we omit. Dividing the product $X_{2,3}(u)H_{2,3}(u)$ by $F_3(u)$, we obtain the remainder

$$\begin{aligned}R_{2,3}(u) =& \\ &[(x_0 - x_2 + x_4 - x_6)(h_0 - h_2 + h_4 - h_6) - (x_1 - x_3 + x_5 - x_6)(h_1 - h_3 + h_5 - h_6)] \\ &+[(x_0 - x_2 + x_4 - x_6)(h_1 - h_3 + h_5 - h_6) + (x_1 - x_3 + x_5 - x_6)(h_0 - h_2 + h_4 - h_6)]u \\ =& (m_5 - m_7) + (m_5 - m_6 + m_7)u,\end{aligned} \qquad (2.49)$$

where

$$m_5 = (x_0 - x_2 + x_4 - x_6)(h_0 - h_2 + h_4 - h_6), \qquad (2.50)$$

$$m_6 = (x_0 - x_1 - x_2 + x_3 + x_4 - x_5)(h_0 - h_1 - h_2 + h_3 + h_4 - h_5), \qquad (2.51)$$

and

$$m_7 = (x_1 - x_3 + x_5 - x_6)(h_1 - h_3 + h_5 - h_6). \qquad (2.52)$$

We also have

$$R_{2,4}(u) = (m_8 - m_{10}) + (m_8 - m_9)u, \qquad (2.53)$$

where

$$m_8 = (x_0 - x_2 + x_3 - x_5)(h_0 - h_2 + h_3 - h_5), \qquad (2.54)$$

$$m_9 = (x_0 - x_1 + x_3 - x_4)(h_0 - h_1 + h_3 - h_4), \qquad (2.55)$$

and

$$m_{10} = (x_1 - x_2 + x_4 - x_5)(h_1 - h_2 + h_4 - h_5); \qquad (2.56)$$

and we have

$$R_{2,5}(u) = (m_{11} - m_{13}) + (-m_{11} + m_{12})u, \qquad (2.57)$$

where

$$m_{11} = (x_0 - x_2 - x_3 + x_5)(h_0 - h_2 - h_3 + h_5), \qquad (2.58)$$

$$m_{12} = (x_0 + x_1 - x_3 - x_4)(h_0 + h_1 - h_3 - h_4), \qquad (2.59)$$

and

$$m_{13} = (x_1 + x_2 - x_4 - x_5)(h_1 + h_2 - h_4 - h_5). \qquad (2.60)$$

Finally, let us express $Q(u)$ in terms of multiplication steps. It is clear that $Q(u)$ depends only on the terms of $X_2(u)H_2(u)$ of degree at least nine. Therefore, we may write

$$\begin{aligned}X_2(u)H_2(u) =& (x_5 - x_6)(h_5 - h_6)u^{10} + [(x_4 - x_6)(h_5 - h_6) + (x_5 - x_6)(h_4 - h_6)]u^9 + \cdots \\ =& m_{16}u^{10} + (m_{14} - m_{15} + m_{16})u^9 + \cdots,\end{aligned} \qquad (2.61)$$

where

$$m_{14} = (x_4 - x_6)(h_4 - h_6), \qquad (2.62)$$

$$m_{15} = (x_4 - x_5)(h_4 - h_5), \qquad (2.63)$$

and

$$m_{16} = (x_5 - x_6)(h_5 - h_6). \qquad (2.64)$$

Therefore, since

$$F_1(u) \ldots F_5(u) = u^9 + u^8 + \cdots, \quad (2.65)$$

we find by performing division of polynomials (see equation (2.24)) that

$$Q(u) = m_{16}u + (m_{14} - m_{15}). \quad (2.66)$$

To compute the matrix A, we substitute these expressions for $R_{2,1}(u), \ldots, R_{2,5}(u)$, and $Q(u)$ into formulas (2.23) - (2,26). We expand the resulting expression for $X(u)H(u)$ as a sixth degree polynomial in u whose coefficients are linear forms in the variables $m_0, \ldots, m_{16}$. We compute the i-th row of the 7-by-17 matrix C by taking the coefficient of u^i and forming the vector of coefficients of the linear form in $m_0, \ldots, m_{16}$, $0 \leq i \leq 6$.

The above calculations may, with difficulty, be carried out by hand. We have instead made use of SCRATCHPAD programs to obtain these matrices.

The 17-by-7 matrix B is the matrix whose i-th row consists of the vector of coefficients of $x_0, \ldots, x_6$ in the x-factor of m_i, $0 \leq i \leq 16$. Similarly, the matrix A is the matrix whose i-th row consists of the vector of coefficients of $h_0, \ldots, h_6$ in the h-factor of m_i, $0 \leq i \leq 16$. Obviously, $A = B$. The resulting algorithm is referred to as C7CR17 in Table 1.

Each iterative algorithm requires nineteen rational multiplications. For one of them, which we call C7LI19, we express $X_2(u)H_2(u)$ as a product of linear polynomials in u^3 whose coefficients are quadratic polynomials in u. For the other we express $X_2(u)H_2(u)$ as a product of quadratic polynomials in u^2 whose coefficients are linear polynomials in u. This is denoted by C7QI19 in Table 1. In the last two cases, we obtain an iteratively stable algorithm requiring nineteen rational multiplications.

2.3.4 The case $p = 11$

For this prime we give five algorithms, the first two based on the CRT and the last three mixed or iterative algorithms. All but the first will be iteratively stable algorithms and the first will fail iterative stability by the presence of exactly one "2" in each 35-by-11 coefficient matrix; the first CRT algorithm requires 35 rational multiplications and the second 36, so we need exactly one additional multiplication to eliminate the 2's of the first algorithm. These are called C11CR35 and C11CR36, respectively.

The two CRT algorithms obviously differ in the polynomials Modulo which we compute $X_2(u)H_2(u)$. For the first of these algorithms we use the polynomial

$$F(u) = (u+1)u^2(u^2+1)(u^2+u+1)(u^2-u+1)(u^4+1)(u^4+u^3+u^2+u+1)(u-\infty)^2 \quad (2.67)$$

and for the second

$$F(u) = (u+1)(u^2+1)(u^2+u+1)u^3(u^4+1)(u^4+u^3+u^2+u+1)(u-\infty)^3. \quad (2.68)$$

Two of our last three algorithms for $p=11$ involve iteration combined with the CRT for computing the coefficients of $X_2(u)H_2(u)$. In one we multiply linear expressions in u^5 whose coefficients are quartics in u, while in the other we multiply quartics in u^2 whose coefficients are linear in u. Both of these algorithms require thirty-seven rational multiplications. These are referred to as C11LI37 and C11QI37, respectively.

The final algorithm for $p=11$, which requires forty-six rational multiplications, may be regarded as a "pure" iterative algorithm, at least in the computation of the coefficients of $X_2(u)H_2(u)$. We write $X_2(u) = \sum_{j=0}^{4} \alpha_j u^{2j}$ and $H_2(u) = \sum_{j=0}^{4} \beta_j u^{2j}$, where α_j and β_j are, for each j, linear polynomials in u. Then

VLSI Signal Processing

$$X_2(u)H_2(u) = \sum_{j=0}^{8} \gamma_j u^{2j}, \qquad (2.69)$$

where

$$\gamma_j = \sum_{k=0}^{j} \alpha_k \beta_{j-k}, \qquad (2.70)$$

with each term on the right being a quadratic in u. We compute the five products $\alpha_k \beta_k$, $0 \leq k \leq 4$, and the ten terms of the form $\alpha_k \beta_{j-k} + \beta_k \alpha_{j-k}$, $j \neq k$, by computing the ten products of the form $(\alpha_k - \alpha_{j-k})(\beta_k - \beta_{j-k})$, $k < j - k$, and using the relation

$$\begin{aligned}\alpha_k \beta_{j-k} + \beta_k \alpha_{j-k} = \\ \alpha_k \beta_k - (\alpha_k - \alpha_{j-k})(\beta_k - \beta_{j-k}) + \alpha_{j-k} \beta_{j-k}.\end{aligned} \qquad (2.71)$$

Since each of the above fifteen products of linear polynomials requires three rational multiplications, we see that the computation of the coefficients of $X_2(u)H_2(u)$ by this algorithm requires forty-five rational multiplications. Counting the multiplication m_0, we have an algorithm requiring forty-six rational multiplications. It will be interesting to see how economically this algorithm may be implemented on a vector processor. This algorithm is denoted by C11LI46 in Table 1.

2.4 The Prime Power Case

With this paper we include only one algorithm on N points, where N is a power of a prime number, namely the case $N=9$. This algorithm has received a rather detailed discussion in [4], so we shall not discuss it here. It is listed as C9CR19 in Table 1 below. In the more complete version of this paper referred to above we expect to include several more examples of algorithms for the prime power case and to discuss the methods and problems of designing such algorithms in much more detail.

2.5 The Composite Case

We include only one algorithm below in Section 4 for the case in which N has distinct prime factors, the case $N=30$. The only algorithm we discuss is a tensor product algorithm. We emphasize that other algorithms and methods of constructing them may be of equal interest. In any case, we take an algorithm on two points, an algorithm on three points, and an iteratively stable algorithm on five points, and combine them to to obtain an iteratively stable algorithm on thirty points. Since the algorithm on two points requires two rational multiplications, the algorithm on three points four multiplications, and the algorithm on five points ten rational multiplications, the algorithm on thirty points which we construct by tensoring them requires eighty rational multiplications.

3. Multidimensional Technique for Digital Convolution

3.1 Mapping of a Convolution into a Three-Dimensional Convolution

The convolution which we consider here is defined by

$$y_j = \sum_{k=0}^{N-1} h_{j-k} x_k, \qquad (3.1)$$

where all indices are to be taken Mod N. Thus, the convolution is circular, i.e., when the index $j - k$ of h_{j-k} is negative, it is replaced by $N + j - k$. As discussed above in Section 1, an RT algorithm for computing the convolution (3.1) may be written

$$H_n = \sum_{k=0}^{N-1} A_{n,k} h_k \qquad (3.2)$$

VLSI Signal Processing

$$X_n = \sum_{k=0}^{N-1} B_{n,k} x_k \tag{3.3}$$

$$Y_n = H_n X_n \tag{3.4}$$

$$y_j = \sum_{n=0}^{M-1} C_{j,n} Y_n \tag{3.5}$$

where $n = 0, 1, \ldots, M-1$, M being the number of multiplications required by the algorithm. Algorithms of the form (3.2) - (3.5) are the result of the methods described in the previous section. Table 1 below gives the theoretical minimum number of multiplications, $2N - K$, where K is the number of factors of N, including 1 and N, as proven by Winograd [14]. The numbers of multiplications M, input additions A_I, output additions A_O, and total additions A required for the algorithms of [3] and for the algorithms derived for this report are also listed in Table 1.

Table 1.
Summary of Operations Count
for Convolution Algorithms.

Algorithm	N	2N − K	M	A_I	A_O	A
Ref.[3]	2	2	2	2	2	4
Ref.[3]	3	4	4	5	6	11
Ref.[3]	5	8	10	13	22	35
Ref.[3]	5	8	10	13	18	31
C5CR10	5	8	10	13	19	32
C5I10	5	8	10	13	18	31
Ref.[3]	7	12	19	23	49	72
Ref.[3]	7	12	19	23	35	58
C7CR17	7	12	17	28	39	57
C7QI19	7	12	19	26	38	64
C7LI19	7	12	19	24	36	60
Ref.[3]	9	15	22	29	69	98
Ref.[3]	9	15	22	29	42	71
C9CR19	9	15	19	36	54	90
C11CR35	11	20	35	75	98	173
C11CR36	11	20	36	71	93	164
C11LI37	11	20	37	59	85	144
C11QI37	11	20	37	63	89	152
C11LI46	11	20	46	55	90	145

For large N-values one can perform a mapping of the arrays into multidimensional arrays and do a multidimensional convolution using the algorithms discussed above. Agarwal and Burrus [1,2] first suggested this idea and showed that if the mapping were defined by assigning the elements in sequential order to the indices of the multidimensional array taken in lexicographical order, one could reduce the number of multiplications. This was in spite of the fact that the resulting multidimensional convolution was non-cyclic in all but one dimension. The convolutions in the non-cyclic dimensions had to be done by padding with zeros thereby making the convolution in those dimensions twice as long. Ref. [3] shows how to perform a CRT mapping of the indices which converts the one-dimensional convolution (3.1) into a multi-dimensional convolution which is cyclic in all dimensions. For this, however, one must require that the factors of N be mutually prime. The result is that the dimensions of the array are the mutually prime factors of N. Assuming, for the sake of the following discussion, that we use three relatively prime factors of N such that

$$N = N_1 N_2 N_3,$$

VLSI Signal Processing

the mapping of the single index j onto the triple of indices j_1, j_2, and j_3, is defined by the equation,

$$j = j_1\tau_1 + j_2\tau_2 + j_3\tau_3 \text{ Mod } N. \tag{3.6}$$

where $\tau_\nu = N/N_\nu$. In [3] it was suggested that the mapping be defined by the CRT. The mapping (3.6) is a simple permutation of the CRT mapping and is used here because it produces a more efficient computer program. If we define the same mapping for k, the convolution (3.1) can be written as a three-dimensional convolution

$$y_{j_1,j_2,j_3} = \sum_{j_1=0}^{N_1-1} \sum_{j_2=0}^{N_2-1} \sum_{j_3=0}^{N_3-1} h_{j_1-k_1,j_2-k_2,j_3-k_3} x_{k_1,k_2,k_3}. \tag{3.7}$$

After mapping the one- dimensional arrays h_k and x_k into the three- dimensional arrays h_{k_1,k_2,k_3} and x_{k_1,k_2,k_3}, one may operate with algorithms of the form (3.2) - (3.5) on each of the three dimensions. This may be described in matrix vector notation with subscripts on the operators A, B, and C corresponding to the operations in (3.2) (3.3) and (3.5) for the respective sequence lengths N_ν, $\nu = 1,2,$ and 3 as follows,

$$H = A_3 A_2 A_1 h, \tag{3.8}$$

$$X = B_3 B_2 B_1 x, \tag{3.9}$$

$$Y = H \times X., \tag{3.10}$$

$$y = N^{-1} C_1 C_2 C_3 Y. \tag{3.11}$$

3.2 Count of Arithmetic Operations

In (3.8) h may be thought of as the 3-dimensional $N_1 \times N_2 \times N_3$ array of elements h_{j_1,j_2,j_3} resulting from the mapping of the original sequence h_j according to (3.6). The result of the operation, $A_1 h$, is the 3-dimensional array obtained by applying the A_1-transform (3.2) to the first index j_1 of h. It is to be noted that, since the matrix of coefficients in (3.2) is a rectangular matrix with M_1 rows M_1 being the number of multiplications for the N_1-point algorithm, the length of the result $A_1 h$ in the first dimension is increased to M_1. The operators A_2 and A_3 have a similar effect on the second and third dimensions. A similar transformation is performed on x. The transformed arrays are multiplied, element-by-element in (3.10) and in (3.11) the result is transformed back, one dimension at a time, to yield the three-dimensional array y. This final array is permuted according to (3.6) into the one-dimensional array of values of the convolution (3.1). The total number of elements in each of the arrays H and X is

$$M = M_1 M_2 M_3 \tag{3.12}$$

which is the total number of multiplications required by the algorithm, with M_ν being the number of multiplications required for the N_ν-point algorithms, for $\nu = 1, 2,$ and 3, respectively.

It is seen that by nesting the algorithms for the factors of N in this manner, one can construct a program by first writing and debugging convolution algorithms for each of the factors of N, N_ν, $\nu = 1,2,$ and 3, and then inserting each of them in a program with loops for repeating the calculation for all values of the other two indices. It is to be noted that the H and X arrays are of length M which is, in general, larger than N. As described in [3], the number of additions depends upon the ordering of the factors. This is due to the fact that each application of the operators A_ν and B_ν enlarges the array. It is also shown in [3] that if A_ν, $\nu = 1, 2, 3$ is the numbers of additions required for the N_ν-point algorithm, then the whole calculation should take

$$A_{tot} = A_1 N_2 N_3 + M_1 A_2 N_3 + M_1 M_2 A_3. \tag{3.13}$$

additions. One can derive from this the fact that the factors of N should be placed in the order with increasing values of the quotient

$$T(N_\nu) = \frac{M_\nu - N_\nu}{A_\nu} \tag{3.14}$$

given in Table 2. The $N = 7, 9$, and 11 algorithm in Table 1 are those with the fewest multiplications given in this paper. The remainder are from Ref. 3.

Table 2.

Table of Values of

$T(N_\nu) = (M_\nu - N_\nu)/A_\nu$

Algorithm	N	$T(N_\nu)$
Ref. [3]	2	.000
Ref. [3]	3	.091
Ref. [3]	4	.067
Ref. [3]	5	.161
Ref. [3]	6	.045
C7CR17	7	.175
Ref. [3]	8	.130
C9CR19	9	.111
C11CR35	11	.139

It can be seen here that reducing one M_ν reduces the total number of multiplications (3.12) by the factor by which M_ν is reduced. On the other hand, a change in the number of additions in one of the algorithms makes a less significant change in the total number of operations since it appears in only one of the terms of the sum in (3.13). This is the reason that in the multi-dimensional method, as one uses more factors, it is better to reduce the number of multiplications even at the expense of increasing the number of additions to obtain an over-all reduction in computation. Therefore, as shown in [2] and [3] the multi-dimensional rectangular transform method yields faster algorithms even in machines where multiplication is as fast as addition.

4. Vectorized Convolution Algorithms

4.1 Parallel Computation

The counting of operations in previous sections gives an overall estimation of computational effort which, in a sequential machine, translates directly into speed of throughput; a fundamental consideration in many large calculations, notably in digital signal processing. With a given amount of required arithmetic, overall throughput can be increased by the simultaneous execution of arithmetic operations. In this section, we consider an implementation on a vector machine where each machine operation is performed on an array of operands.

4.2 Theory

The description of the vectorized rectangular transform (VRT) algorithms is best given in terms of polynomial multiplications as in Section 2 above. Let us describe the 30-point circular convolution which we write in the form

$$Y(u) = H(u) X(u) \text{ Mod } (u^{30} - 1), \tag{4.1}$$

where

$$H(u) = h_0 + h_1 u + \ldots + h_{29} u^{29},$$

with similar definitions for $X(u)$ and $Y(u)$. Calculations with polynomials Mod $(u^{30} - 1)$ simply means that in the course of the calculation, $u^{30} \cong 1$ so that we simply replace u^{30} by 1 wherever it

VLSI Signal Processing

appears. Here, and in what follows, "convolution" and "polynomial multiplication" will be considered to be equivalent and, unless stated otherwise, "convolution" will mean "circular convolution".

For the factorization of the field of polynomials, select any mutually prime factors of the convolution length N. For example, let $N = 30 = 2 \times 15$, and form the mapping

$$j \Longleftrightarrow j_1, j_2, \quad j_1 = 0, 1, \quad j_2 = 0, \ldots, 14,$$

defined by

$$j = 15j_1 + 2j_2 \text{ Mod } 30. \tag{4.2}$$

This is actually a permutation of the CRT mapping. The present form is slightly simpler to explain and to program. We then let

$$u^j = u^{15j_1} u^{2j_2} \tag{4.3}$$

The factoring of the field of polynomials is done by making the replacement $\mu = u^2$ and $\eta = u^{15}$ and expressing all polynomials as 1-st degree polynomials in η with coefficients which are 15-th degree polynomials in μ. Thus, we let

$$H(u) = H_0(\mu) + H_1(\mu)\eta, \tag{4.4}$$

with similar representations for $X(u)$ and $Y(u)$, and express the convolution in the form,

$$Y(u) = Y_0(\mu) + Y_1(\mu)\eta = \tag{4.5}$$

$$= [H_0(\mu) + H_1(\mu)\eta][X_0(\mu) + X_1(\mu)\eta].$$

4.3 Vector form of the 30-Point Convolution Algorithm

Choosing the order of factors as in the previous section prescribes using the factor 2 first. Therefore, one starts by computing the 2-point convolution algorithm with the calculation in the field of polynomials Mod ($\mu^{15} - 1$). Thus, the algorithm for the 30-point transform is essentially a two-point convolution algorithm in polynomial arithmetic which can be described as follows:

1. Permute the coefficients of $H(u)$ and $X(u)$ so as to yield the coefficients of $H_0(\mu)$, $H_1(\mu)$, $X_0(\mu)$, and $X_1(\mu)$.

2. Let

$$A_0(\mu) = .5[H_0(\mu) + H_1(\mu)] \tag{4.6}$$

$$A_1(\mu) = .5[H_0(\mu) - H_1(\mu)]$$

3. Let

$$B_0(\mu) = X_0(\mu) + X_1(\mu) \tag{4.7}$$

$$B_1(\mu) = X_0(\mu) - X_1(\mu)$$

4. Let

$$M_j(\mu) = A_j(\mu) B_j(\mu), \quad j = 0,1 \text{ Mod } (\mu^{15} - 1) \tag{4.8}$$

5. Let

$$Y_0(\mu) = M_0(\mu) + M_1(\mu) \tag{4.9}$$

$$Y_1(\mu) = M_0(\mu) - M_1(\mu)$$

6. Do the inverse of the permutation in Step 1 on the coefficients of $Y_0(\mu)$ and $Y_1(\mu)$ to get the coefficients of $Y(u)$.

VLSI Signal Processing

In Step 4 of the above algorithm, it is seen that polynomial products Mod ($\mu^{15} - 1$) must be computed. These may, in turn, be computed by a 15-point algorithm, using two mutually prime factors of 15. The 15-point algorithm has the same general structure as that given above. Table 2 suggests using the factorization $N = 15 = 3 \times 5$, so that the 15-point algorithm would use the mapping defined by

$$j = 5j_1 + 3j_2 \text{ Mod } 15 \tag{4.10}$$

giving a mapping into a set of 3 polynomials with 5-th degree polynomials in some variable ξ as coefficients. Thus, the initial permutation in Step 1 results in three X and three H polynomials and the 3-point convolution algorithm in polynomial arithmetic Mod ($\xi^5 - 1$) is used. This calls upon a 5-point algorithm to do the polynomial multiplication. We describe the 15-point algorithm referring to the dummy variables $X(\mu)$ and $H(\mu)$ as input and the dummy variable $Y(\mu)$ as output. Thus, the algorithm for the 15-point transform is essentially a 3-point convolution algorithm in polynomial arithmetic which may be described as follows:

1. Permute the coefficients of $H(\mu)$ and $X(\mu)$ so as to yield the coefficients of

    $$H_0(\xi), H_1(\xi), H_2(\xi), X_0(\xi), X_1(\xi), \text{ and } X_2(\xi)$$

2. Let

    $$A_0(\xi) = [H_0(\xi) + H_1(\xi) + H_2(\xi)]/3 \tag{4.11}$$

    $$A_1(\xi) = H_0(\xi) - H_2(\xi)$$

    $$A_2(\xi) = H_1(\xi) - H_2(\xi)$$

    $$A_3(\xi) = (A_1(\xi) + A_2(\xi))/3$$

3. Let

    $$B_0(\xi) = X_0(\xi) + X_1(\xi) + X_2(\xi) \tag{4.12}$$

    $$B_1(\xi) = X_0(\xi) - X_2(\xi)$$

    $$B_2(\xi) = X_1(\xi) - X_2(\xi)$$

    $$B_3(\xi) = B_1(\xi) + B_2(\xi)$$

4. Let

    $$M_j(\xi) = A_j(\xi)B_j(\xi), \; j = 0,1,2, \text{ and } 3. \tag{4.13}$$

5. Let

    $$V_0(\xi) = M_1(\xi) - M_3(\xi) \tag{4.14}$$

    $$V_1(\xi) = M_2(\xi) - M_3(\xi)$$

    $$Y_0(\xi) = M_0(\xi) + V_0(\xi)$$

    $$Y_1(\xi) = M_0(\xi) - V_0(\xi)$$

    $$Y_3(\xi) = M_0(\xi) + V_1(\xi)$$

6. Do the inverse of the permutation in Step 1 on the coefficients of $Y_j(\xi), j = 0, 1, 2,$ and 3, so as to yield the coefficients of $Y(\mu)$.

It is important to note here that there is an increase in dimensionality. Three pairs of $H_j(\xi)$ and $X_j(\xi)$ polynomials are put into the algorithm and four pairs of $A_j(\xi)$ and $B_j(\xi)$ polynomials are generated. This increase occurs for all RT convolution algorithms except for the 2-point algorithm. Thus, with each factor of N, the amount of intermediate storage required is increased resulting in a total of $M = M_1 \times M_2 \times \cdots \times M_r$ where the M_i's are the numbers of multiplications for each

VLSI Signal Processing

factor of N. A further consequence of this, as noted in Section 3 with reference to Equation (3.13), is that an expansion due to early factors in the ordering produces more data and therefore more operations for subsequent factors.

There is no factorization of 5, so the 5-point algorithm, unlike the previous two, operates in the field of rationals and performs a multiplication of scalars in its Step 4. Using this general structure, a collection of program modules for various convolution lengths may be used to do convolutions for a very large selection of composite convolution lengths N.

4.4 General Comments on the Vectorized RT Algorithms

The vector structure of the above algorithm is obvious. Within each module as described in Steps 1 to 6 above, all polynomial additions are simple element-by-element vector additions. In Step 4, a set of identical polynomial multiplications may be performed simultaneously by a call to an appropriate convolution subroutine which operates on a set of vector representations of polynomials. Thus, the vectors representing the polynomials with $j=0,1,...$ in Step 4 can be concatenated so that the subroutine operates on long vectors. Further details of the manipulation of these vectors is somewhat technology-dependent and will not be given here.

5. Error Analysis

As mentioned earlier in this paper, we have a vague idea of the notion of "stability" as meaning that the transform matrices have the nice property of being sparse and containing very simple non-zero numbers, such as ± 1. To associate a quantitative measure of stability in terms of the transform matrices, we analyze the floating point error generated by the arithmetic in the transform and produce variance estimates which may be computed, under broad assumptions, and which depend only on the matrix elements. It is hoped that this may lead to an objective measurement of "goodness" which can be used in a system for the automatic generation of convolution algorithms.

The error analysis considered here follows Wilkinson's methods [10] for floating point operations. A good description of these methods can also be found in McCracken and Dorn [9]. The present analysis is not to be considered an accurate predictor of errors, but only a rough method for comparing the relative merits of the algorithms considered here. We estimate the variance of the errors produced during the "input" transform of x to b and the "output" transform from m to the result y. Of course, the multiplication stage of the algorithm, which computes the element by element product $m = a \times b$, generates error, but, since complications arise due to the fact that this is data-dependent, it is omitted. We have also neglected to consider the sensitivity of the result to errors in the input data. Both of these sources of error should of course be considered in a good analysis and prediction of errors. In any case, the analyses given below are simply for the purpose of comparing the algorithms with each other.

Due to the increased use of high-speed floating point devices, there has, for a time, been less interest in fixed-point calculations in digital signal processing. However, in VLSI applications, floating point calculations generally require more time and space on the silicon. Therefore, there has been a revived interest in the performance of algorithms in fixed point arithmetic. For those who are interested in fixed point calculations, it will only be pointed out here that the variance figures for b and y given below are, except in rare situations, good predictors of the ranges of values required for fixed point calculations.

Following Wilkinson, the assumption made is that when a floating point addition,

$$v_3 = v_2 + v_1, \tag{5.1}$$

is performed, the result is truncated, or rounded, to produce a result,

$$fl(v_3) = v_3(1 + \varepsilon) \tag{5.2}$$

where ε depends upon the number of significant bits in the fractional part of the floating point representation of numbers and upon the type of rounding used. Thus, if v_1 and v_2 contain errors e_1 and e_2, the error in v_3 will be, neglecting products of errors,

$$e_3 = v_3\varepsilon + e_1 + e_2 \tag{5.3}$$

Letting (5.1) represent any addition occurring during the course of the input or output stages of the algorithm and assuming all quantities to be random variables, the variance of v_3 is

$$\sigma_{v_3}^2 = \sigma_{v_1}^2 + \sigma_{v_2}^2 \tag{5.4}$$

if v_1 and v_2 are independent. If $v_1 = v_2$ then

$$\sigma_{v_3}^2 = 4\sigma_{v_2}^2 \tag{5.5}$$

It is not quite true that intermediate quantities generated during the execution of the algorithms are really independent, but, it will be assumed so here for simplicity. Since relative performance of algorithms is being evaluated, it will be assumed that the variance of all input values is 1,

$$\sigma_{x_j}^2 = 1 \tag{5.6}$$

The variance of the errors, again assuming independence of errors and data, is, from (5.3),

$$\sigma_{e_3}^2 = \sigma_{v_3}^2 \sigma_\varepsilon^2 + \sigma_{e_1}^2 + \sigma_{e_2}^2 \tag{5.7}$$

Since all input data is assumed to have no error, all σ_e^2's will be proportional to σ_ε^2. Therefore, if the σ_e^2's are re-defined so that they are given in units of σ_ε^2, (5.7) can be re-written in the form

$$\sigma_{e_3}^2 = \sigma_{v_3}^2 + \sigma_{e_1}^2 + \sigma_{e_2}^2 \tag{5.8}$$

The variance of variables and errors were calculated according to the above formulas by relatively simple editing of the programs for the algorithms. By setting all input variables in the input and output transformation algorithms to 1 and by changing all "$-$" signs to "$+$" a program was obtained which computed the variances of all b and y variables. Then, by changing the names of all variables and inserting an extra term, expressions of the form (5.4) were converted to expressions of the form (5.8) giving the variances of the errors with initial errors assumed to be 0. For expressions of the form

$$v_4 = v_1 + v_2 + v_3 \tag{5.9}$$

it was assumed that the machine will add from left to right so that the variance of the error would be given by

$$\sigma_{e_4}^2 = \sigma_{v_4}^2 + \sigma_{v_1}^2 + \sigma_{v_2}^2 + \sigma_{e_1}^2 + \sigma_{e_2}^2 + \sigma_{e_3}^2 \tag{5.10}$$

This is easily generalized to formulas with any number of terms.

VLSI Signal Processing

Table 3. Variance of input transform b_j

Algorithm	Max $\sigma^2_{b_j}$	$\sigma^2_{b_j}$											
C3CR4	4	3	2	2	4								
C5CR10	6	5	2	4	2	2	4	4	2	6	2		
C5I10	6	5	2	2	4	2	4	6	2	2	4		
C7CR17	8	7	2	2	4	6	4	4	8	4	4	8	4 ...
C7LI19	7	7	2	2	4	2	4	6	2	2	4	2	2 ...
C7QI19	7	7	2	4	2	2	4	2	2	6	4	2	6 ...
C9CR19	12	9	6	6	12	2	2	4	6	6	12	6	6 ...
C11CR35	16	11	10	2	2	4	6	10	16	8	8	16	8 ...
C11CR36	14	11	10	6	6	12	8	6	14	2	2	4	2 ...
C11LI37	12	11	2	2	4	6	6	12	4	4	8	6	6 ...
C11QI37	12	11	2	6	4	2	6	4	4	8	4	4	8 ...
C11LI46	11	11	2	2	4	2	2	4	2	2	4	2	2 ...

Table 4. Variance of output transform y_j

Algorithm	Max $\sigma^2_{y_j}$	$\sigma^2_{y_j}$										
C3CR4	5	5	3	3								
C5CR10	8	7	5	7	8	6						
C5I10	9	9	5	5	9	5						
C7CR17	16	16	9	10	12	12	8	10				
C7LI19	11	11	7	7	9	7	11	7				
C7QI19	11	9	11	9	7	9	9	7				
C9CR19	21	21	13	11	13	13	11	13	15	11		
C11CR35	38	38	20	25	32	23	26	23	26	28	24	20
C11CR36	24	23	22	18	24	21	18	19	16	18	24	4
C11LI37	33	19	33	17	19	15	15	15	15	15	15	17
C11QI37	21	21	19	17	17	19	15	19	17	19	19	15
C11LI46	13	13	11	11	13	11	13	11	13	11	13	11

Table 5. Variance of Error ε_{b_j} in input transform

Algorithm	Max $\sigma^2_{\varepsilon_{b_j}}$	$\sigma^2_{\varepsilon_{b_j}}$											
C3CR4	8	5	2	2	8								
C5CR10	16	12	2	8	2	2	8	8	2	16	2		
C5I10	16	14	2	2	8	2	8	16	2	2	8		
C7CR17	25	19	2	2	8	15	8	8	24	8	8	24	8 ...
C7LI19	27	27	2	2	8	2	8	16	2	2	8	2	2 ...
C7QI19	21	21	2	8	2	2	8	2	2	16	8	2	16 ...
C9CR19	41	22	14	15	41	4	4	12	12	12	36	14	12 ...
C11CR35	67	41	38	2	2	8	15	36	67	26	22	64	28 ...
C11CR36	55	33	28	15	16	43	24	17	55	2	2	8	2 ...
C11LI37	45	45	2	2	8	16	16	44	6	8	22	14	14 ...
C11QI37	43	41	2	14	10	2	18	10	8	26	10	8	26 ...
C11LI46	65	65	2	2	8	2	2	8	2	2	8	2	2 ...

Table 6. Variance of Error ϵ_{yj} in output transform

Algorithm	Max $\sigma^2_{\epsilon_{yi}}$	$\sigma^2_{\epsilon_{yj}}$									
C3CR4	12	12	5	5							
C5CR10	25	20	12	20	25	16					
C5I10	30	30	12	12	30	12					
C7CR17	68	68	29	34	46	46	24	38			
C7LI19	42	42	21	21	32	20	42	20			
C7QI19	51	30	51	30	21	30	30	21			
C9CR19	109	109	51	39	52	51	39	52	65	39	
C11CR35	240	240	92	138	198	117	132	119	144	151	128	94
C11CR36	124	121	107	80	124	96	78	87	75	83	123	66
C11LI37	204	88	204	76	84	59	62	65	63	59	60	76
C11QI37	101	101	99	82	73	84	60	93	76	85	83	60
C11LI46	60	60	45	45	60	45	60	45	60	45	60	45

6. Concluding Remarks

The experience gained in the work described above yielded considerable insight and skill in devising strategies for deriving efficient convolution algorithms. These involve the many choices of reducing polynomials and the sequencing of operations, all with a view towards some hoped-for properties in the final implementation of the algorithms. The results are somewhat revealing.

Of the three $N=7$ algorithms, C7CR17, using the CRT, gave the smallest number of multiplications and additions. However, it seemed obvious that the C7LI19 linear iterative algorithm would give simpler matrices. This was shown to be true in the variance calculations in Tables 3-6 where the variance of generated error in C7CR17 is 50% greater than in C7LI19. The $N=11$ algorithm, C11CR35, with the fewest multiplications saves 1 multiplication at the cost of 5 additions when compared with C11CR36. If $N=11$ is the last of the factors in the sequence length, these are roughly equivalent. However, the latter has simpler matrix elements in its output transform matrix and an error variance only half as great. Therefore, for many applications it may be preferable.

In the above, it is seen that various strategies can lead to algorithms with great differences in computation and in error generation. However, no really "hard" results in terms of theorems and objective methods were found which could be automated. Furthermore, it has not yet been possible to characterize the properties of "good" algorithms from the point of view of their implementation. The present work carries us in that direction but there is much yet to be done. In future work, we intend to implement some of these algorithms in various types of computers, hopefully in cooperation with VLSI experts and computer architects to establish some firm criteria which can be formulated and incorporated in the algorithm development process.

VLSI Signal Processing
References

[1] R. C. Agarwal and C. S. Burrus, "Fast Convolution Using Fermat Number Transforms With Applications to Digital Filtering," *IEEE Trans. Acoust., Speech, Signal Processing,* vol. ASSP-22, No. 2, pp. 87-99, Apr. 1974. Also, *Proc. IEEE,* vol. 63, No.4, pp 550-560, Apr. 1975.

[2.] R. C. Agarwal and C. S. Burrus, "Fast One-Dimensional Digital Convolution by Multidimensional Techniques," *IEEE Trans. Acoust., Speech, Signal Processing,* vol. ASSP-22, No. 1, pp. 1-10, Feb. 1974.

[3] R. C. Agarwal and J. W. Cooley, "New Algorithms for Digital Convolution," *IEEE Trans. Acoust., Speech, Signal Processing,* vol. ASSP-25, pp. 392-410, Oct. 1977.
Also, Research Report RC 6446, IBM Watson Research Center, P. O. Box 218, Yorktown Hts., New York, 10598. Mar. 23, 1977

[4] R. K. Brayton, C. L. Chen, C. T. McMullen, R. H. J. M. Otten, Y. J. Yamour, "Automated Implementation of Switch Functions as Dynamic CMOS Circuits", *Proceedings of the 1984 Custom Integrated Circuit Conference,* pp. 346-350.

[5] L. Auslander and A. Silberger, "On the Use of SCRATCHPAD in the Construction of Convolution Algorithms," Research Report RC 10554 IBM Watson Research Center, P. O. Box 218, Yorktown Hts., New York, 10598., May 1984.

[6] J. W. Cooley, "Some Applications of Computational Complexity Theory to Digital Signal Processing", *Proceedings, 1981 Joint Automatic Control Conference,* University of Virginia, Charlottesville, Va. June 17-19, 1981 Also in Research Report No. RC8805, Center, P. O. Box 218, Yorktown Heights, New York, 10598, April 1981.

[7] J. W. Cooley, "Rectangular Transforms for Digital Convolution on the Research Signal Processor", *IBM J. Res. Develop.* 26, No. 4. pp. 424-430, July 1982

[8] J. H. Griesmer, R. D. Jenks, and D. Y. Y. Yun, "SCRATCHPAD User's Manual" *I.B.M. Research Report, RA 70,* I.B.M. Watson Research Center, P.O. Box 218, Yorktown Hts. New York, 10598, June 1975.

[9] D. D. McCracken and W. S. Dorn, *Numerical Methods and Fortran Programming,* Wiley, New York 1964.

[10] J.H.Wilkinson, *Rounding Errors in Algebraic Processes,* Prentice-Hall, Englewood Cliffs, New Jersey, 1963.

[11] S. Winograd, "Some Bilinear Forms Whose Multiplicative Complexity Depends on the Field of Constants," *I.B.M. Research Report, RC 5669,* I.B.M.Watson Research Center, P.O. Box 218, Yorktown Hts. New York, 10598, Oct. 10,1975.

[12] S. Winograd, "On Computing the Discrete Fourier Transform,"*Proc. Nat. Acad. Sci. USA,* Vol. 73, No. 4, pp 1005-1006, April 1976.

[13] S. Winograd, "On Computing the Discrete Fourier Transform," *Math. of Computation,* Vol. 32, No. 141, pp 175-199, January 1978.

[14] S. Winograd, Arithmetic Complexity of Computations, *CBMS-NSF Conference Series in Applied Mathematics,* No. 33, SIAM, 1980.

VLSI Signal Processing

Appendix.

The following is a listing of some of the algorithms derived for this paper. In all cases, it is assumed that the h sequence is fixed and the additions involved in its computation are not counted. The values of $M=$ the number of multiplications, $A_I=$ the number of input additions, $A_O=$ the number of output additions and $A = A_I + A_O$ are given. The RT matrices, A and C are given explicitly. For all algorithms considered here, the B matrix is the same as the A matrix. Note, however, that this is not necessarily always true for RT algorithms.

The transposition of the A and C matrices, the optimization of the additions, and the automatic production of PLI program statements which execute the algorithm was done by computer programs employing various strategies and heuristics but lacking the firm theoretical framework of the theory for reducing multiplications. Where optimization of additions was done by hand, the program did it equally well or better and, of course, did it much faster. One may eventually consider developing this program into a system which automatically produces program statements in a form suitable for direct input to a computer-driven VLSI design system such as that produced by Brayton [4]

C3CR4- $N = 3$, $M = 4$, Using the Chinese Remainder Theorem.
$A_I = 5$, $A_O = 6$, $A = 11$.

```
       A                    3 C'
   1   1   1            1   1   1
   1   0  -1            1  -2   1
   1  -1   0            1   1  -2
   0   1  -1           -2   1   1
```

The transform of x (Input transform)
```
B(0) = X(0)+X(1)+X(2);
B(1) = X(0)-X(2);
B(2) = X(0)-X(1);
B(3) = B(1)-B(2);
```

The multiplication:

$$m_j = a_j \times b_j, \ j = 0, 1, 2, \text{ and } 3$$

The transform of m (Output transform)
```
V(0) = C(1)+C(3);
V(1) = C(2)-C(3);
Y(0) = C(0)+V(0)+V(1);
Y(1) = C(0)-V(0);
Y(2) = C(0)-V(1);
```

C5CR10- $N = 5$, $M = 10$, Using the Chinese Remainder Theorem.
$A_I = 13$, $A_O = 19$, $A = 32$.

```
         A                       10 C'
  1   1   1   1   1        2   2   2   2   2
  1   0   0   0  -1       -4  -4   6   6  -4
  1  -1   1  -1   0        1   1  -4   1   1
  1   0  -1   0   0       12   2  -8  -8   2
  0   1   0  -1   0        2  -8  -8   2  12
  1  -1  -1   1   0       -7   3   8   3  -7
  1   0  -1   1  -1       -2  -2  -2  -2   8
  0   1  -1   0   0       -6   4   4   4  -6
  1  -1   0   1  -1        8  -2  -2  -2  -2
  0   0   0   1  -1       -4   6   6  -4  -4
```

The transform of x (Input transform)
```
V(0) = +X(1)  +X(3);
V(1) = +X(0)  +X(2);
V(2) = +X(0)  -X(2);
V(3) = +X(3)  -X(4);
B(0) = +X(4)  +V(0)  +V(1);
B(1) = +X(0)  -X(4);
```

VLSI Signal Processing

```
B(2)= -V(0) +V(1);
B(3)= +V(2);
B(4)= +X(1) -X(3);
B(6)= +V(2) +V(3);
B(7)= +X(1) -X(2);
B(9)= +V(3);
B(5)=   B(3) -B(4);
B(8)=   B(6) -B(7);
```

The multiplication:

$$m_j = a_j \times b_j, \; j = 0, 1, 2, \ldots, 9$$

The transform of **m** (Output transform)
```
V(0)= +C(6) +C(8);
V(1)= +C(7) -C(8);
V(2)= +C(3) +C(5);
V(3)= +C(4) -C(5);
V(4)= +C(2) -V(1);
V(5)= +C(0) +V(0);
V(6)= +C(0) -V(0);
Y(0)= +C(1) +C(2) +V(2) +V(5);
Y(1)= -C(1) -C(9) +V(6);
Y(2)= -C(2) +C(9) -V(3) +V(5);
Y(3)= -V(2) +V(4) +V(6);
Y(4)= +C(0) +V(3) -V(4);
```

C5I10- $N = 5, M = 10$, Using the Iterative method.

$A_I = 13, \; A_O = 18, \; A = 31.$

```
              A                        5 C'
   1  1  1  1  1           1  1  1  1  1
   1  0  0  0 -1           1 -4  1  1  1
   1 -1  0  0  0           2  2 -3  2 -3
   0  1  0  0 -1          -4  1  1  1  1
   1  0 -1  0  0           2  2 -3 -3  2
   1 -1 -1  1  0          -1 -1  4 -1 -1
   0  1  0 -1  0           2 -3 -3  2  2
   0  0  1  0 -1           1  1  1  1 -4
   0  0  1 -1  0          -3  2 -3  2  2
   0  0  0  1 -1           1  1  1 -4  1
```

The transform of **x** (Input transform)
```
V(0)= +X(0) -X(1);
V(1)= +X(2) -X(3);
B(0)= +X(0) +X(1) +X(2) +X(3) +X(4);
B(1)= +X(0) -X(4);
B(2)= +V(0);
B(4)= +X(0) -X(2);
B(5)= +V(0) -V(1);
B(7)= +X(2) -X(4);
B(8)= +V(1);
B(3)=   B(1) -B(2);
B(6)=   B(4) -B(5);
B(9)=   B(7) -B(8);
```

The multiplication:

$$m_j = a_j \times b_j, \; j = 0, 1, 2, \ldots, 9$$

The transform of **m** (Output transform)
```
V(0)= +C(7) +C(9);
V(1)= +C(8) -C(9);
V(2)= +C(4) +C(6);
V(3)= +C(5) -C(6);
V(4)= +C(1) +C(3);
V(5)= +C(2) -C(3);
V(6)= +C(0) +V(3);
V(7)= +C(0) -V(3);
```

```
Y(0) = +V(2) +V(4) +V(5) +V(6);
Y(1) = +C(0) -V(0) -V(4);
Y(2) = -V(1) +V(6);
Y(3) = +V(0) +V(1) -V(2) +V(7);
Y(4) = -V(5) +V(7);
```

C7CR17- $N = 7$, $M = 17$, Using the Chinese Remainder Theorem

$A_I = 28$, $A_O = 39$, $A = 57$.

<pre>
              A   42 C'
  1 1 1 1 1 1 1   6   6   6   6   6   6   6
  1 0 0 0 0 0 -1   12 -30   12   12   12   12 -30
  1 -1 0 0 0 0 0   30 -12 -12 -12 -12   30 -12
  0 1 0 0 0 0 -1   -30   12   12   12   12 -30   12
  1 -1 1 -1 1 -1 0   -6   8   -6   8   -6   1   1
  1 0 -1 0 1 0 -1   -6   -6   -6   -6   -6   -6   36
  0 1 0 -1 0 1 -1   -6   -6   -6   -6   -6   36   -6
  1 -1 -1 1 1 -1 0   6   6   6   6   6 -15 -15
  1 0 -1 1 0 -1 0   3   3 -18   3   3   3   3
  1 -1 0 1 -1 0 0   -12   9   9 -12   9 -12   9
  0 1 -1 0 1 -1 0   9 -12   9   9 -12   9 -12
  1 0 -1 -1 0 1 0   -3   11   18   11   -3 -17 -17
  1 1 0 -1 -1 0 0   12   5   -9 -16   -9   -2   19
  0 1 1 0 -1 -1 0   -9 -16   -9   5   12   19   -2
  0 0 0 0 1 0 -1   12   12   12   12 -30   12 -30
  0 0 0 0 1 -1 0   -12 -12 -12 -12   30 -12   30
  0 0 0 0 0 1 -1   12   12   12 -30   12 -30   12
</pre>

The transform of x (Input transform)
```
V(0) = +X(1) +X(5);
V(1) = +X(0) -X(2);
V(2) = +X(0) -X(1);
V(3) = +X(0) +X(2);
V(4) = +X(4) +V(3);
V(5) = +X(4) -X(6);
V(6) = +X(3) -X(5);
V(7) = +X(3) +X(6);
B(0) = +V(0) +V(4) +V(7);
B(1) = +X(0) -X(6);
B(2) = +V(2);
B(4) = -X(3) -V(0) +V(4);
B(5) = +V(1) +V(5);
B(6) = +V(0) -V(7);
B(8) = +V(1) +V(6);
B(9) = +X(3) -X(4) +V(2);
B(11) = +V(1) -V(6);
B(12) = +X(0) +X(1) -X(3) -X(4);
B(14) = +V(5);
B(15) = +X(4) -X(5);
B(3) =  B(1) -B(2);
B(7) =  B(5) -B(6);
B(10) = B(8) -B(9);
B(13) = B(12) -B(11);
B(16) = B(14) -B(15);
```

The multiplication:

$$m_j = a_j \times b_j, \; j = 0, 1, 2, \ldots, 16$$

The transform of m (Output transform)
```
V(0) = +C(14) +C(16);
V(1) = +C(15) -C(16);
V(2) = +C(12) +C(13);
V(3) = +C(11) -C(13);
V(4) = +C(8) +C(10);
V(5) = +C(9) -C(10);
V(6) = +C(5) +C(7);
V(7) = +C(6) -C(7);
V(8) = +C(1) +C(3);
V(9) = +C(2) -C(3);
```

VLSI Signal Processing

```
V(10)= +C(0)  +C(4);
V(11)= +C(0)  -C(4);
V(12)= +V(0)  +V(6);
V(13)= +V(5)  +V(9);
V(14)= +V(7)  +V(11);
V(15)= +V(2)  +V(3);
V(16)= +V(4)  +V(6);
Y(0) = +V(8)  +V(10) +V(13) +V(15) +V(16);
Y(1) = +C(0)  -V(7)  -V(8)  -V(12);
Y(2) = -V(1)  +V(3)  -V(4)  +V(14);
Y(3) = +V(1)  -V(2)  -V(5)  +V(10)  +V(12);
Y(4) = +V(4)  +V(5)  -V(7)  +V(11)  -V(15);
Y(5) = -V(3)  +V(10) -V(16);
Y(6) = +V(2)  -V(13) +V(14);
```

C7LI19- $N = 7, M = 19$, Using the Iterative Method with Linear Polynomials

$A_I = 24$, $A_O = 36$, $A = 60$.

A

1	1	1	1	1	1	1
1	0	0	0	0	0	-1
1	-1	0	0	0	0	0
0	1	0	0	0	0	-1
1	0	-1	0	0	0	0
1	-1	-1	1	0	0	0
0	1	0	-1	0	0	0
0	0	1	0	0	0	-1
0	0	1	-1	0	0	0
0	0	0	1	0	0	-1
1	0	0	0	-1	0	0
0	1	0	0	-1	0	0
1	-1	0	0	-1	1	0
0	0	0	0	1	0	-1
0	0	0	0	1	-1	0
0	0	0	0	0	1	-1
0	0	1	0	-1	0	0
0	0	0	1	0	-1	0
0	0	1	-1	-1	1	0

$7\ C$

1	1	1	1	1	1	1
1	-6	1	1	1	1	1
3	3	-4	3	-4	3	-4
-6	1	1	1	1	1	1
2	2	2	2	-5	-5	2
-1	-1	-1	-1	6	-1	-1
2	2	2	-5	-5	2	2
1	1	1	1	1	1	-6
-4	3	-4	3	-4	3	3
1	1	1	1	1	-6	1
2	2	-5	-5	2	2	2
2	-5	-5	2	2	2	2
-1	-1	6	-1	-1	-1	-1
1	1	1	1	-6	1	1
-4	3	-4	3	3	-4	3
1	1	1	-6	1	1	1
-5	-5	2	2	2	2	2
-5	2	2	2	2	2	-5
6	-1	-1	-1	-1	-1	-1

The transform of x (Input transform)

```
V(0) = +X(0)  -X(1);
V(1) = +X(2)  -X(3);
B(0) = +X(0)  +X(1) +X(2) +X(3) +X(4) +X(5) +X(6);
B(1) = +X(0)  -X(6);
B(2) = +V(0);
B(4) = +X(0)  -X(2);
B(5) = +V(0)  -V(1);
B(7) = +X(2)  -X(6);
B(8) = +V(1);
B(10)= +X(0)  -X(4);
B(11)= +X(1)  -X(5);
B(13)= +X(4)  -X(6);
B(14)= +X(4)  -X(5);
B(16)= +X(2)  -X(4);
B(17)= +X(3)  -X(5);
B(3) =  B(1)  -B(2);
B(6) =  B(4)  -B(5);
B(9) =  B(7)  -B(8);
B(12)=  B(10) -B(11);
B(15)=  B(13) -B(14);
B(18)=  B(16) -B(17);
```

The multiplication:

$m_j = a_j \times b_j$, $j = 0, 1, 2, \ldots, 18$

The transform of m (Output transform)

```
V(0) = +C(16) +C(18);
V(1) = +C(17) -C(18);
V(2) = +C(13) +C(15);
```

VLSI Signal Processing

```
V(3)=  +C(14) -C(15);
V(4)=  +C(10) +C(12);
V(5)=  +C(11) -C(12);
V(6)=  +C(7)  +C(9);
V(7)=  +C(8)  -C(9);
V(8)=  +C(4)  +C(6);
V(9)=  +C(5)  -C(6);
V(10)= +C(1)  +C(3);
V(11)= +C(2)  -C(3);
V(12)= +C(0)  +V(9);
V(13)= +C(0)  -V(9);
Y(0)=  +V(4)  +V(8)  +V(10) +V(11) +V(12);
Y(1)=  +C(0)  -V(2)  -V(6)  -V(10);
Y(2)=  +C(0)  -V(1)  -V(3)  -V(5);
Y(3)=  +C(0)  -V(0)  +V(2)  +V(3)  -V(4);
Y(4)=  +V(1)  -V(7)  +V(12);
Y(5)=  +V(0)  +V(6)  +V(7)  -V(8)  +V(13);
Y(6)=  +V(5)  -V(11) +V(13);
```

C7QI19- $N = 7$, $M = 19$, Using the Iterative Method with Quadratic Polynomials

$A_I = 26$, $A_O = 38$, $A = 64$.

A							42 C'						
1	1	1	1	1	1	1	2	2	2	2	2	2	2
1	0	0	0	0	0	-1	-2	-2	-2	-2	12	-2	-2
1	-1	1	0	0	0	-1	4	-3	4	-3	-3	4	-3
1	0	-1	0	0	0	0	8	-6	-6	8	-6	-6	8
0	1	0	0	0	0	-1	-8	-8	6	6	-8	6	6
1	-1	-1	0	0	0	1	0	7	0	-7	7	0	-7
0	0	1	0	0	0	-1	-2	-2	-2	12	-2	-2	-2
1	0	0	-1	0	0	0	8	-6	-6	-6	8	8
1	-1	0	-1	1	-1	0	-2	5	-2	5	-2	-2	-2
1	0	-1	-1	0	1	0	-4	10	10	-4	-4	-4	-4
0	1	0	0	-1	0	0	4	4	-10	-10	4	4	4
1	-1	-1	-1	1	1	0	0	-7	0	7	0	0	0
0	0	1	0	0	-1	0	-6	-6	-6	-6	8	8	8
0	0	0	1	0	0	-1	-2	12	-2	-2	-2	-2	-2
0	0	0	1	-1	1	-1	-3	-3	4	-3	4	-3	4
0	0	0	1	0	-1	0	8	-6	-6	8	8	-6	-6
0	0	0	0	1	0	-1	6	-8	6	6	-8	-8	6
0	0	0	1	-1	-1	1	-7	7	0	-7	0	7	0
0	0	0	0	0	1	-1	12	-2	-2	-2	-2	-2	-2

The transform of x (Input transform)

```
V(0)=  +X(0)  +X(2);
V(1)=  +X(3)  +X(5);
V(2)=  +X(0)  -X(2);
V(3)=  +X(3)  -X(5);
V(4)=  +X(1)  +X(6);
V(5)=  +X(1)  -X(4);
B(0)=  +X(4)  +V(0)  +V(1)  +V(4);
B(1)=  +X(0)  -X(6);
B(2)=  +V(0)  -V(4);
B(3)=  +V(2);
B(4)=  +X(1)  -X(6);
B(6)=  +X(2)  -X(6);
B(7)=  +X(0)  -X(3);
B(8)=  +V(0)  -V(1)  -V(5);
B(9)=  +V(2)  -V(3);
B(10)= +V(5);
B(12)= +X(2)  -X(5);
B(13)= +X(3)  -X(6);
B(14)= -X(4)  -X(6)  +V(1);
B(15)= +V(3);
B(16)= +X(4)  -X(6);
B(18)= +X(5)  -X(6);
B(5)=   B(3)  -B(4);
B(11)=  B(9)  -B(10);
B(17)=  B(15) -B(16);
```

VLSI Signal Processing

The multiplication:

$$m_j = a_j \times b_j, \ j = 0, 1, 2, \ldots, 18$$

The transform of *m* (Output transform)
```
V(0)= +C(15) +C(17);
V(1)= +C(16) -C(17);
V(2)= +C(9)  +C(11);
V(3)= +C(10) -C(11);
V(4)= +C(3)  +C(5);
V(5)= +C(4)  -C(5);
V(6)= +C(2)  +C(8);
V(7)= +C(8)  -C(14);
V(8)= +C(0)  +V(6);
V(9)= +C(0)  -V(7);
V(10)= +C(7) +V(2);
V(11)= +C(12) -V(2);
Y(0)= +C(1)  +V(4)  +V(8)  +V(10);
Y(1)= +C(0)  -C(1)  -C(2)  -C(6)  -C(13) -C(14) -C(18) -V(1) -V(5);
Y(2)= +C(18) -V(0)  +V(9)  -V(11);
Y(3)= +C(0)  +V(1)  -V(3)  +V(7);
Y(4)= +C(13) +V(0)  +V(9)  -V(10);
Y(5)= +C(6)  -V(4)  +V(8)  +V(11);
Y(6)= +C(0)  +V(3)  +V(5)  -V(6);
```

C9CR19- N = 9, M = 19, Using the Chinese Remainder Theorem.

$A_I = 36$, $A_O = 54$, $A = 90$.

```
                    A                                           9 C'
 1  1  1  1  1  1  1  1  1       1  1  1  1  1  1  1  1  1
 1  0 -1  1  0 -1  1  0 -1       1 -2  1  1 -2  1  1 -2  1
 1 -1  0  1 -1  0  1 -1  0       1  1 -2  1  1 -2  1  1 -2
 0  1 -1  0  1 -1  0  1 -1      -2  1  1 -2  1  1 -2  1  1
 1  0  0  0  0 -1  0  0  0       9  0  0 -9  0  0  0  0  0
 1  0  0 -1  0  0  0  0  0       0  0  0  9  0  0 -9  0  0
 0  0  0  1  0  0 -1  0  0      -9  0  0  0  0  0  9  0  0
 1  1  1  0  0  0 -1 -1 -1      -2 -2 -2  1  1  1  1  1  1
 1  1  1 -1 -1 -1  0  0  0       1  1  1 -2 -2 -2  1  1  1
 0  0  0  1  1  1 -1 -1 -1       1  1  1  1  1  1 -2 -2 -2
 1  0 -1  0 -1  1 -1  1  0      -2  1  1  1  1 -2  1 -2  1
 1 -1  0 -1  0  1  0  1 -1       1 -2  1 -2  1  1  1  1 -2
 0  1 -1  1 -1  0 -1  0  1      -2  1  1 -2  1  1  1  1 -2
 1 -1  0  0  1 -1 -1  0  1      -2  1  1  1 -2  1  1  1 -2
 1  0 -1 -1  1  0  0 -1  1       1 -2 -2  1  1  1  1 -2  1
 0 -1  1  1  0 -1 -1  1  0       1 -2  1  1 -2  1  1  1  1
 0  0  1  0  0  0  0  0 -1       0  0  9  0  0  0  0  0 -9
 0  0  1  0  0 -1  0  0  0       0  0 -9  0  0  9  0  0  0
 0  0  0  0  1  0  0  0 -1       0  0  0  0  0 -9  0  0  9
```

The transform of *x* (Input transform)
```
V(0)= +X(0)  -X(3);
V(1)= +X(0)  -X(6);
V(2)= +X(0)  +X(3);
V(3)= +X(6)  +V(2);
V(4)= +X(2)  -X(5);
V(5)= +X(2)  -X(8);
V(6)= +X(2)  +X(5);
V(7)= +X(5)  -X(8);
V(8)= +X(8)  +V(6);
V(9)= +X(1)  +X(4);
V(10)= +X(1) -X(4);
V(11)= +X(1) -X(7);
V(12)= +X(4) -X(7);
V(13)= +X(7) +V(9);
B(0)= +V(3)  +V(8)  +V(13);
B(1)= +V(3)  -V(8);
B(2)= +V(3)  -V(13);
B(4)= +V(1);
```

```
B(5) = +V(0);
B(7) = +V(1)  +V(5)  +V(11);
B(8) = +V(0)  +V(4)  +V(10);
B(10)= +V(1)  -V(4)  -V(12);
B(11)= +V(0)  +V(7)  -V(11);
B(13)= +V(1)  -V(7)  -V(10);
B(14)= +V(0)  -V(5)  +V(12);
B(16)= +V(5);
B(17)= +V(4);
B(3) =   B(1)  -B(2);
B(6) =   B(4)  -B(5);
B(9) =   B(7)  -B(8);
B(12)=   B(10) -B(11);
B(15)=   B(13) -B(14);
B(18)=   B(16) -B(17);
```

The multiplication:

$$m_j = a_j \times b_j, \; j = 0, 1, 2, \ldots, 18$$

The transform of **m** (Output transform)
```
V(0) = +C(16) +C(18);
V(1) = +C(17) -C(18);
V(2) = +C(13) +C(15);
V(3) = +C(14) -C(15);
V(4) = +C(10) +C(12);
V(5) = +C(11) -C(12);
V(6) = +C(7)  +C(9);
V(7) = +C(8)  -C(9);
V(8) = +C(4)  +C(6);
V(9) = +C(5)  -C(6);
V(10)= +C(1)  +C(3);
V(11)= +C(2)  -C(3);
V(12)= +V(10) +V(11);
V(13)= +C(0)  -V(10);
V(14)= +C(0)  -V(11);
V(15)= +C(0)  +V(12);
V(16)= +V(2)  +V(3);
V(17)= +V(4)  +V(5);
V(18)= +V(6)  +V(7);
Y(0) = +V(8)  +V(9)  +V(15) +V(16) +V(17) +V(18);
Y(1) = -V(0)  -V(5)  -V(6)  +V(13) +V(16);
Y(2) = -V(3)  -V(6)  +V(14) +V(17);
Y(3) = -V(2)  -V(4)  -V(6)  -V(8)  +V(15);
Y(4) = -V(1)  -V(2)  -V(7)  +V(13) +V(17);
Y(5) = -V(4)  -V(7)  +V(14) +V(16);
Y(6) = -V(3)  -V(5)  -V(7)  -V(9)  +V(15);
Y(7) = +V(0)  +V(1)  -V(3)  -V(4)  +V(13) +V(18);
Y(8) = -V(2)  -V(5)  +V(14) +V(18);
```

C11CR35- $N = 11$, $M = 35$, Using the Chinese Remainder Theorem

$A_I = 75, A_O = 98, A = 173.$

A

1	1	1	1	1	1	1	1	1	1	1
1	-1	1	-1	1	-1	1	-1	1	-1	0
1	0	0	0	0	0	0	0	0	0	-1
1	-1	0	0	0	0	0	0	0	0	0
0	1	0	0	0	0	0	0	0	0	-1
1	0	-1	0	1	0	-1	0	1	0	-1
1	-1	-1	1	1	-1	-1	1	1	-1	0
0	1	0	-1	0	1	0	-1	0	1	-1
1	0	-1	1	0	-1	1	0	-1	1	-1
1	-1	0	1	-1	0	1	-1	0	1	-1
0	1	-1	0	1	-1	0	1	-1	0	0
1	0	-1	-1	0	1	1	0	-1	-1	1
1	1	0	-1	-1	0	1	1	0	-1	-1
0	1	1	0	-1	-1	0	1	1	0	-2
1	0	0	0	-1	0	0	0	1	0	-1
1	-1	1	-1	-1	1	-1	1	1	-1	0
1	0	-1	0	-1	0	1	0	1	0	-1

VLSI Signal Processing

```
 1 -1 -1  1 -1  1  1 -1  1 -1  0
 0  1  0 -1  0 -1  0  1  0  1 -1
 1  0 -1  1 -1  0  1 -1  1  0 -1
 1 -1  0  1 -1  1  0 -1  1 -1  0
 0  1 -1  0  0 -1  0  0  1 -1
 0  0  0  1  0  0  0 -1  0  0  0
 1  0  0  0 -1  1  0  0  0 -1  0
 1 -1  1 -1  0  1 -1  1 -1  0  0
 1  0 -1  0  0  1  0 -1  0  0  0
 1 -1 -1  1  0  1 -1 -1  1  0  0
 0  1  0 -1  0  0  1  0 -1  0  0
 1  0 -1  1 -1  1  0 -1  1 -1  0
 1 -1  0  1 -1  1 -1  0  1 -1  0
 0  1 -1  0  0  1 -1  0  0  0
 0  0  0  1 -1  0  0  0  1 -1  0
 0  0  0  0  0  0  0  1  0 -1
 0  0  0  0  0  0  0  1 -1  0
 0  0  0  0  0  0  0  1 -1
```

132 C′

```
   12    12    12    12    12    12    12    12    12    12    12    12    12
   -5    28     6    28    -5     6   -16   -16   -16   -16     6   -16     6
 -132  -264  -132     0   132   264   132   264     0     0  -264     0  -264
  -12  -276  -408  -408  -276   -12   120   384   384   384   120   384   120
   12   276   408   408   276    12  -120  -384  -384  -384  -120   384  -120
  -36  -102  -102  -102  -102   -36    96    96    96    96    96    96    96
   69   102   102   102    69   -30   -96   -96   -96   -96   -30   -96   -30
 -102  -102  -102  -102   -36    96    96    96    96    96   -36    96   -36
   24    90    90    24    24   -42   -42  -108   -42   -42    24   -42    24
   24    24    90    90    24    24   -42   -42  -108   -42   -42   -42   -42
  -48  -114  -180  -114   -48    18    84   150   150    84    18    84    18
  -20    46    90   112   112    90     2   -64  -130  -130  -108   130  -108
  112   112    90    46   -20  -108  -130  -130   -64     2    90     2    90
  -92  -158  -180  -158   -92    18   128   194   194   128    18   128    18
  468   468   336    72  -192  -456  -456  -456  -192    72   336    72   336
  -99   -66     0    66    99   132    66    66   -66  -132   -66  -132
 -264  -198   -66    66   198   264   264   132     0  -132  -264   132  -264
  231   132     0  -132  -231  -264  -198   -66    66   198   264   198   264
 -198   -66    66   198   264   264   132     0  -132  -264  -264   264  -264
 -204  -204  -204   -72    60   192   192   192   192   -72   -72   -72   -72
  -60    72   204   204   204    72   -72  -192  -192  -192  -192   192  -192
  264   132     0  -132  -264  -264  -264     0     0   264   264   264   264
  192   -72  -336  -468  -468  -336   -72   192   456   456   456   456   456
 -216  -216  -216   -84    48   180   180   180    48   -84    48   -84
   54    54    54   -12   -12   -78   -12   -78   -12   -12    54   -12    54
  240   240   108   -24  -156  -288  -156  -156   -24   -24   240   -24   240
 -102    30   162   228   162   -36  -102  -168  -168  -234   168  -234
  -36  -300  -432  -432  -300   -36   228   360   360   360   228   360   228
   24  -108  -108  -108  -108    24    24   156    24   156    24   156    24
   24    24  -108  -108  -108   -108   24   156    24   156    24    24   156
  -48    84   216   216   216    84   -48  -180  -180  -180  -180   180  -180
   60   192   324   324   192    60  -204  -204  -336  -204  -204   204  -204
  276   408   408   276    12  -120  -384  -384  -384  -120    12   120    12
 -276  -408  -408  -276   -12   120   384   384   384   120   -12   120   -12
  132     0  -132  -264  -132  -264     0     0   264   132   264   132   264
```

The transform of *x* (Input transform)
```
V(0)  = +X(0)  +X(8);
V(1)  = +X(3)  -X(7);
V(2)  = +X(4)  +X(9);
V(3)  = +X(1)  -V(0);
V(4)  = +X(2)  -X(6);
V(5)  = +X(5)  -V(2);
V(6)  = +X(2)  +X(6);
V(7)  = +V(3)  -V(5);
V(8)  = +V(1)  -V(4);
V(9)  = +X(10) -V(0);
V(10) = +X(0)  -X(10);
V(11) = +X(4)  +V(9);
V(12) = +X(3)  +X(7);
V(13) = +X(4)  -X(9);
V(14) = +X(0)  +X(5);
V(15) = +X(1)  -V(1);
```

VLSI Signal Processing

```
V(16) = +X(6)  +V(10);
V(17) = +X(3)  +X(9);
V(18) = +X(8)  -X(10);
V(19) = +V(3)  -V(13);
V(20) = +X(5)  +V(6);
V(21) = +X(5)  -V(6);
B(0)  = +X(1)  +X(10) +V(0)  +V(2)  +V(12) +V(20);
B(1)  = -V(12) -V(19) -V(21);
B(2)  = +V(10);
B(3)  = +X(0)  -X(1);
B(5)  = +X(4)  -V(6)  -V(9);
B(6)  = +V(12) -V(19) -V(20);
B(8)  = -X(5)  -X(8)  -V(4)  +V(10) +V(17);
B(9)  = -V(13) -V(15) +V(16);
B(11) = -V(4)  +V(14) -V(17) -V(18);
B(12) = -V(2)  +V(15) +V(16);
B(14) = -V(11);
B(15) = -V(7)  -V(8);
B(16) = -V(4)  -V(11);
B(17) = -V(7)  +V(8);
B(19) = +V(8)  -V(11);
B(20) = +V(1)  -V(7);
B(22) = +V(1);
B(23) = +X(0)  +V(5);
B(24) = -X(1)  -X(8)  -V(8)  +V(14);
B(25) = -X(2)  -X(7)  +V(14);
B(26) = +V(1)  -V(3)  +V(21);
B(28) = -X(2)  +V(0)  +V(1)  +V(5);
B(29) = +X(3)  -X(6)  -V(7);
B(31) = +X(3)  +X(8)  -V(2);
B(32) = +V(18);
B(33) = +X(8)  -X(9);
B(4)  =  B(2)  -B(3);
B(7)  =  B(5)  -B(6);
B(10) =  B(8)  -B(9);
B(13) = -B(11) +B(12);
B(18) =  B(16) -B(17);
B(21) =  B(19) -B(20);
B(27) =  B(25) -B(26);
B(30) =  B(28) -B(29);
B(34) =  B(32) -B(33);
```

The multiplication:

$$m_j = a_j \times b_j, \; j = 0, 1, 2, \ldots, 34$$

The transform of **m** (Output transform)
```
V(0)  = +C(32) +C(34);
V(1)  = +C(33) -C(34);
V(2)  = +C(28) +C(30);
V(3)  = +C(29) -C(30);
V(4)  = +C(25) +C(27);
V(5)  = +C(26) -C(27);
V(6)  = +C(19) +C(21);
V(7)  = +C(20) -C(21);
V(8)  = +C(16) +C(18);
V(9)  = +C(17) -C(18);
V(10) = +C(11) -C(13);
V(11) = +C(12) +C(13);
V(12) = +C(8)  +C(10);
V(13) = +C(9)  -C(10);
V(14) = +C(5)  +C(7);
V(15) = +C(6)  -C(7);
V(16) = +C(2)  +C(4);
V(17) = +C(3)  -C(4);
V(18) = +V(7)  +V(9);
V(19) = +V(3)  +V(18);
V(20) = +C(15) +C(24);
V(21) = +V(6)  +V(8);
V(22) = +V(2)  +V(19);
V(23) = +V(5)  +V(15);
V(24) = +V(12) +V(13 );
```

VLSI Signal Processing

```
V(25)= +C(0)  +C(1);
V(26)= +C(0)  -C(1);
V(27)= +C(31) +V(22);
V(28)= +V(14) +V(23 );
V(29)= +C(14) +V(21);
V(30)= +V(10) +V(11 );
V(31)= +C(15) +V(27);
V(32)= +C(22) +V(6);
V(33)= +V(1)  +V(31);
V(34)= +V(9)  +V(21);
V(35)= +V(11) -V(13 );
V(36)= +V(5)  -V(15);
V(37)= +V(26) +V(35 );
V(38)= +V(0)  -V(10);
V(39)= +C(23) +V(22);
V(40)= +V(11) +V(29 );
V(41)= +V(16) +V(24 );
V(42)= +V(24) +V(26 );
V(43)= +V(30) -V(42 );
V(44)= +V(25) +V(30 );
V(45)= +V(12) -V(25 );
V(46)= +V(28) +V(29 );
V(47)= +V(4)  +V(10);
V(48)= +V(2)  -V(20);
V(49)= +V(17) +V(20 );
Y(0)= +V(4)  +V(39) +V(41) +V(44) +V(46) +V(49);
Y(1)= +C(0)  -V(14) -V(38) -V(40) -V(41);
Y(2)= -C(23) -V(15) -V(33) -V(43);
Y(3)= -C(24) +V(33) +V(38) -V(45) +V(46);
Y(4)= -V(4)  -V(18) -V(32) -V(36) +V(37) -V(48);
Y(5)= -V(3)  -V(20) +V(24) -V(28) +V(34) +V(44);
Y(6)= -V(12) +V(20) +V(26) +V(36) +V(39) +V(47);
Y(7)= -C(23) -V(13) +V(14) +V(15) +V(25)  -V(31)   -V( 40);   ;
Y(8)= -V(20) +V(23) +V(27) +V(32) -V(43);
Y(9)= -V(28) -V(34) -V(45) -V(47) -V(48);
Y(10)= -V(19) -V(23) +V(37) -V(49);
```

C11CR36- N = 11, M = 36, Using the Chinese Remainder Theorem

$A_I = 71, A_O = 93, A = 164.$

$$A$$

1	1	1	1	1	1	1	1	1	1	1
1	-1	1	-1	1	-1	1	-1	1	-1	0
1	0	-1	0	1	0	-1	0	1	0	-1
1	-1	-1	1	1	-1	-1	1	1	-1	0
0	1	0	-1	0	1	0	-1	0	1	-1
1	0	-1	1	0	-1	1	0	-1	1	-1
1	-1	0	1	-1	0	1	-1	0	1	-1
0	1	-1	0	1	-1	0	1	-1	0	0
1	0	0	0	0	0	0	0	0	0	-1
1	-1	0	0	0	0	0	0	0	0	0
0	1	0	0	0	0	0	0	0	0	-1
1	0	-1	0	0	0	0	0	0	0	0
0	0	1	0	0	0	0	0	0	0	-1
1	0	0	0	-1	0	0	0	1	0	-1
1	-1	1	-1	1	-1	1	1	1	-1	0
1	0	-1	0	-1	1	0	1	1	0	-1
1	-1	-1	1	-1	1	1	-1	1	-1	0
0	1	0	-1	0	-1	0	1	0	1	-1
1	0	-1	1	-1	0	1	-1	1	0	-1
1	-1	0	1	-1	1	0	-1	1	-1	0
0	1	-1	0	0	-1	1	0	0	1	-1
0	0	0	1	0	0	0	-1	0	0	0
1	0	0	0	-1	1	0	0	0	-1	0
1	-1	1	-1	0	1	-1	1	-1	0	0
1	0	-1	0	1	0	-1	0	0	0	0
1	-1	-1	1	0	1	-1	-1	1	0	0
0	1	0	-1	0	0	1	0	-1	0	0
1	0	-1	1	-1	1	0	-1	1	-1	0
1	-1	0	1	-1	1	-1	0	1	-1	0
0	1	-1	0	0	0	1	-1	0	0	0
0	0	0	-1	0	0	0	1	-1	0

```
0   0   0   0   0   0   0   1   0   0  -1
0   0   0   0   0   0   0   1   0  -1   0
0   0   0   0   0   0   0   0   1   0  -1
0   0   0   0   0   0   0   0   1  -1   0
0   0   0   0   0   0   0   0   0   1  -1
```

132 C'

```
   4    4    4    4    4    4    4    4    4    4    4
 -17  -17  -28  -17  -17    5   16   27   27   16    5
  12   34   56   56   34   12  -10  -54  -76  -54  -10
 -23  -45  -56  -45  -23   -1   32   65   65   32   -1
  34   56   56   34   12  -10  -54  -76  -54  -10   12
 -60  -60  -60  -60  -16   28   72   72   72   28  -16
 -16  -60  -60  -60  -60  -16   28   72   72   72   28
  76  120  120  120   76  -12 -100 -144 -144 -100  -12
  44  -44  -44  -44  -44  -44  -44   44   44   88   44
 228  360  404  360  228    8 -300 -432 -476 -300  -80
-180 -224 -224 -180  -92   40  216  260  260  128   -4
 -48 -136 -180 -180 -136  -48   84  172  216  172   84
  48  136  180  180  136   48  -84 -172 -216 -172  -84
  20  -24  -68 -112 -112  -68   20   64  108  108   64
 -11  -11    0   11   11   11    0   11  -11    0  -11
-112 -134 -112  -68   -2   64  130  130  108   42  -46
  55   33    0  -33  -55  -55  -44  -11   11   44   55
   2   68  112  134  112   46  -42 -108 -130 -130  -64
 116  248  292  292  204   28 -192 -324 -324 -280  -60
-204 -292 -292 -248 -116   60  280  324  324  192  -28
  88   44    0  -44  -88  -88  -88    0    0   88   88
 112  112   68   24  -20  -64 -108 -108  -64  -20   68
  64  108  108  108   64  -24 -112 -156 -112  -68   20
 -50  -72  -72  -72  -28   16   82   82   82   38   -6
  -8   36   36   36   36   36   36   -8  -52  -96  -52
 -14  -36  -36  -36  -36  -36  -14   30   74   74   30
  36   36   36   36   36   36   -8  -52  -96  -52   -8
   8  -36  -36  -36  -36    8    8   52    8   52    8
  56  144  144  144  100   12  -76 -164 -164 -164  -32
 -64 -108 -108 -108  -64  -20   68  112  156  112   24
 -28  -72  -72  -72  -72  -28   16  104  104  104   16
 136  180  180  136   48  -84 -172 -216 -172  -84   48
-136 -180 -180 -136  -48   84  172  216  172   84  -48
 -92 -180 -224 -224 -180   -4  128  260  260  216   40
 228  360  404  360  228  -80 -300 -476 -432 -300    8
 -44  -44  -44  -44   44   44   88   44   44  -44  -44
```

The transform of *x* (Input transform)

```
V( 0)=+X( 0)   +X( 8);
V( 1)=+X( 2)   -X( 6);
V( 2)=+X( 1)   +X( 9);
V( 3)=+X( 3)   -X( 7);
V( 4)=+X( 4)   -V(0);
V( 5)=+X(10)   +V(1);
V( 6)=+X( 8)   -X( 9);
V( 7)=+X( 4)   +V(0);
V( 8)=+X( 2)   +X( 6);
V( 9)=+X( 3)   +X( 7);
V(10)=+X( 5)   +V(2);
V(11)=+X( 5)   -V(2);
V(12)=+X( 0)   +X( 5);
V(13)=+X( 1)   -X( 8);
V(14)=+X( 1)   -V(1);
V(15)=+V(3)    -V(4);
V(16)=+V(7)    +V(8);
V(17)=+X(10)   +V(9);
B( 0)=+V(10)   +V(16)   +V(17);
B( 1)=-V(9)    -V(10)   +V(16);
B( 2)=-X(10)   +V(7)    -V(8);
B( 3)=+V(10)   -V(17);
B( 5)=+X( 0)   +X( 3)   -X( 5)   -V(5)  -V(6);
B( 6)=-X( 2)   +X( 4)   -X( 5)   +X( 7) +V(13);
B( 8)=+X( 0)   -X(10);
B( 9)=+X( 0)   -X( 1);
B(11)=+X( 0)   -X( 2);
B(12)=+X( 2)   -X(10);
```

VLSI Signal Processing

```
B(13)=-X(10)   -V(4);
B(14)=+V(1)    -V(3)    -V(4)    +V(11);
B(15)=-V(4)    -V(5);
B(16)=-X(10)   -V(3)    -V(11);
B(18)=-V(5)    +V(15);
B(19)=-V(5)    -V(11);
B(21)=+V(3);
B(22)=-X( 4)   -X( 9)   +V(12);
B(23)=-X( 8)   -V(3)    +V(12)   -V(14);
B(24)=-X( 2)   -X( 7)   +V(12);
B(25)=-X( 3)   +X( 6)   +V(13);
B(27)=-X( 2)   +X( 5)   -X( 9)   +V(15);
B(28)=-X( 7)   +V(14);
B(30)=+X( 3)   -X( 4)   +V(6);
B(31)=+X( 7)   -X(10);
B(32)=+X( 7)   -X( 9);
B(33)=+X( 8)   -X(10);
B(34)=+V(6);
B( 4)= B( 2)   -B( 3);
B( 7)= B( 5)   -B( 6);
B(10)= B( 8)   -B( 9);
B(17)= B(15)   -B(16);
B(20)= B(18)   -B(19);
B(26)= B(24)   -B(25);
B(29)= B(27)   -B(28);
```

The multiplication:

$$m_j = a_j \times b_j, \; j = 0, 1, 2, \ldots, 35$$

The transform of m (Output transform)
```
V(0) =+C(33)   +C(35);
V(1) =+C(34)   -C(35);
V(2) =+C(27)   +C(29);
V(3) =+C(28)   -C(29);
V(4) =+C(24)   +C(26);
V(5) =+C(25)   -C(26);
V(6) =+C(18)   +C(20);
V(7) =+C(19)   -C(20);
V(8) =+C(15)   +C(17);
V(9) =+C(16)   -C(17);
V(10)=+C( 8)   +C(10);
V(11)=+C( 9)   -C(10);
V(12)=+C( 5)   +C( 7);
V(13)=+C( 6)   -C( 7);
V(14)=+C( 2)   +C( 4);
V(15)=+C( 3)   -C( 4);
V(16)=+C(14)   +C(23);
V(17)=+V(6)    +V(8);
V(18)=+V(7)    +V(12);
V(19)=+C( 0)   +C( 1);
V(20)=+C( 0)   -C( 1);
V(21)=+V(3)    -V(16);
V(22)=+V(9)    +V(15);
V(23)=+C(13)   +V(17);
V(24)=+C(22)   +V(2);
V(25)=+C(14)   +C(30);
V(26)=+V(14)   +V(19);
V(27)=+V(14)   -V(19);
V(28)=+C(21)   +V(6);
V(29)=+V(1)    +V(25);
V(30)=+V(20)   +V(22);
V(31)=+V(0)    +V(23);
V(32)=+V(2)    +V(28);
V(33)=+V(9)    -V(15);
V(34)=+V(17)   +V(18);
V(35)=+V(21)   +V(34);
V(36)=+V(5)    +V(13);
V(37)=+V(13)   +V(33);
V(38)=+C(11)   +V(4);
V(39)=+V(4)    -V(37);
V(40)=+V(16)   +V(24);
```

```
Y( 0)=+V(10)+V(11)  +V(12)   +V(23)  +V(26)  +V(38)  +V(40);
Y( 1)=+C( 0)  -C(12)  -C(31)  -V(10)  -V(14)  -V(18)  -V(22)  -V(31);
Y( 2)=-C(32)  +V(18)  -V(24)  -V(29)  +V(30);
Y( 3)=-C(23)  +V(2)   -V(12)  +V(26)  +V(29)  +V(31)  -V(36);
Y( 4)=+C(31)  +C(32)  +V(20)  -V(21)  -V(32)  -V(39);
Y( 5)=+V(5)   -V(27)  +V(35);
Y( 6)=-V(18)  +V(20)  +V(39)  +V(40);
Y( 7)=+V(13)  -V(23)  -V(24)  -V(25)  +V(26);
Y( 8)=+C(30)  -V(5)   +V(12)  -V(16)  +V(20)  -V(22)  +V(32);
Y( 9)=+C(12)  -V(2)   -V(13)  -V(27)  -V(35)  -V(38);
Y(10)=+V(7)   -V(11)  +V(21)  +V(30)  +V(36);
```

C11LI37- $N = 11, M = 37$, Using Linear Polynomials

$A_I = 59, A_O = 85, A = 144.$

A

```
 1   1   1   1   1   1   1   1   1   1   1
 1   0   0   0   0   0   0   0   0   0  -1
 1  -1   0   0   0   0   0   0   0   0   0
 0   1   0   0   0   0   0   0   0   0  -1
 1   0  -1   0   1   0  -1   0   1   0  -1
 0   1   0  -1   0   1   0  -1   0   1  -1
 1  -1  -1   1   1  -1  -1   1   1  -1   0
 1   0   0   0  -1   0   0   0   1   0  -1
 0   1   0   0   0  -1   0   0   0   1  -1
 1  -1   0   0  -1   1   0   0   1  -1   0
 1   0  -1   0  -1   0   1   0   1   0  -1
 0   1   0  -1   0  -1   0   1   0   1  -1
 1  -1  -1   1  -1   1   1  -1   1  -1   0
 0   0   1   0   0   0  -1   0   0   0   0
 0   0   0   1   0   0   0  -1   0   0   0
 0   0   1  -1   0   0  -1   1   0   0   0
 1   0   0   0  -1   0   1   0   0   0  -1
 0   1   0   0   0  -1   0   1   0   0  -1
 1  -1   0   0  -1   1  -1   0   0   0   0
 1   0  -1   0   0   1   0  -1   0   0   0
 0   1   0  -1   0   0   1   0  -1   0   0
 1  -1  -1   1   0   1  -1  -1   1   0   0
 0   0   1   0  -1   0   0   0   1   0  -1
 0   0   0   1   0  -1   0   0   0   1  -1
 0   0   1  -1  -1   1   0   0   1  -1   0
 1   0   0   0  -1   0  -1   0   0   0   1
 0   1   0   0   0  -1   0  -1   0   0   1
 1  -1   0   0  -1   1   1  -1   0   0   0
 1   0  -1   0   0   0  -1   0  -1   0   0
 0   1   0   1   0   0   0  -1   0  -1   0
 1  -1   1  -1   0   0  -1   1  -1   1   0
 0   0   1   0   1   0   0   0  -1   0  -1
 0   0   0   1   0   1   0   0   0  -1  -1
 0   0   1  -1   1  -1   0   0  -1   1   0
 0   0   0   0   0   0   0   1   0  -1   0
 0   0   0   0   0   0   0   0   1  -1   0
 0   0   0   0   0   0   0   0   1  -1
```

66 C'

```
  6    6    6    6    6    6    6    6    6    6    6
 54  -12  -12  -12  -12  -12  -12   54  -12  -12  -12
-60   72  -60   72  -60   72  -60    6    6    6    6
-12  -12  -12  -12  -12  -12   54  -12  -12  -12   54
-10  -10   12   12  -10  -10   12    1  -10    1   12
-10   12   12  -10  -10   12    1  -10    1   12  -10
 16   -6   -6   -6   16   -6   -6    5    5   -6   -6
  6    6    6    6    6    6    6  -60    6    6    6
  6    6    6    6    6    6  -60    6    6    6    6
 30  -36   30  -36   30  -36   30   30  -36   30  -36
  0    0    0    0    0    0    0   33    0  -33    0
  0    0    0    0    0    0   33    0  -33    0    0
  0    0    0    0    0    0    0  -33   33    0    0
 -6   -6   -6   -6   -6   -6   -6   -6   60   -6
 -6   -6   -6   -6   -6   -6   -6   -6   60   -6   -6
-30   36  -30   36  -30   36  -30   36  -30  -30   36
-24  -24    9    9    9    9  -24  -24    9   42    9
```

VLSI Signal Processing

```
 -24    9    9    9    9  -24  -24    9   42    9  -24
  12   12  -21   12  -21   12   12   12  -21  -21   12
  15   15  -18  -18   15   15   15  -18  -18  -18   15
  15  -18  -18   15   15   15  -18  -18  -18   15   15
   9  -24   42  -24    9  -24    9    9    9    9  -24
   9    9    9    9  -24  -24    9   42    9  -24  -24
   9    9    9  -24  -24    9   42    9  -24  -24    9
 -21   12  -21   12   12   12  -21  -21   12   12   12
  20   20    9    9  -13  -13  -24   -2  -13   -2    9
  20    9    9  -13  -13  -24   -2  -13   -2    9   20
 -32   12  -21   12    1   12   12  -10   23  -21   12
  -7   -7  -18  -18   -7   -7   15    4   26    4   15
  -7  -18  -18   -7   -7   15    4   26    4   15   -7
  31  -24   42  -24   31  -24    9  -13  -13    9  -24
 -13  -13    9    9   20   20    9   -2  -13   -2  -24
 -13    9    9   20   20    9   -2  -13   -2  -24  -13
   1   12  -21   12  -32   12  -21   23  -10   12   12
 -12  -12  -12  -12  -12   54  -12  -12  -12   54  -12
 -60   72  -60   72  -60    6    6    6    6  -60   72
 -12  -12  -12  -12   54  -12  -12  -12   54  -12  -12
```

The transform of x (Input transform)

```
V(  0)= +X(  4) +X(10);
V(  1)= +X(  2) -X(  6);
V(  2)= +X(  0) +X(  8);
V(  3)= +X(  1) +X(  7);
V(  4)= +X(  3) -X(  9);
V(  5)= +X(  5) +X(10);
V(  6)= +X(  1) -X(  7);
V(  7)= +X(  3) +X(  9);
V(  8)= +X(  4) -X(10);
V(  9)= +X(  5) -X(10);
V( 10)= +X(  0) -X(  8);
V( 11)= +X(  2) +X(  6);
V( 12)= +V(  0) -V(  2);
V( 13)= +V(  3) -V(  5);
B(  0)= +X(  5) +V(  0) +V(  2) +V(  3) +V(  7) +V( 11);
B(  1)= +X(  0) -X( 10);
B(  2)= +X(  0) -X(  1);
B(  4)= +V(  2) +V(  8) -V( 11);
B(  5)= -V(  4) +V(  6) +V(  9);
B(  7)= -V( 12);
B(  8)= +X(  1) +X(  9) -V(  5);
B( 10)= -V(  1) -V( 12);
B( 11)= -V(  4) +V( 13);
B( 13)= +V(  1);
B( 14)= +X(  3) -X(  7);
B( 16)= +X(  0) +X(  6) -V(  0);
B( 17)= +V( 13);
B( 19)= -V(  1) +V( 10);
B( 20)= +V(  3) -V(  7);
B( 22)= +X(  2) +X(  8) -V(  0);
B( 23)= -V(  5) +V(  7);
B( 25)= +X(  0) -X(  6) -V(  8);
B( 26)= +V(  6) -V(  9);
B( 28)= +V(  1) +V( 10);
B( 29)= +V(  4) +V(  6);
B( 31)= +X(  2) -X(  8) +V(  8);
B( 32)= +V(  4) +V(  9);
B( 34)= +X(  8) -X( 10);
B( 35)= +X(  8) -X(  9);
B(  3)=  B(  1) -B(  2);
B(  6)=  B(  4) -B(  5);
B(  9)=  B(  7) -B(  8);
B( 12)=  B( 10) -B( 11);
B( 15)=  B( 13) -B( 14);
B( 18)=  B( 16) -B( 17);
B( 21)=  B( 19) -B( 20);
B( 24)=  B( 22) -B( 23);
B( 27)=  B( 25) -B( 26);
B( 30)=  B( 28) -B( 29);
B( 33)=  B( 31) -B( 32);
B( 36)=  B( 34) -B( 35);
```

VLSI Signal Processing

The multiplication:

$$m_j = a_j \times b_j, \; j = 0, 1, 2, \ldots, 36$$

The transform of **m** (Output transform)
```
V( 0)= +C(34) +C(36);
V( 1)= +C(35) -C(36);
V( 2)= +C(31) +C(33);
V( 3)= +C(32) -C(33);
V( 4)= +C(28) +C(30);
V( 5)= +C(29) -C(30);
V( 6)= +C(25) +C(27);
V( 7)= +C(26) -C(27);
V( 8)= +C(22) +C(24);
V( 9)= +C(23) -C(24);
V(10)= +C(19) +C(21);
V(11)= +C(20) -C(21);
V(12)= +C(16) +C(18);
V(13)= +C(17) -C(18);
V(14)= +C(13) +C(15);
V(15)= +C(14) -C(15);
V(16)= +C(10) +C(12);
V(17)= +C(11) -C(12);
V(18)= +C( 7) +C( 9);
V(19)= +C( 8) -C( 9);
V(20)= +C( 4) +C( 6);
V(21)= +C( 5) -C( 6);
V(22)= +C( 1) +C( 3);
V(23)= +C( 2) -C( 3);
V(24)= +V(17) +V(19);
V(25)= +V(12) +V(16);
V(26)= +V( 9) +V(24);
V(27)= +V( 8) +V(18);
V(28)= +C( 0) +V(21);
V(29)= +C( 0) -V(21);
V(30)= +C( 0) +V(20);
V(31)= +C( 0) -V(20);
V(32)= +V( 0) +V(27);
V(33)= +V( 2) -V(16);
V(34)= +V( 2) -V( 6);
V(35)= +V(22) +V(25);
V(36)= +V( 3) -V(26);
V(37)= +V( 7) +V(13);
V(38)= +V( 7) -V(13);
V(39)= +V(10) -V(14);
V(40)= +V( 4) +V( 6);
V(41)= +V( 4) +V(33);
V(42)= +V( 5) +V(11);
V(43)= +V( 5) -V(11);
V(44)= +V(15) -V(17);
V(45)= +V(43) +V(44);
Y( 0)= +V(10) +V(18) +V(23) +V(30) +V(35) +V(40);
Y( 1)= -V( 3) -V(20) -V(26) +V(29) -V(32) -V(34) -V(35) +V(38);
Y( 2)= -V( 1) +V(28) -V(36) -V(42);
Y( 3)= +V( 1) -V(10) +V(30) +V(32) -V(41);
Y( 4)= +V(29) -V(38) -V(45);
Y( 5)= +V(25) +V(31) +V(39) -V(40);
Y( 6)= +V(28) +V(36) -V(37);
Y( 7)= -V(25) -V(27) +V(30) +V(34);
Y( 8)= +V( 3) +V( 9) +V(29) +V(45);
Y( 9)= +V( 8) +V(31) -V(39) +V(41);
Y(10)= -V(23) +V(24) +V(28) +V(37) +V(42);
```

C11QI37- N = 11, M = 37, Using the Iterative Method with Quadratic Polynomials

$A_I = 63, A_O = 89, A = 152$

$$A$$

```
1 1 1 1 1 1 1 1 1 1 1
1 0 0 0 0 0 0 0 0 0 -1
```

VLSI Signal Processing

```
1 -1  1 -1  1  0  0  0  0  0 -1
1  0 -1  0  1  0  0  0  0  0 -1
0  1  0 -1  0  0  0  0  0  0  0
1 -1 -1  1  1  0  0  0  0  0 -1
1  0 -1  1  0  0  0  0  0  0 -1
1 -1  0  1 -1  0  0  0  0  0  0
0  1 -1  0  1  0  0  0  0  0 -1
1  0 -1 -1  0  0  0  0  0  0  1
1  1  0 -1 -1  0  0  0  0  0  0
0  1  1  0 -1  0  0  0  0  0 -1
0  0  0  1  0  0  0  0  0  0 -1
1  0  0  0  0 -1  0  0  0  0  0
1 -1  1 -1  1 -1  1 -1  1 -1  0
1  0 -1  0  1 -1  0  1  0 -1  0
0  1  0 -1  0  0 -1  0  1  0  0
1 -1 -1  1  1 -1  1  1 -1 -1  0
1  0 -1  1  0 -1  0  1 -1  0  0
0  1 -1  0  1  0 -1  1  0 -1  0
1 -1  0  1 -1 -1  1  0 -1  1  0
1  0 -1 -1  0 -1  0  1  1  0  0
0  1  1  0 -1  0 -1 -1  0  1  0
1  1  0 -1 -1 -1 -1  0  1  1  0
0  0  0  0  1  0  0  0  0 -1  0
0  0  0  0  0  1  0  0  0  0 -1
0  0  0  0  1 -1  1 -1  1 -1
0  0  0  0  1  0 -1  0  1 -1
0  0  0  0  0  1  0 -1  0  0
0  0  0  0  1 -1 -1  1  1 -1
0  0  0  0  1  0 -1  1  0 -1
0  0  0  0  1 -1  0  1 -1  0
0  0  0  0  0  1 -1  0  1 -1
0  0  0  0  1  0 -1 -1  0  1
0  0  0  0  1  1  0 -1 -1  0
0  0  0  0  0  1  1  0 -1 -1
0  0  0  0  0  0  0  0  1 -1
```

11 **C'**

```
  6    6    6    6    6    6    6    6    6    6    6
-12  -12  -12  -12   54  -12   54  -12  -12  -12  -12
 12  -10   12  -10    1    1  -10   12  -10   12  -10
  6    6    6    6  -60    6    6    6    6    6    6
 -6   -6   -6   -6   -6   60   -6   -6   -6   -6   -6
  0    0    0    0   33  -33    0    0    0    0    0
  9  -24    9    9  -24   42  -24    9    9  -24    9
 15   15  -18   15  -18  -18   15  -18   15   15  -18
-24    9    9  -24   42  -24    9    9  -24    9    9
  9   20    9  -13   -2   -2   20    9  -13  -24  -13
 15   -7  -18   -7    4    4   -7  -18   -7   15   26
-24  -13    9   20   -2   -2  -13    9   20    9  -13
-12  -12  -12   54  -12   54  -12  -12  -12  -12  -12
  6  -60  -60  -60  -60    6    6   72   72   72    6
 -6   16   -6   16   -6    5   -6   -6   -6   -6    5
 30   30   30   30   30  -36  -36  -36  -36  -36   30
-30  -30  -30  -30  -30  -30   36   36   36   36   36
  0    0    0    0    0   33    0    0    0    0  -33
-21   12  -21  -21   12  -21   12   12   12   12   12
 12  -21  -21   12  -21   12   12   12   12   12  -21
  9    9   42    9    9    9  -24  -24  -24  -24    9
-21  -32  -21    1   12   23   12   12   12   12  -10
 12    1  -21  -32  -21  -10   12   12   12   12   23
  9   31   42   31    9  -13  -24  -24  -24  -24  -13
-60  -60  -60  -60    6    6   72   72   72    6    6
-12   54  -12  -12  -12  -12  -12  -12  -12  -12   54
  1  -10   12  -10   12  -10   12  -10   12  -10    1
  6    6    6    6    6    6    6    6    6    6  -60
 60   -6   -6   -6   -6   -6   -6   -6   -6   -6   -6
-33    0    0    0    0    0    0    0    0    0   33
 42  -24    9    9  -24    9    9  -24    9    9  -24
-18   15  -18   15   15  -18   15   15  -18   15  -18
-24    9    9  -24    9    9  -24    9    9  -24   42
 -2   20    9  -13  -24  -13    9   20    9  -13   -2
  4   -7  -18   -7   15   26   15   -7  -18   -7    4
 -2  -13    9   20    9  -13  -24  -13    9   20   -2
```

VLSI Signal Processing

54 −12 −12 −12 −12 −12 −12 −12 54 −12

The transform of x (Input transform)
```
V( 0)= +X( 0) −X( 2);
V( 1)= +X( 5) −X( 7);
V( 2)= +X( 6) −X( 9);
V( 3)= +X( 1) −X( 3);
V( 4)= +X( 6) +X( 9);
V( 5)= +V( 0) −V( 1);
V( 6)= +X( 5) −X( 8);
V( 7)= +X( 0) +X( 2);
V( 8)= +X( 4) +V( 7);
V( 9)= +X( 1) +X( 3);
V(10)= +V( 8) −V( 9);
V(11)= +X( 5) +X( 8);
V(12)= +X( 6) −X( 8);
V(13)= +X( 3) −X( 8);
V(14)= +V( 2) −V( 6);
V(15)= +X( 0) −X( 4);
V(16)= +X( 2) −X( 4);
V(17)= +X( 4) −X( 9);
V(18)= +X( 7) +V( 2);
V(19)= +X( 7) −V(14);
V(20)= +X( 3) −X(10);
V(21)= +X( 4) −X(10);
V(22)= +X( 8) −X(10);
V(23)= +X( 9) −X(10);
B( 0)= +X( 7) +X(10) +V( 4) +V( 8) +V( 9) +V(11);
B( 1)= +X( 0) −X(10);
B( 2)= −X(10) +V(10);
B( 3)= +V( 0) +V(21);
B( 4)= +V( 3);
B( 6)= +V( 0) +V(20);
B( 7)= −V( 3) +V(15);
B( 9)= +V( 0) −V(20);
B(10)= +V( 3) +V(15);
B(12)= +V(21);
B(13)= +X( 0) −X( 5);
B(14)= +V(10) −V(19);
B(15)= +V( 5) +V(17);
B(16)= +V( 3) −V(12);
B(18)= +V( 5) +V(13);
B(19)= +X( 1) +X( 7) −V( 4) −V(16);
B(21)= +V( 5) −V(13);
B(22)= +X( 1) +V(16) −V(18);
B(24)= +V(17);
B(25)= +X( 5) −X(10);
B(26)= −X(10) +V(19);
B(27)= +V( 1) +V(23);
B(28)= +V(12);
B(30)= +V( 1) +V(22);
B(31)= −V( 4) +V(11);
B(33)= +V( 1) −V(22);
B(34)= +V( 2) +V( 6);
B(35)= −X(10) +V(18);
B(36)= +V(23);
B( 5)= B( 3) −B( 4);
B( 8)= B( 6) −B( 7);
B(11)= B(10) −B( 9);
B(17)= B(15) −B(16);
B(20)= B(18) −B(19);
B(23)= B(22) +B(21);
B(29)= B(27) −B(28);
B(32)= B(30) −B(31);
```

The multiplication:

$$m_j = a_j \times b_j,\ j = 0, 1, 2, \ldots, 36$$

The transform of m (Output transform)
```
V( 0)= +C(30) +C(32);
V( 1)= +C(31) −C(32);
```

VLSI Signal Processing

```
V( 2)= +C(27) +C(29);
V( 3)= +C(28) -C(29);
V( 4)= +C(22) +C(23);
V( 5)= +C(21) +C(23);
V( 6)= +C(18) +C(20);
V( 7)= +C(19) -C(20);
V( 8)= +C(15) +C(17);
V( 9)= +C(16) -C(17);
V(10)= +C(10) +C(11);
V(11)= +C( 9) -C(11);
V(12)= +C( 6) +C( 8);
V(13)= +C( 7) -C( 8);
V(14)= +C( 3) +C( 5);
V(15)= +C( 4) -C( 5);
V(16)= +C( 2) +C(14);
V(17)= +C(14) -C(26);
V(18)= +V( 4) -V( 7);
V(19)= +V( 8) -V(18);
V(20)= +V( 5) +V( 6);
V(21)= +V(12) +V(14);
V(22)= +V( 0) +V( 2);
V(23)= +C( 0) +V(16);
V(24)= +C( 0) -V(17);
V(25)= +C( 1) +V(21);
V(26)= +C(13) +V( 8);
V(27)= +C(24) +V(19);
V(28)= +C(25) +V(22);
V(29)= +V( 4) +V( 7);
V(30)= +V( 5) -V( 6);
V(31)= +C( 0) -V(16);
V(32)= +C( 0) +V(17);
V(33)= +C(33) -C(35);
V(34)= +V(19) +V(20);
V(35)= +V(20) +V(26);
V(36)= +V(10) +V(11);
V(37)= +V(13) -V(15);
V(38)= +C(33) +C(34);
V(39)= +C(34) +C(35);
V(40)= +V( 1) -V( 3);
V(41)= +V( 9) +V(29);
V(42)= +V( 9) +V(30);
Y( 0)= +V(13) +V(23) +V(25) +V(35) +V(36);
Y( 1)= +C( 0) -C( 2) -C(12) -C(26)
       -C(36) +V(11) -V(25) -V(28) +V(33);
Y( 2)= +C(36) -V( 1) +V( 2) +V(24) -V(27) -V(39);
Y( 3)= +V( 0) +V(32) -V(38) +V(40) +V(42);
Y( 4)= -V(22) +V(24) -V(33) +V(34);
Y( 5)= +V(32) +V(39) -V(40) -V(41);
Y( 6)= +V( 1) +V(24) +V(28) -V(35) +V(38);
Y( 7)= +C(12) -V(10) -V(13) +V(14) +V(23) +V(27);
Y( 8)= +V(12) +V(31) -V(36) +V(37) -V(42);
Y( 9)= -V(11) -V(21) +V(23) -V(34);
Y(10)= +V(10) +V(31) -V(37) +V(41);
```

C11L146- N = 11, M = 46, Using the Iterative Method with Linear Polynomials

$A_I = 55, A_O = 90, A = 145.$

$$A$$

1	1	1	1	1	1	1	1	1	1	1
1	0	0	0	0	0	0	0	0	0	-1
1	-1	0	0	0	0	0	0	0	0	0
0	1	0	0	0	0	0	0	0	0	-1
0	0	1	0	0	0	0	0	0	0	-1
0	0	1	-1	0	0	0	0	0	0	0
0	0	0	1	0	0	0	0	0	0	-1
0	0	0	0	1	0	0	0	0	0	-1
0	0	0	0	1	-1	0	0	0	0	0
0	0	0	0	0	1	0	0	0	0	-1
0	0	0	0	0	0	1	0	0	0	-1
0	0	0	0	0	0	1	-1	0	0	0
0	0	0	0	0	0	0	1	0	0	-1
0	0	0	0	0	0	0	0	1	0	-1

```
0  0  0  0  0  0  0  1 -1  0
0  0  0  0  0  0  0  0  1 -1
1  0 -1  0  0  0  0  0  0  0
0  1  0 -1  0  0  0  0  0  0
1 -1 -1  1  0  0  0  0  0  0
1  0  0  0 -1  0  0  0  0  0
0  1  0  0  0 -1  0  0  0  0
1 -1  0 -1  1  0  0  0  0  0
1  0  0  0  0 -1  0  0  0  0
0  1  0  0  0  0 -1  0  0  0
1 -1  0  0  0 -1  1  0  0  0
1  0  0  0  0  0 -1  0  0  0
0  1  0  0  0  0  0 -1  0  0
1 -1  0  0  0  0 -1  1  0  0
0  0  1  0 -1  0  0  0  0  0
0  0  0  1  0 -1  0  0  0  0
0  0  1 -1 -1  1  0  0  0  0
0  0  1  0  0 -1  0  0  0  0
0  0  0  1  0  0 -1  0  0  0
0  0  1 -1  0 -1  1  0  0  0
0  0  1  0  0  0 -1  0  0  0
0  0  0  1  0  0  0 -1  0  0
0  0  1 -1  0  0 -1  1  0  0
0  0  0  0  1  0 -1  0  0  0
0  0  0  0  0  1  0 -1  0  0
0  0  0  0  1 -1 -1  1  0  0
0  0  0  0  1  0  0 -1  0  0
0  0  0  0  0  1  0  0 -1  0
0  0  0  0  1 -1  0 -1  1  0
0  0  0  0  0  0  1  0 -1  0
0  0  0  0  0  0  0  1  0 -1
0  0  0  0  0  0  1 -1 -1  1  0
```

11 *C*′

```
  1    1    1    1    1    1    1    1    1    1
  1  -10    1    1    1    1    1    1    1    1
  5    5   -6    5   -6    5   -6    5   -6    5   -6
-10    1    1    1    1    1    1    1    1    1
  1    1    1    1    1    1    1    1    1  -10
 -6    5   -6    5   -6    5   -6    5   -6    5    5
  1    1    1    1    1    1    1    1    1  -10    1
  1    1    1    1    1    1    1  -10    1    1
 -6    5   -6    5   -6    5   -6    5    5   -6    5
  1    1    1    1    1    1    1  -10    1    1    1
  1    1    1    1    1    1  -10    1    1    1    1
 -6    5   -6    5   -6    5    5   -6    5   -6    5
  1    1    1    1    1  -10    1    1    1    1    1
  1    1    1    1  -10    1    1    1    1    1    1
 -6    5   -6    5    5   -6    5   -6    5   -6    5
  1    1    1  -10    1    1    1    1    1    1    1
  2    2    2    2    2    2    2   -9   -9    2
  2    2    2    2    2    2   -9   -9    2    2
 -1   -1   -1   -1   -1   -1   -1   -1   10   -1   -1
  2    2    2    2    2   -9   -9    2    2    2
  2    2    2    2   -9   -9    2    2    2    2
 -1   -1   -1   -1   -1   -1   10   -1   -1   -1   -1
  2    2    2    2   -9   -9    2    2    2    2
  2    2    2   -9   -9    2    2    2    2    2
 -1   -1   -1   -1   10   -1   -1   -1   -1   -1
  2    2   -9   -9    2    2    2    2    2    2
  2   -9   -9    2    2    2    2    2    2    2
 -1   -1   10   -1   -1   -1   -1   -1   -1   -1
  2    2    2    2   -9   -9    2    2    2    2
  2    2    2   -9   -9    2    2    2    2    2
 -1   -1   -1   -1   10   -1   -1   -1   -1   -1
  2    2   -9   -9    2    2    2    2    2    2
  2   -9   -9    2    2    2    2    2    2    2
 -1   -1   10   -1   -1   -1   -1   -1   -1   -1
 -9   -9    2    2    2    2    2    2    2    2
 -9    2    2    2    2    2    2    2    2   -9
 10   -1   -1   -1   -1   -1   -1   -1   -1   -1
 -9   -9    2    2    2    2    2    2    2    2
 -9    2    2    2    2    2    2    2    2   -9
```

VLSI Signal Processing

```
10  -1  -1  -1  -1  -1  -1  -1  -1  -1  -1
 2   2   2   2   2   2   2   2   2  -9  -9
 2   2   2   2   2   2   2   2  -9  -9   2
-1  -1  -1  -1  -1  -1  -1  -1  -1  10  -1
 2   2   2   2   2   2   2  -9  -9   2   2
 2   2   2   2   2   2  -9  -9   2   2   2
-1  -1  -1  -1  -1  -1  -1  10  -1  -1  -1
```

The transform of x (Input transform)
```
B( 0)= +X( 0) +X( 1) +X( 2) +X( 3)
       +X( 4) +X( 5) +X( 6) +X( 7) +X( 8) +X( 9) +X(10);
B( 1)= +X( 0) -X(10);
B( 2)= +X( 0) -X( 1);
B( 4)= +X( 2) -X(10);
B( 5)= +X( 2) -X( 3);
B( 7)= +X( 4) -X(10);
B( 8)= +X( 4) -X( 5);
B(10)= +X( 6) -X(10);
B(11)= +X( 6) -X( 7);
B(13)= +X( 8) -X(10);
B(14)= +X( 8) -X( 9);
B(16)= +X( 0) -X( 2);
B(17)= +X( 1) -X( 3);
B(19)= +X( 0) -X( 4);
B(20)= +X( 1) -X( 5);
B(22)= +X( 0) -X( 6);
B(23)= +X( 1) -X( 7);
B(25)= +X( 0) -X( 8);
B(26)= +X( 1) -X( 9);
B(28)= +X( 2) -X( 4);
B(29)= +X( 3) -X( 5);
B(31)= +X( 2) -X( 6);
B(32)= +X( 3) -X( 7);
B(34)= +X( 2) -X( 8);
B(35)= +X( 3) -X( 9);
B(37)= +X( 4) -X( 6);
B(38)= +X( 5) -X( 7);
B(40)= +X( 4) -X( 8);
B(41)= +X( 5) -X( 9);
B(43)= +X( 6) -X( 8);
B(44)= +X( 7) -X( 9);
B( 3)= B( 1) -B( 2);
B( 6)= B( 4) -B( 5);
B( 9)= B( 7) -B( 8);
B(12)= B(10) -B(11);
B(15)= B(13) -B(14);
B(18)= B(16) -B(17);
B(21)= B(19) -B(20);
B(24)= B(22) -B(23);
B(27)= B(25) -B(26);
B(30)= B(28) -B(29);
B(33)= B(31) -B(32);
B(36)= B(34) -B(35);
B(39)= B(37) -B(38);
B(42)= B(40) -B(41);
B(45)= B(43) -B(44);
```

The multiplication:

$$m_j = a_j \times b_j, \; j = 0, 1, 2, \ldots, 45$$

The transform of m (Output transform)
```
V( 0)= +C(43) +C(45);
V( 1)= +C(44) -C(45);
V( 2)= +C(40) +C(42);
V( 3)= +C(41) -C(42);
V( 4)= +C(37) +C(39);
V( 5)= +C(38) -C(39);
V( 6)= +C(34) +C(36);
V( 7)= +C(35) -C(36);
V( 8)= +C(31) +C(33);
V( 9)= +C(32) -C(33);
```

```
V(10)= +C(28) +C(30);
V(11)= +C(29) -C(30);
V(12)= +C(25) +C(27);
V(13)= +C(26) -C(27);
V(14)= +C(22) +C(24);
V(15)= +C(23) -C(24);
V(16)= +C(19) +C(21);
V(17)= +C(20) -C(21);
V(18)= +C(16) +C(18);
V(19)= +C(17) -C(18);
V(20)= +C(13) +C(15);
V(21)= +C(14) -C(15);
V(22)= +C(10) +C(12);
V(23)= +C(11) -C(12);
V(24)= +C( 7) +C( 9);
V(25)= +C( 8) -C( 9);
V(26)= +C( 4) +C( 6);
V(27)= +C( 5) -C( 6);
V(28)= +C( 1) +C( 3);
V(29)= +C( 2) -C( 3);
Y( 0)= +C( 0) +V(12) +V(14) +V(16) +V(18) +V(28) +V(29);
Y( 1)= +C( 0) -V(20) -V(22) -V(24) -V(26) -V(28);
Y( 2)= +C( 0) -V( 1) -V( 3) -V( 7) -V(13) -V(21);
Y( 3)= +C( 0) -V( 0) -V( 2) -V( 6) -V(12) +V(20) +V(21);
Y( 4)= +C( 0) +V( 1) -V( 5) -V( 9) -V(15) -V(23);
Y( 5)= +C( 0) +V( 0) -V( 4) -V( 8) -V(14) +V(22) +V(23);
Y( 6)= +C( 0) +V( 3) +V( 5) -V(11) -V(17) -V(25);
Y( 7)= +C( 0) +V( 2) +V( 4) -V(10) -V(16) +V(24) +V(25);
Y( 8)= +C( 0) +V( 7) +V( 9) +V(11) -V(19) -V(27);
Y( 9)= +C( 0) +V( 6) +V( 8) +V(10) -V(18) +V(26) +V(27);
Y(10)= +C( 0) +V(13) +V(15) +V(17) +V(19) -V(29);
```

Part VII
DIGITAL SIGNAL PROCESSING, CAD

Edited by:
Dr. Richard Lyon
Fairchild Research Center

ALGIC - A FLEXIBLE SILICON COMPILER SYSTEM FOR DIGITAL SIGNAL PROCESSING

J. SCHUCK, M. GLESNER, H. JOEPEN

Technical University Darmstadt, FR Germany

ABSTRACT

This paper presents a flexible concept of a silicon compiler for the full custom realization of digital signal processing circuits. Essential features of the silicon compiler system are a flexible technology-independent design language, a strict separation of the silicon compiler software from technology and process dependent parameters, efficient layout generation of the signal path supported by a hierarchically structured cell design methodology and a powerful cell library.

1. INTRODUCTION

The appearance of VLSI-technology offers many advantages for digital signal processing (DSP) in telecommunication. Key elements are temperature stability, definite signal to noise ratio, any precesion of transfer functions, long time stability, insensitivity to process irregulation, etc..

For the implementation of filters and other signal processing functions there are important demands concerning computation rate, power consumption and silicon area. Therefore DSP circuits for field of special applications cannot always be realized by using general purpose processors. In these cases the layout of DSP circuits has to be implemented for custom oriented applications. The advantages of custom VLSI design, however, are restricted by the high costs caused by long time development and multiple design iterations. To reduce the design costs some interesting approaches for an automatic generation of the layout of DSP circuits have been presented in the past (Refs.1-4).

When implementing a silicon compiler system serious problems are involved concerning an efficient generation of the signal and control path. Examples for these problems are the specification of the architecture, the layout generation of DSP cells in the signal path, etc..

Overcoming these problems we have developed a flexible concept for a silicon compiler system with an efficient layout generation of customized circuits. This paper presents a silicon compiler for VLSI digital signal processing circuits.

2. FEATURES OF OUR SILICON COMPILER SYSTEM

A new concept for a flexible silicon compiler system called ALGIC (Automated Layout Generation for Integrated Circuits) has been developed as a powerful design environment for an efficient layout generation of custom VLSI circuits (Ref.5). The ALGIC-system is highly suitable for two different approaches:

a) A general approach for high throughput systems which are described in a behavioural manner by an universal block structured input language. In contrast to earlier approaches in silicon compilation we are not using a fixed target architecture. Instead of this an efficient architecture is constructed using data-flow analysis and optimization techniques (Ref.6).

b) A special approach, presented here, for a full custom realization of VLSI digital signal processing circuits. A survey of this part of the ALGIC silicon compiler system is shown in figure 1.
The design system processes a flexible technology-independent language which is suitable for a structural and functional description of signal and control path of a DSP circuit. A hierarchically structured cell design methodology supports an efficient generation of the signal path. Based on this cell design technique the cell layout of DSP functions can automatically be generated according to declared signal operation (e. g. multiplication, addition), to the specific circuit implementation (e. g. piplining) and to the wordwidth of signals to be processed. Thus a minimization of silicon area, a reduction of power consumption and an increase of computing rate can be achieved. Furthermore an increase of performance is supported by the component layout generators which generate the layout of controllers like PLA, finite-state-machine, Weinberger-array and multiplexer using complex algorithms to minimize the silicon area.
After general cell placement and routing with successive floorplan compaction the result will be an efficient geometric layout of a DSP circuit.

3. THE FLEXIBLE DESIGN LANGUAGE

The silicon compiler system processes a flexible technology-independent design language which is highly suitable for a structural and functional description of custom VLSI digital signal processing circuits. Among the interesting characteristics of the language are Pascal-like structures to facilitate the learning of the language, block structuring concepts to support a hierarchical dissection of the DSP circuit into smaller sub-systems by the user and the possibilities to declare DSP cells according to their specific circuit implementation and to the wordwidth of the signals to be processed.

3.1 The Declaration Part

All elements used in a program or procedure must be declared in the declaration part of the block to support type-checking mechanisms and plausibility checks. In the block declaration part we can differ between constants, ports, terminals, procedures and functions.
The CONSTANT declaration specifies integer constants. Constant lists declare the values of coefficients for digital filters.
The PORT declaration defines the interface to the outside of the DSP circuit or sub-system. The attributes IN, OUT and INOUT specify whether the signal is running to or from the circuit or in both directions. The signals are declared by their names and their wordwidth.
The TERMINAL declaration declares local single lines or vectors in the circuit. The declaration of PROCEDURE or FUNCTION supports a modularization and hierarchical dissection of the DSP circuit into small sub-circuits. Procedures and functions must be declared by using the keyword PROCEDURE or FUNCTION followed by the name of the subsystem and its parameters which specify the interface to the outside. After this declaration header the sub-circuit can be described in the procedure or function body.

Furthermore procedure and function mechanisms are highly suitable to specify certain DSP cells like multiplier, adder, arithmetic unit, etc. which are to be generated. The cell declaration uses the keyword PROCEDURE or FUNCTION

VLSI Signal Processing

followed by a cell operator and its parameters. Each cell operator specifies an arithmetic or logical function which is to be performed (e. g. addition, multiplication) and refers to a function cell in the cell library. The parameters of each cell operator define the specific circuit implementation (e.g. piplining) and the wordwidth of the signals to be processed.
E. g. a cell declaration is given by

FUNCTION Par_Mul (PORT IN A [0..7] , B [0..11]) : Product [0..19] ;
LIB ;

With this declaration a 8 x 12 bit parallel multiplier processing the two input vectors A and B is declared. The layout generation methods of the specified cell are outlined in chapter 4 and 6. The keyword LIB is assigned for a reference to the cell library.

3.2 Statements

The DSP circuit is described by using statements which define the functional interconnections of the signals. Two classes of statements can be used:

a) Simple statements for the structural description of the signal path (procedure, function and assignment statements).
   A procedure or function can be called by its name followed by a list of actual parameters corresponding to the formal parameter list. The actual parameters may not conflict with the PORT or TERMINAL attributes or indexranges of formal parameters. The result of a function must be assigned to a signal of type PORT or TERMINAL. E. g. the mechanism of a function call of a parallel multiplier described above is then as follows:

   Result := Par_Mul (Signal, Coefficient) ;

   The assignment statements denote signal definitions and connections. Examples for assignments are:

   A 0..7   := F;
   D 0,2,4 := B'101';
   K := B AND C;

b) Conditional statements for the functional description of the control path (IF..THEN..ELSE, ON..DO statements).
   The well-known IF..THEN..ELSE statement is a perfect mechanism for specifying the synchronisation of the system or for describing multiplexers or demultiplexers. The statement allows only the description of transmission gates. E. g. the statement IF a THEN b := c specifies a transmission gate with source c, drain b and gate a.
   As an example the description of a 4:1 multiplexer is given in figure 2.

```
IF a
  THEN
    IF b
      THEN
        Out := In4
      ELSE
        Out := In3
  ELSE
    IF b
      THEN
        Out := In2
      ELSE
        Out := In1;
```

Fig.2: Description of a 4:1 multiplexer

VLSI Signal Processing

The ON..DO statement can be used if more than two alternatives must be specified. The ON statement allowes the specification of a list of alternatives, each alternative marked by one or more ON-selectors. The mechanism of the statement works as follows:

```
ON list_of_operands DO
    BEGIN
        condition 1, condition 2,...condition i : action 1;
              .                          .                .
              .                          .                .
              .                          .                .
        condition j,..............condition k : action n;
    END
ELSE
    action (n+1);
```

The ON-statement is highly suitable for describing ROM, PLA and finite-state-machine (FSM). As an example the description of a FSM realisation of a RS-Flip Flop is shown in figure 3.

```
ON (R, S, State_one) /B DO
    BEGIN
        001, 011, 101, 111 : BEGIN
                                Q:=1;
                                QN:=1;
                             END;
        000, 010, 100, 110 : BEGIN
                                Q:=0;
                                QN:=1;
                             END;
    END;
ON (R, S, State_one) /B DO
        001, 010, 011, 110 : NEXT (State_one, Phi_1, Phi_2):=1
ELSE
        NEXT (State_one, Phi_1, Phi_2):=0;
```

Fig.3: Description of a FSM realisation of a RS-Flip Flop

The NEXT-clause of the ON-statement indicates the new state of state_one. The parameters Phi_1 and Phi_2 specify the clock cycle of the feedback register R (figure 4).

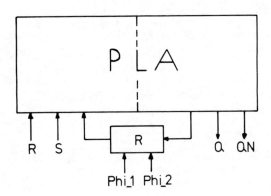

Fig.4: FSM realisation of a RS-Flip Flop

VLSI Signal Processing

The statements can be grouped in compound statements by enclosing them in BEGIN .... END blocks.

The scanner and parser for the language have been generated from the language specification in an extended Bacchus-Nauer-Form using a compiler generator tool (Ref.7).

4. THE HIERARCHICALLY STRUCTURED CELL DESIGN METHODOLOGY

In digital signal processing the wordwidth of signals to be processed varies according to the special application. Serious problems will arise when autocally generating a signal path of customized circuits.

All cells in a signal path should be available for any desired wordwidth and special application to achieve a minimum of silicon area, a reduction of power consumption and an increase of computing rate. Furthermore the inputs, outputs and the VDD and GND lines have to be positioned with respect to a regular bus structure and architecture of the signal path.

To meet these demands a flexible silicon compiler system has been developed its important feature being a hierarchically structured cell design methodology (Ref.8). Based on this cell design methodology complex cells like adder, multiplier, arithmetic unit can automatically be generated for desired wordwidth of the signal and for special application only by abutting simple handcrafted cells in a certain manner.

The cell design methodology uses a top-down technique for a regular and hierarchical dissection of complex cells into simple elementary cells. In this respect it is important to specify an efficient and regular architecture of the complex cell to achieve a hierarchical dissection into only few elementary cells.

Figure 5 demonstrates these hierarchical design methodology which consists of 5 hierarchy levels of cell representation. The hierarchy level 2 consists of elementary cells like 1 bit full adder, superbuffer, clockgenerator etc., which are fixed in size and function. Using a bottom up design one ore more of these elementary cells have to be grouped to sub-cells of level 3 to achieve a simple structure for the automatic generation of cells of a higher hierarchy. This must be done with respect to size and line connections to neighboured sub-cells. Parametrised cells of hierarchy level 4 consist of several sub-cells. In contrast to fixed cells, which are not variable and changeable in size and function, the floorplan of the parametrised cells is flexible according to special application and desired wordwidth. Therefore these cells can automatically be generated without using placement or routing techniques by abutting sub-cells according to simple construction rules. Parametrised cells can be grouped to macro-cells of the highest hierarchy level, which now realize complex functions like an arithmetic unit.

4.1 Parameterised Multipliers In Digital Signal Processing

The most complex DSP cell in high performance custom design circuits is the multiplier. Multipliers are required for bit-serial and bit-parallel processing, for any precions and speeds depending on the application. Other variables are the desired wordwidth of signals, piplining processing, rounding or truncation of the product. Based on the hierarchical cell design methodology parameterised multipliers for the automatic layout generation have been designed for various applications (Ref.9):

a) For bit-serial processing a parameterised fully piplined bit-serial multiplier has been developed. Its most important features are fully pipelined processing, bit-parallel input of the coefficient, bit-serial processing of signal, wordwidth n of signal independent of wordwidth k of the

VLSI Signal Processing

coefficient (n=k, n<k, n>k), rounding of the product during processing, result and input signal represented in two's complement, in-to-out delay equal (k+1) clock cycles, computation of the product in n clock cycles (Ref.10).

b) For bit-parallel processing a fixed coefficient multiplier, a pipelined guild array multiplier (Ref.11) and a bit-parallel multiplier have been designed as parameterised cells:

- The first one multiplies a two's complement input signal with a fixed coefficient by adding and hardware-shifting the input signal. CSD coding is used to reduce the number of additions.

- The most important features of the parameterised piplined guild array multiplier are a very efficient floorplan for the realization of bit-parallel pipelining multiplication, two's complement representation of signal and coefficient, various amount of pipelining (latch layers) in the array, optionally addition of other functions (e. g. pipelined adder) and a small number of elementary cells.

- The last one, the two's complement bit-parallel multiplier, has been developed with respect to the following demands: Selection of an efficient multiplication algorithm for two's complement representation of input vectors (Ref.12), selection of a suitable type of parallel array multiplier (Ref.13-16), definition of an efficient architecture and a minimization of the number of elementary cells.
As an example figure 6 shows the building block of a 6 x 4 bit-parallel multiplier. It consists of only 5 elementary cells (exor-gate, super-buffer, half- and full-adder,and-gate) which are assembled to the sub-cells PMi and PMKj (1≤i,j≤12) to achieve an efficient parameterised multiplier structure for a simple layout generation. The sub-cell representation is shown in figure 7. The general construction rules for the automatic layout generation of a n x k bit-parallel multiplier can be described line by line (Figure 8).

```
PMK1        (n-4)PM2                                    PM3
- - - - - - - - - - - - - - - - - - - - - - - - - - - - - - -
        for i = 1 to k - 3 do
PM4         (i-1)PM5      PMK5       (n-i-4)PM5         PM6
P̄M̄7̄ - - - -(k̄-3)P̄M̄8̄ - - - P̄M̄K̄1̄0̄ - - (n̄-k̄-1̄)P̄M̄8̄ - - - P̄M̄K̄X̄
```

Fig.8: General construction rules of a n x k bit parallel multiplier

By abbutting the sub-cells according to the construction rules the layout of two's complement bit-parallel multipliers can automatically be generated for any desired wordwidth of input signals without using placement and routing techniques (Ref.10). There is no restriction with regard to wordwidth of multiplicand or multiplier (n=k, n≠k; n,k=wordwidth of the two input signals). Additionally the product can be rounded to the wordwidth n or k respectively. After generation the final layout will be a good result with regard to silicon area, power consumption and computing rate.

5. CELL LIBRARY AND PROCESS DEFINITION FILE

One essential idea of the ALGIC-system is based on the fact that the geometric layout of the elementary or sub-cells and all parameters depending on the technology and actual process are strictly seperated from the ALGIC-compiler tools like the cell layout generation, verification and floorplanning tools. This seperation has the evident advantage that new designs or redesigns of cells and any change of the electrical process parameters, geometric design rules etc., does not lead to a modification and recompiling of the implemented

software. The separation has been realized by the
- implementation of a process definition file (PDF)
- implementation of a powerful cell library.

In the process definition file all necessary parameters of the actual process are stored like e. g. the number of mask levels and the geometric design rules. The floorplanning tools and the timing verifier have access to the PDF.

In the cell library (figure 9) the whole geometric and electrical data of cells of all hierarchy levels are stored to provide the cell generation and verification tools with the necessary information. In the cell library we can distinguish between 4 types of data:

- geometric layout of the elementary and sub-cells
- construction rules for the generation of the parameterised and macro-cells
- electrical properties of the cells
- cell organisation

The geometric layout is represented in a simple CIF-like format. The construction rules for the automatic generation of the parameterised and macrocells are described using the SELLAV-language (Ref.17). Part of the electrical properties of cells are data on the delay time of the cell, power consumption, output drive capability, input capacitance etc..

The cell organisation includes cell type, size, pin connection, clocking etc..

All the cell data are available to the cell operator handler, the component layout generators, the service routines and the timing verifier. All these tools communicate with the cell library via the library interface.

6. THE LAYOUT GENERATION PROCEDURE

Starting from the specification of the desired signal processing function in the input language the cell operator handler receives the following information:

- name of the cell operator
- parameters for the definition of the interface to the outside
- parameters for the definition of the wordlength of the signal to be processed
- parameters for the definition of the specific circuit implementation (e.g. pipelining)

The cell operator specifies the declared signal operation (e.g. addition, multiplication) and refers to a functional cell in the cell library. Via the library interface the cell operator handler provides the cell library with the cell operator and its parameters. As a response the cell operator handler receives from the library the following informations:

- type of the function cell (elementary, parameterised or macro-cell)
- construction rules for the automatic layout generation of the parameterise and macro-cells
- the geometric layout of the elementary or sub-cells
- cell organisation (size, pin connection positions and layer of the I/O, process reference etc.)

In contrast to the cell operator handler, which generates the cell layout only by abutting simple handcrafted cells the component layout generators use complex algorithms for the layout generation to increase performance and to minimize the silicon area. E. G., the PLA/FSM generator uses advanced column and row folding techniques for the minimization of the silicon area. This tool only needs a certain number of elementary cells (input and output registers, pull-up and pull-down transistors, contacts etc.) which are stored in the library.

7. THE TIMING ANALYSIS OF THE DSP SYSTEM

The timing analysis of the DSP circuit supports the user in locating problem timing in the clocked system. The timing verifier which is under development will be able to validate the path delays (primary input or storage element to primary output or storage element). It is based on a block-oriented algorithm which leads to running time essentially proportional to the number of blocks (cells). Instead of worst-case technique which usually leads to unrealistic worst-case criteria the path delays will be calculated by using formalized statistical methods (Ref. 18-19).

To calculate the path delays in the DSP system the following basic data are needed:
- A complete netlist of the system
  The netlist is represented by a graph, in which nodes indicate cells and edges specify the signal connections between the cells.
- The geometric position of all cells (available after general cell placement)
- Shortest, typical and longest in-to-out delay of each cell
- Optionally the meantime delay m and variance σ of all cells
  The meantime delay and variance can be calculated by timing simulations using a large number of input patterns.
- The input capacitance of all cell inputs
- The output drive capability of all cell outputs
- The specific resistance R and capacitance C of all layers (metall, poly, diffusion)

If the meantime delay m and variance σ of a cell is unknown they must be calculated. The values of m and σ can be approximated by the following formulae (Ref. 20):

$$m = (\text{shortest delay} + 4 \times \text{typical delay} + \text{longest delay}) / 6 \quad (1)$$

$$\sigma^2 = (\frac{\text{longest delay} - \text{shortest delay}}{6})^2 \quad (2)$$

Additionally the path delays of all line connections between the cells must be involved in the timing analysis. The solution of this problem is as follows:
- Computation of the Manhattan distances between cell outputs and the corresponding cell inputs
- Substitution of each line by a simple RC network
- Extension of each RC network by the input capacitance C_{IN} of the cell input and by a linear resistance R_{OUT} which approximates the non-linear pull-up or pull-down of the cell output
- Computation of the meantime delay m_L and variance σ_L of all lines
- Substitution of all lines in the netlist by pseudo-cells. The meantime delay

VLSI Signal Processing

m_L and variance σ_L of each line are assigned to the corresponding pseudo-cells.

Under the assumption that the delay distribution of all cells and pseudo-cells are Gaussian the timing verifier calculates now the expected or most propable delays, the critical timing paths and the timing slacks.

8. THE FLOORPLANNING TOOLS

After running the layout generation procedure the floorplanning tools will create the final layout of the DSP circuit.

The placement sub-system is qualified for general cell assemblies combining the advantages of constructive and mincut methods. The cell-set is devided into groups according to some constraints defined by the user or the ALGIC-system respectively, e. g. netweights. After placing the cells inside a group the groups themselves are placed. Finally the overall placement is improved by rotating and squeezing single cells. The placement process can be controlled by the user or system according to a set of parameters.

After routing a floorplan compaction will minimize the silicon area. During compaction the geometric layout and the size of the cells are not changed. The process definition file (PDF) provides the compactor with the geometric design rules. The result is a correct and compact layout of the DSP circuit.

9. CONCLUSION

A flexible concept of a silicon compiler system has been presented. At the moment important parts of the silicon compiler are completed, e. g. scanner and parser, PLA/FSM generator, library interface, first version of the floorplan compactor, design of parameterised cells like adder, multipliers etc.. These components have to be integrated in the system in the next future. Other components like general cell placement, routing and timing verification are under development.

To test the functions and to check the specifications automatic generated cells like Dual-Port RAM, two's complement bit-parallel multiplier, bit-serial fully pipelined multiplier and Finite-State-Machine have been fabricated on the 5 μm NMOS-Technology line of the Technical University Darmstadt.

This investigation is partly supported by the European Economic Community within the project "CAD-VLSI for Telecommunication".

10. REFERENCES

/1/ J.Siskind, J.R. Southard, K. W. Crouch:
   "Generating Custom High Performance VLSI-Designs from Succinct Algorithmic Description", Proc. Conference on Advanced Research in VLSI, MIT, January 1982, pp. 28-40

/2/ P. Denyer, D. Renshaw, N. Bergmann:
   "A Silicon Compiler for VLSI Signal Processors", ESSCIRC Proceedings, Brussels, 1982, pp. 215-218

/3/ D. Gajski:
   "The Structure of a Silicon Compiler", IEEE Proceedings on Design Automation, 1982, pp. 272-276

/4/ D. L. Johannsen:
   "Bristle Blocks: A Silicon Compiler", Proc. 16th Design Automation Conference, June 1979, pp. 310-313

/5/ M. Glesner, J. Schuck, H. Joepen:
"A Flexible Silicon Compiler for Digital Signal Processing Circuits", ICCD, October 1984, New York

/6/ H. Joepen, M. Glesner:
"Preliminary ALGIC-Language Reference and Report", CVT-Report, THD, Institut für Halbleitertechnik, February 1984

/7/ B. Kreling, A. Kappas:
"TASD-Toolmaker's Aid for Syntax Development", Reference Manual, Institut für Datentechnik, 1983

/8/ J. Schuck, M. Glesner:
"Custom Design and Synthesis of VLSI Digital Signal Processing Circuits", CVT-Report, THD, Institut für Halbleitertechnik, July 1983

/9/ J. Schuck, M. Glesner:
"Layout Generation for Multipliers in VLSI Digital Signal Processing", ESSCIRC, September 1984, Edinburgh

/10/ S. Meier, H. Schlappner:
"Entwurf von Multiplizierern für die Digitale Signalverarbeitung", Studienarbeit THD, Institut für Halbleitertechnik, December 1983

/11/ H. H. Guild:
"Fully Iterative Fast Array for Binary Multiplication and Addition", Electron. Lett., Vol 5, June 1969, p. 263

/12/ C. R. Baugh, B. A. Wooley:
"A Two's Complement Parallel Array Multiplication Algorithm", IEEE Transactions on Computers, Vol C-22, pp. 1045-1047, 1973

/13/ C. S. Wallace:
"A Suggestion for a Fast Multiplier", IEEE Transactions on Electronic Computers, Vol. EC-13, pp. 14-17, 1964

/14/ A. D. Booth:
"A Signed Multiplication Technique", Quart. Journal of Mech. Appl. Math., Vol 4, pp. 236-240, 1951

/15/ S. D. Pezaris:
"A 4ons 17-bit Array Multiplier", IEEE Transactions on Computers, Vol C-20, pp. 442-447, April 1971

/16/ N. F. Benschop:
"Layout Compiler for Variable Array-Multipliers", IEEE, Custom Integrated Circuits Conference, 1983, pp. 336-339

/17/ G. Schäfer, M. Glesner:
"Sellav Report", CVT-Report, THD, Institut für Halbleitertechnik, February 1984

/18/ R. B. Hitchcock, G. L. Smith, D. L. Cheng:
"Timing Analysis of Computer Hardware", IBM I. Res. Develop., Vol 26, No. 1, January 1982, pp. 100-105

/19/ R. B. Hitchcock:
"Timing Verification and the Timing Analysis Program", 19th Design Automation Conference, 1982, pp. 594-604

/20/ T. I. Kirkpatrick, N. R. Clark:
"PERT as an Aid to Logic Design", IBM Journal, March 1966, pp. 135-141

VLSI Signal Processing

Author's Biographies:

Johannes Schuck was born in Würzburg, FR Gremany, in 1955. He received the Diplom in electrical engineering from the Technical University Munich in 1982. Since 1982 he is research assistant of the Technical University Darmstadt and is working on full custom design of DSP circuits and on silicon compilation for DSP.

Manfred Glesner was born in Saalouis, FR Gremany, in 1943. He received the Diplom in physics and electrical engineering in 1968 and his Dr.rer.nat. degree in 1975, both from the Saarland University. From 1976 to 1980 he was lecturer at the Saarland University. Since 1980 he is full professor of electrical engineering at the Technical University Darmstadt, where he is currently working on the development of CAD-tools for VLSI custom design, silicon compilation techniques, full custom design of DSP circuits and switch-level simulation.

Horst Joepen was born in Bad Homburg, FR Gremany, in 1957. He received the Diplom in electrical engineering from the Technical University Darmstadt in 1982. Since 1983 he is research assistant of the Technical University Darmstadt and is working on silicon compilation for highly throughput digital systems.

Mailing address of authors:

Technical University Darmstadt
Institut für Halbleitertechnik
Fachgebiet Halbleiterschaltungstechnik
Schloßgartenstr. 8
D 6100 Darmstadt, FR Germany

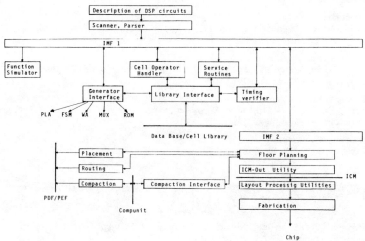

Fig.1: The Silicon Compilation System for Digital Signal Processing

VLSI Signal Processing

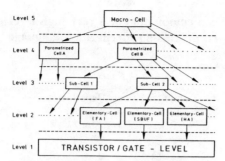

Fig.5: Hierarchy Levels of Cells

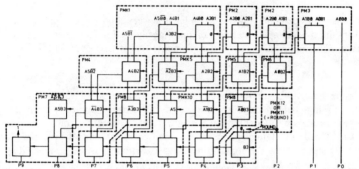

Fig.6.: Building Block of a Two's Complement 6 x 4 Bit Parallel Multiplier

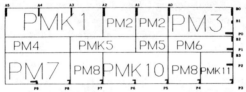

Fig.7: Sub-Cell Representation of a Two's Complement 6 x 4 Bit Parallel Multiplier

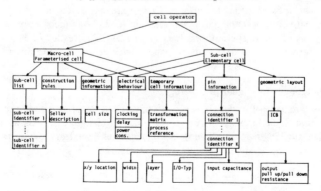

Fig.9: Structure of the Cell Library

25 Silicon compilation of a very high level signal processing specification language

Mark Kahrs†

Computer Science Department
University of Rochester
Rochester, NY 14627

Abstract:

SILI is a language independent silicon compiler for very high level languages. CLASP is a very high level specification language designed for expressing signal processing tasks. This paper discusses the implementation of CLASP in SILI.

1. Introduction

Signal processing presents a unique challenge to the designer of VLSI design tools. First, signal processing lags behind other areas in the use of existing compiler technology. Second, digital signal processing algorithms are "conceptually different" from many other programming tasks. Third, the languages presently used in digital signal processing are algorithmic; they fully specify how the signal processing algorithm is to be performed.

This paper presents a different perspective on how to create VLSI circuits from programs. First, the input language CLASP[1] is a *specification language,* i.e., it specifies what the chip *must do* rather than *how* it does it. One definition of the difference between specification languages and implementation languages is as follows:

> "implementation languages are for describing efficient algorithms, and specification languages are for describing behaviors." (Ref. 1) (pp. 91)

The task of a specification language compiler is to try to choose the most efficient implementation for a given behavior. The second difference is that SILI[2], the compiling system, attempts to construct a machine out of predefined library modules.

This paper will begin by reviewing some past work in programming languages for signal processing. The most unusual features of CLASP are described in the subsequent section. Next, the SILI silicon compiler system is outlined. A much more complete description of SILI is available in the author's thesis (Ref. 2). The results of SILI's implementation of CLASP are presented next and the paper concludes with some directions for further research.

2. Programming languages for signal processing

Special purpose languages for signal processing have been developed over the last 20 or so years. The first such languages were block diagram languages such as BLODI (Ref. 3). Block diagram languages can handle a great deal of signal processing tasks but lack the ability to express "system" level tasks such as the Fast Fourier Transform.

---

† This work was performed at the University of Rochester. Author's present address: AT&T Bell Laboratories, 600 Mountain Avenue, Murray Hill, NJ 07974

[1] Complex Language for Attacking Signal Processing

[2] Short for SILIcon

However, recently there has been work in taking some of the more current programming language concepts such as data abstraction and object oriented programming and creating new languages based on these concepts. SIPROL (Ref. 4) and SPL (Ref. 5) are examples of this trend. However, these languages are not specification languages since they dictate *exactly* how the algorithm is to be performed. For example, the code for a second order filter section would explicitly state the computation to be performed. This level of description is acceptable to someone who knows the design of the algorithm down to the lowest level. However, it is possible to describe these algorithms at a higher level; that is, to express the behavior of the algorithm, not the algorithm itself. In particular, filters have many performance criteria such as bandwidth, Q and noise bounds. A signal processing specification language could use these specifications to help choose implementations much the way a designer of a signal processing circuit would use these specifications when designing a circuit. CLASP uses performance specifications to describe the filters and other operators. It is the task of systems like SILI to make the actual implementation selection.

3. CLASP

This section will give a short overview of CLASP. The salient features of CLASP include very high level types, filters, nonlinear operators and a time specific looping construct. These will be detailed in the sections that follow. A full specification of CLASP can be found in Ref. 6.

3.1. Types

CLASP has very high level types that hide implementations from the user of CLASP. In particular, CLASP has both sets and tuples. Tuples are unordered multisets of base types (integers, floats, complex and characters). Sets and tuples can be nested to any level, i.e., sets of tuples of sets of integers. Sets and tuples also have properties that can be specified in the type declaration. These declarations include the range of the base type and the maximum size of the sets and tuples. Note that these declarations (or *specifications*) are a replacement for detailed program analysis that could've been performed (had it been implemented) before selection.

3.2. Functions

CLASP uses function calling as a syntactic device to describe connections between various modules. The connections between operators (expressed as functions) are derived by the data flow analysis procedure. Nonlinear operators such as rectification can be expressed by writing the operators as functions. Note that this method of denoting connections (implicitly by function calling) is different from the style of SIGNAL (Ref. 7) where ports are bound explicitly by name.

3.2.1. Filters

Filters are such ubiquitous elements in signal processing that they form an integral part of CLASP. Filters are expressed directly in the syntax of CLASP as either expressions or in variable declarations. For example, a low pass filter at A440 with a a stopband attenuation of 56 db could be written in an expression (shown here in an assignment) as:

```
... := filter input from DC to 440 with stopband attenuation of 56 db;
```

or it could be declared like a variable as:

```
declare filter from DC to 440 with stopband attenuation of 56 db : a440filter;
```

and then used further on in an expression by name (in this case, a440filter):

VLSI Signal Processing

```
... := a440filter(input);
```

Of course, since filters are functions, filters can be nested to any level (a filter of a filter of a filter ...). This nesting forms a cascade of filters. In fact, the nesting calling of functions produces a cascade of functions. The filter parameters in the declaration form the specification of the filter used by the compiler. These parameters are used by the selection phase to choose the appropriate filter and the appropriate filter order. CLASP currently permits the following specifications to be attached to any filter declaration or use:

```
              stopband attenuation of ... db [down]
              passband attenuation of ... db [down]
                        Q of ...
```

3.2.2. Nonlinear operators

Nonlinear operators (such as a halfwave rectification and limiting) can be directly expressed as functions. Therefore, a rectified limited (at `limitpoint`) with a filtered input would be written:

```
              HWR(limit(filter ..., limitpoint))
```

Functions can have any number of inputs but currently only one output. This limits the expressibility of CLASP but the range of available functions is still high.

3.3. Time specific looping

"Vanilla" looping constructs are not sufficiently expressive for the signal processing domain. To be more specific, consider the case when a new sample must be (re)computed within a certain timeframe (specifically the reciprocal of the sampling frequency). For this specific case, CLASP has a new iteration construct every (the old "vanilla" ones are included too). The every construct specifies that the loop body must be performed within the time limit. Note that this is an infinite loop and it does *not* have an exit. For example, if the sampling rate is 20 kilosamples per second, then the specification would be written:

```
              every 50 microseconds do ...
```

This timing restriction dictates that the length of the microcode store multipled by the microcycle time must not exceed this bound. This is a *global resource bound* (global constraint) and can be used to call optimization operators.

4. A silicon compiler for very high level languages

4.1. SILI

SILI is a table driven language independent silicon compiler. The philosophy of SILI is simple: data flow → data path, control flow → control section. This philosophy generates a classical horizontally microprogrammed machine. The "generic" machine generated by SILI is illustrated on the next page.

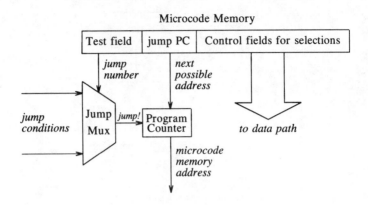

Figure 1. Generic machine architecture

The data path of this machine is constructed from the library modules by the data flow analysis routines. The function calling style of CLASP is converted by the flow analysis routines into a cascade of operators. The jump conditions are output from various tests (such as the result of if statement boolean expressions).

4.2. SILI implementation

SILI is organized along the following lines of the figure shown on the next page. The input program, written in the appropriate language (CLASP in this case), are given to the **scanner** and the **parser.** The parser output (a parse tree) is used by the flow analysis routines to create the flow graphs. The output from the data flow analysis phase is a directed (*not* acyclic) graph of nodes. These nodes correspond to the variables and operators in the input program. The output from the control flow analysis phase is a similar graph of nodes that represent the control flow of the input program, i.e., one node per statement plus the additional "overhead" nodes. The **control flow compression** phase compresses the control flow graph by eliminating branches to branches and the overhead nodes. The data flow graph is given to the **matcher** which matches the data flow subgraphs in the library (written in the same language as the data flow analysis routines) with the data flow graph. The results of the matcher are used by the **binder** to assign values to the instances of the parameterized cells in the library. The output from the binder is used by the **search** phase to select implementations. After selections of parts have been made, then the **control unit generator** takes over and creates the appropriate control unit for the compressed control flow graph.

4.3. CLASP in SILI

The files needed for the specification of a language in the SILI system are shown in the previous figure (the labels on the dashed lines are the names of input specifications). Given these data files, the SILI system will accept a program written in CLASP and generate the generic machine mentioned above. Space does not permit the inclusion of the

VLSI Signal Processing

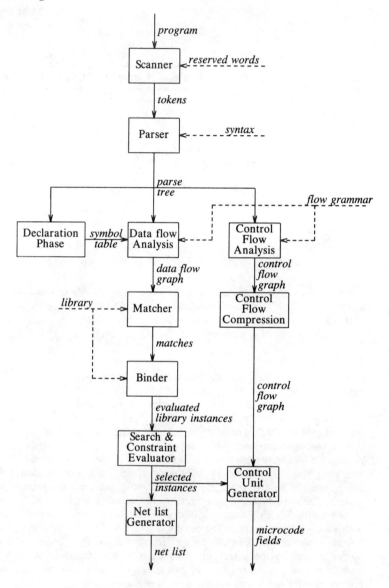

Figure 2. System block diagram

full specification of CLASP in this paper, therefore only a sample of the full library specification can be found in Appendix A.

4.4. An example : Touch Tone Receiver

Consider the touch tone receiver (decoder) of Jackson, et. al. (Ref. 8). The CLASP program that implements this decoder is given in Appendix B. The data path derived by SILI's data flow analysis is shown in Appendix C. Notice how the filters are not shared and how the bandpass filter bounds are fed to the filter blocks (the implicit assumption being that frequencies are fed to the filters, not coefficients).

4.5. Practical Experience with CLASP in SILI

CLASP was implemented within the SILI system which in turn was implemented in Lisp. The CLASP library was derived from Lyon's NMOS library (contained in Ref. 9) and included basic second order filter sections along with both serial and parallel arithmetic operators.

Unfortunately, the circuits generated by SILI are quite large. This is due to the lack of busing and functional unit sharing. This is directly attributable to the lack of optimization operators. Of course, this is not an insurmountable problem; it can be solved by either adding an optimizer or by calling *critics* (specialized bodies of code) to solve the resource allocation problems when the resource bounds (global constraints) are exceeded.

Another problem is CLASP specific. When a filter is selected, the filter coefficients must be computed. There is presently no mechanism for performing these calculations within the confines of SILI. This is because SILI doesn't really have a sophisticated mechanism for dealing with types. A similar problem with SILI concerns filter banks. Take the following CLASP specification:

```
foreach lowerBound in [ 440, 880, 1320 ] do
    filter input from DC to lowerBound;
```

This is a straightforward low pass filter bank. The obvious implementation is a tuple of integers "feeding" the lower bound input of a filter. However, this implementation is not reasonable. In particular, the frequencies of the tuple must be converted to filter coefficients in order for the implementation to be reasonable. This problem also arises in the case of a variable filter, i.e.,

```
filter input from lower to upper do ...
```

where lower and upper are variables. Here, the variable values must be converted into coefficients. The common problem with these two examples is that since filter coefficient calculation is extremely difficult, it is an unreasonable strategy to convert integer frequencies to coefficients in hardware because the calculations are quite complex. The first case could be resolved if SILI was aware of the conversion of integer filter bounds to coefficients at compile time. The resolution of the second case depends on adding type inference mechanisms.

This version of CLASP also had syntactic support for the FFT. This permitted the following syntactic construct:

```
... := transform input from frequency domain to time domain
```

However nice this "syntactic sugar" looks, in fact this detracts from the stated purpose of CLASP and SILI: to remove implementation knowledge from the language. Perhaps a more reasonable way to *imply* the use of transforms is through the introduction of types that express time and frequency domains. Then the use of a transform is really the same as a type coercion. This remains to be explored.

VLSI Signal Processing

5. Future work

Obviously, there are two areas for improvement: SILI and CLASP. SILI does not implement common subexpression elimination, although such an optimization is well defined and the procedure is well known. Application of common subexpression elimination as well as other optimization operators would result in much more reasonable circuits. Besides optimization, the problems concerning the conversion of filter coefficients must be resolved before the circuits could be considered acceptable to a human designer.

6. Conclusion

Silicon compilation of very high level specification languages like CLASP in SILI has demonstrated that it is possible not only to define reasonable signal processing function within a specification language but also to generate semantically plausible implementations. Hopefully, this work has demonstrated that signal processing functions can be described at a high enough level to make their use by non-experts (and experts) fast and painless.

7. Acknowledgements

This work was partially supported in part by NSF grants IST-8012418, MCS-8104008 and DARPA grant N00014-78-C-0164.

8. Bibliography

1. P. E. London and M. S. Feather, Specification of Implementation freedoms, *Science of Computer Programming 2* ,2 (Nov. 1982), 91-112.

2. M. W. Kahrs, *Silicon Compilation of Very High Level Languages*, PhD thesis, University of Rochester, October 1983.

3. B. J. Karafin, A New Block Diagram Compiler for SImulation of Sampled-Data Systems, *FJCC 1965*, 1965, 55-61.

4. H. Gethoeffer, SIPROL : A High Level Language for Digital Signal Processing, *ICASSP*, 1980, 1056-1059.

5. G. E. Kopec, *The representation of Discrete-time signals and systems in programs*, PhD thesis, MIT, May 1980.

6. M. Kahrs, CLASP - A very high level signal processing specification language, (In preparation), 1984.

7. P. L. Guernic, A. Nenveniste and T. Gautier, SIGNAL: Un language pure le traitement du signal, INRIA Rapport No. 206, May 1983.

8. L. B. Jackson, J. F. Kaiser and H. S. McDonald, An Approach to the Implementation of Digital Filters, *IEEE Transactions on Audio and Electroacoustics AU-16* ,3 (Sep. 1968), 413-421.

9. J. A. Mathews and R. G. Newkirk, *The VLSI Designers Library*, Addison Wesley, Reading, MA, 1983.

VLSI Signal Processing

9. Appendix A: Sample library definition

The following is a sample definition of a serial butterworth filter with variable filter bounds.

```
(SerialVariableButterworthFilter
            (variable butterworth-order)
            (inputs (inputStream 1)
                    (lowerFrequency 1)
                    (upperFrequency 1))
            (outputs (outputStream 1))
            (control (control 1))
            (others resident)
            (properties (inputStream serial)
                        (outputStream serial)
                        (lowerFrequency serial)
                        (upperFrequency serial))
            (area (width (times butterworth-order 100))
                  (height 200)
                  (overhead 0))
            (time (delay (lookup delay)) (period))
            (power (times 200 width))
            (parts (basis (control (control 0))
                          (graph
                            (join (node FILTER)
                                  (port upperFrequency
                                        (node ANY))
                                  (port lowerFrequency
                                        (node ANY))
                                  (port resident
                                        (node IDENTIFIER
                                                FILTER))))
                          (timing
                            (delay butterworth-order))
                          (bind
                            (butterworth-order resident
                                          stopband-attenuation
                                          butterworth-filter-order)))
                   (input (control (control 0))
                          (graph
                            (attach-tail (port inputStream
                                               (node
                                                ANY))
                                         (port resident
                                               (node IDENTIFIER
                                                       FILTER))))
                          (timing
                            (delay butterworth-order)))
                   (use (control (control 1))
                        (graph
                          (attach-head (port resident
                                             (node IDENTIFIER
```

235

VLSI Signal Processing

```
                                            FILTER))
                                (port outputStream
                                    (node
                                        ANY))))
                            (timing (delay 1)))))
```

There are two interestings points:

(1) The binding of the filter parameter `filter-order` uses the function `butterworth-filter-order` defined within SILI. This function computes the filter order required to satisfy the the specification of the filter.

(2) The ports of the filter are declared as input and/or output along with the width of the signal in bits. Properties of the filter ports are declared by naming the port and the properties associated with the port.

10. Appendix B: Example details

```
module TouchToneDecoder

    -- The now classic touch tone decoder, as done originally in 1963 by
    -- a group in Bell, then done again in 1968 by Jackson, et al and
    -- done again by Lyon.

    declare tuple of integer : lowerBand, upperBand;
    declare tuple of integer : lowerBandCenterFrequencies;
    declare tuple of integer : upperBandCenterFrequencies;
    declare tuple of integer : detection;
    declare integer : result; -- output from hum filter (and iteration variable)
    declare integer : bandLimit; -- bandpass (and lowpass) band limit
    declare integer : input; -- input from the A/D
    declare integer : output; -- output from the module
    declare filter from 180 to INFINITY : noHum; -- Line hum filter

    lowerBandCenterFrequencies := [ 697, 770, 852, 941 ] ;
    upperBandCenterFrequencies := [ 1209, 1336, 1447, 1663 ] ;

    result := noHum(input);
    lowerBand := filter result from DC to 1070 : lowerGroupFilter;
    upperBand := filter result from 1070 to INFINITY : upperGroupFilter;

    detection := phi;

    foreach centerFrequency in lowerBandCenterFrequencies do
        detection := detection plus
                    filter HalfWaveRectifier(
                        filter lowerBand
                            from centerFrequency-bandLimit
                            to centerFrequency+bandLimit
                                with Q of 15 and
                                with stopband attenuation of 16 db down
                            : lowerBandPass)
                        from DC
                        to centerFrequency+bandLimit
```

```
                    : detectLowGroup;

foreach centerFrequency in upperBandCenterFrequencies do
      detection := detection plus
                   filter HalfWaveRectifier(
                       filter upperBand
                          from centerFrequency-bandLimit
                          to centerFrequency+bandLimit
                              with Q of 15 and
                              with stopband attenuation of 16 db down
                              : upperBandPass)
                          from DC
                          to centerFrequency+bandLimit
                    : detectUpperGroup;

foreach result in detection do
      output := LevelDetect(result)

end.
```

VLSI Signal Processing

11. Appendix C: Data flow graph for the example

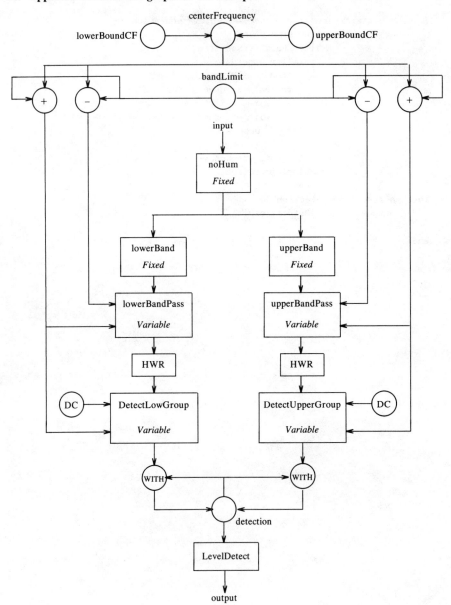

26

VLSI Signal Processing

Automated Design of Signal Processors Using Macrocells

Stephen Pope[†]
Jan Rabaey[‡]
Robert W. Brodersen
Department of Electrical Engineering and Computer Science
University of California, Berkeley CA 94720

Abstract: Design of semi-custom digital signal processing IC's can be performed rapidly and efficiently using a design system based on *macrocells*. Macrocells are large, parameterized blocks of circuitry assembled from a cell library. Software support for the macrocell system consists of an *emulator* and a two-pass *silicon compiler*.

Both the emulator and the first pass of the silicon compiler accept a *design file* (a text file prepared by the designer) as input. Emulation from the design file permits verification of circuit performance prior to fabrication.

The design file has a very readable format. Each IC contains several processors operating concurrently. The designer specifies hardware parameters and symbolic microinstructions for each processor. Local constants and variables are declared. Optionally, a user-defined finite state machine (FSM) is specified. The FSM is used for decision making tasks common to signal processing systems. Interprocessor and off-chip communications are also specified in the design file.

The first pass of the compiler extracts hardware parameters from the design file and assembles the symbolic microcode into binary. Bit-serial communications paths among the processors are set up. Communications between processors and off-chip hosts are also allowed, supported by an interrupt-acknowledge sequence. Signal input/output at the sample rate is supported over a separate interface. The output of the first pass is a hardware description which specifies the configuration of each macrocell and the interconnections among macrocells.

The second pass assembles the macrocells, accessing a cell library. A 3u NMOS cell library is currently in use, with a 3u CMOS cell library under development. Following macrocell assembly, placement and routing subroutines complete the compilation, generating a mask-level description of the IC.

A major architectural feature of the system is the use of multiple, bit-parallel processors. Each processor has an efficient single-accumulator organization. Dedicated data paths for control, address and signal data, along with the use of pipelining, further increase the level of concurrency.

The system is intended to be used in the design of dedicated DSP chips for speech, audio and telecommunications functions. Typical applications are speech coding, modems, adaptive equalizers and filter banks.

1. Introduction

There are several approaches to the design of dedicated integrated circuits for signal processing. Full-custom design methods have the greatest potential for efficient use of the technology, but the magnitude of design effort can be very large. Standard-cell or gate-array methods can shorten the design time, but with a considerable loss in efficiency. Another choice would be the use of a single-chip programmable signal processor such as the Texas Instruments TMS320 [1].

[†] Current Affiliation: Cyclotomics Corp., Berkeley CA 94704
[‡] On Leave from Catholic University of Leuven, Belgium

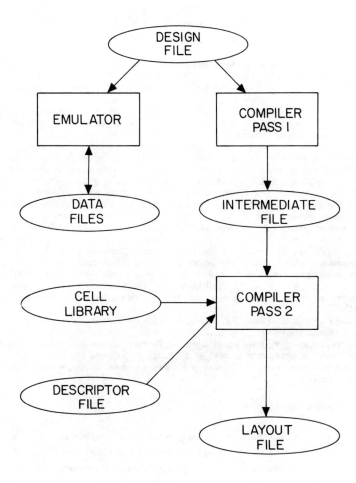

Fig.1 Software package for the macrocell system

VLSI Signal Processing

An alternative to all of these is the *silicon compiler* approach [2,3]. The design is specified procedurally, usually by providing a text file as input to the compiler. The compiler then generates a mask-level description of the circuit layout. This paper describes a compiler developed specifically for the purpose of generating signal processing IC's. Target applications are speech processing (vocoders, noise canceling, speech recognition), telecommunications (modems, line equalizers, echo cancelers), and digital audio (equalization, special effects).

An important aspect of the system is the use of *macrocells*. A macrocell is a large block of circuitry assembled by tiling (arraying) cells from a cell library in two dimensions. This approach is very flexible, allowing the size and function of each macrocell to be customized for the particular design.

The complete software system is shown in Fig.1. All software is written the the C programming language, and runs under the 4.2 BSD version of the UNIX operating system.

The input to the system is the *design file*. A design file for a typical application consists of a few pages of text. Because the silicon compiler generates multiprocessor IC's, the design file contains hardware options and symbolic microcode for each processor in the design. In addition, there is a section which describes interprocessor and off-chip communications.

The emulator enables the designer to debug and verify his design file input. In an interactive mode, the emulator has a full set of debugging commands: breakpoints, tracing, single-step, setting and displaying variables. A batch mode allows non-real-time simulation of the signal processor design. This capability is an important part of the system, since it verifies the algorithm and its implementation from the high-level design file input, rather than from the circuit layout or a lower level description, which would be less efficient.

The compiler is broken into two passes. The first pass extracts hardware parameters and assembles the microcode from the design file. The output of the first pass is an *intermediate file* which describes the macrocells at the level of the smaller library cells. A net list describes how the macrocells are interconnected. The intermediate file is human-readable to aid in program development.

The second pass of the compiler first assembles the macrocells, then interconnects them to form a mask-level description of the IC. In order to make the software technology independent, a special file (the descriptor file) contains dimensional information on the library cells and their terminals. A few additional parameters define the technology for the purposes of macrocell placement and signal routing. With the technology-specific information encapsulated in this way, the compiler may be used for any MOS process by redesigning only the cell library.

The cell library currently in use is for a 3-micron NMOS process, with a 3-micron CMOS cell library under development. The library contains a total of 160 cells. The cells are designed to operate at a 5 MHz maximum clock rate over all possible configurations of the macrocells.

2. Input-output and multiprocessor organization.

Signal processing IC's designed with the compiler will typically be used as peripheral chips to a host microprocessor. For this reason, two separate interfaces to the IC are provided (Fig.2).

Signal data is transferred each sample over a bidirectional signal data bus. Since the system clock rate is much higher then the signal sample rate, many such signals may be transferred each sample interval. These signals are the sampled-data inputs and outputs of the IC. Exactly one of the processors in the multiprocessor IC connects to this bus.

VLSI Signal Processing

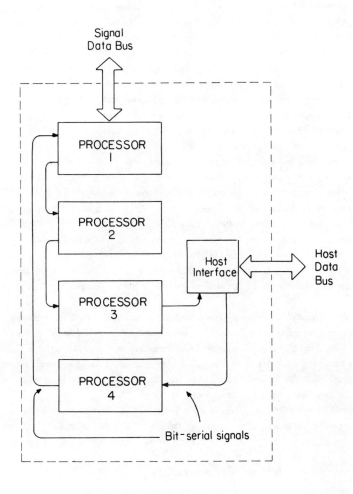

Fig.2 Organization of a four-processor IC

VLSI Signal Processing

A separate host data bus connects the IC to its host microprocessor. Transfers over this bus are performed by the host in response to an interrupt from the signal processing IC. In general, a frame rate is defined, where a frame is an interval many samples in length. A block of data is transferred over the host data bus once each frame. This data is buffered in the IC in a host interface macrocell. Including the host interface hardware in the signal processing IC reduces the hardware costs of systems using the IC.

As shown in Fig.2, the IC contains some number of processors (from one up to perhaps four) and a host interface. The processors themselves contain several different macrocells as described in Section 3. An important aspect of the design of multiprocessor signal processor IC's is that the IC is dedicated to a single algorithm, and each processor is dedicated to a specific part of the algorithm. It is often natural to decompose a signal processing algorithm in this fashion. Because of the flexibility of the macrocell approach, the individual processors can be customized for their particular task.

Interprocessor communication is conducted over bit-serial data paths connecting the processors. The topology of this network is determined by the designer and is expressed in abstract form in the design file. The compiler then translates this topology into the requisite hardware, including parallel-serial and serial-parallel converters only where necessary. Thus it is advantageous to partition the algorithm such that interprocessor communication is minimized, since this reduces the amount of both circuitry and wiring.

3. Processor Architecture

The individual processors of the multiprocessor IC's are characterized by the following:

(1) A control sequencer microprogrammed for a dedicated function
(2) A bit-parallel, single accumulator arithmetic unit
(3) A separate unit to perform address arithmetic
(4) A data memory for local variables and constants
(5) Use of a finite state machine for decision-making.

A study of algorithms within the target range indicated that the same general architecture, with suitable parameterization, can perform the algorithms efficiently. Thus the word length, the size of the data memory, the number of serial I/O ports, and the address arithmetic hardware are all configured by the compiler from declarations contained in the design file. The control sequencer is similarly parameterized, and contains a read-only memory programmed with microcode for the signal processing algorithm.

Each processor is organized into the following macrocells:

PC (program counter)
ROM (microcode read-only memory)
SPC (subprogram counter — optional)
AAU (address arithmetic unit — optional)
FSM (finite state machine — optional)
AUIO (arithmetic unit with I/O ports)
RAM (processor data memory)

VLSI Signal Processing

A fully configured processor is diagrammed in Fig.3.

The processor executes its microprogram (stored in ROM) once per sample interval. The microprogram consist of a *main program*, optionally followed by a fixed number of *subprogram* iterations. Except for this iteration pattern, there are no branches (conditional or unconditional) in the program execution.

The AAU is included only if address indexing is used. The address field of the control ROM is modified by the AAU to create the effective address for the data memory (RAM). Two index counters, IX and IY, may be included in the AAU. Indexing is used in signal processors to allow repeated sections of an algorithm to be multiplexed into a single piece of code.

The IX counter counts the subroutine iterations. IX equals -1 during the main program, 0 during the first subroutine iteration, 1 during the second iteration, and so forth. Indexing by the IX counter allows the subprogram to operate on elements of an array in data memory, accessing a different element each time the subprogram iterates.

IY is a sample counter with a fixed modulus. Indexing by IY allows the main program to operate on elements of an array in data memory, accessing different elements from sample to sample. This permits multiplexed processing of a group of decimated signals.

An additional option allows more general addressing, including pointer arithmetic. In each case, only the hardware needed to perform the types of addressing actually used are compiled into the AAU macrocell.

Viewed together, the ROM, PC, SPC and AAU present a stream of horizontal control and address words to the AUIO, RAM, and (optional) FSM macrocells. Signal data is processed in the arithmetic unit of the AUIO macrocell. RAM is used for data storage. FSM, if included, performs control and decision-making functions.

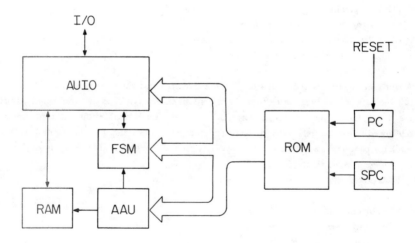

Fig.3 Macrocells forming a single processor

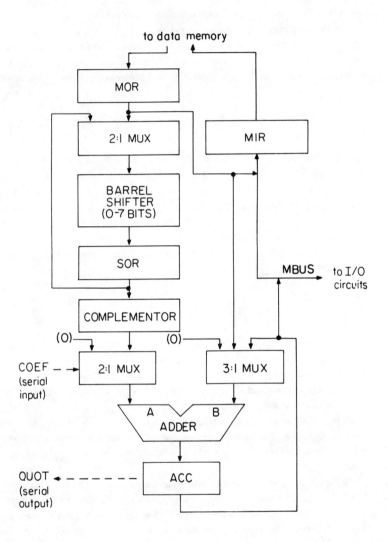

Fig.4 Processor data-path architecture

VLSI Signal Processing

The processor arithmetic unit (Fig.4) consists of a barrel shifter of depth eight, a complementer, a saturating adder, three registers (MOR, SOR and ACC), and a transparent latch (MIR). This processor may be microprogrammed for all common signal processing functions.

Essential to any signal processor architecture is the ability to perform multiply-accumulate operations. Two cases are considered separately: multiplying signal data by a variable coefficient external to the processor; and multiplying signal data by a constant.

In the case of variable coefficients, the coefficient is available externally in a bit-serial, MSB first, fractional two's complement format. Each processor arithmetic unit allows two such serial coefficient inputs. To multiply a signal A by an n-bit coefficient k, the following equation is used:

$$kA = -A + \bar{k}_{n-1}A + k_{n-2}\frac{A}{2} + k_{n-3}\frac{A}{4} + ...$$

Here k_i is the i-th bit of k. Each term on the right-hand side is a partial product, which is non-zero depending on one of the bits of the coefficient k. So to perform these variable-coefficient multiplies, the external bit-serial coefficient must be multiplexed into the control path for the arithmetic unit. This allows the partial products to be summed sequentially, one per clock cycle.

In the case of a fixed coefficient, it is not necessary to input the coefficient into the processor. Instead, a signed-digit representation of the coefficient is embedded in the control stream. This is done by specifying a sequence of shift depths and sign values. Thus, a fixed coefficient g has signed-digit representations of the form

$$g = (-1)^{s_0}2^{e_0} + (-1)^{s_1}2^{e_1} + ...$$

Any binary number has such a representation with the minimum number of digits. This is known as the canonical signed-digit (CSD) representation. The importance of the CSD representation is that it minimizes the number of clock cycles required to perform a multiply by a fixed coefficient. The multiply is performed by sequencing the barrel shifter and complementer so that it presents the desired terms to the accumulator.

In both the above cases, a sequence of multiplies may be performed, with the result of each being accumulated. This is an efficient feature of the single-accumulator architecture. The same accumulator is used for adding partial products, and for accumulating the results of several multiplies.

The arithmetic unit may also be microprogrammed to perform two-quadrant divide operations. Circuitry is included that makes the bit-serial quotient available in two's-complement form.

Generally the result of a multiply-accumulate operation is to be written back into the data memory. The transfer from ACC to the data memory is through the transparent latch MIR. By holding MIR transparent the transfer may be made immediately. Alternatively, MIR may store the data until the next free memory cycle. This use of MIR prevents congestion resulting from too frequent memory accesses.

The data memory itself may contain both read/write and read-only locations. The read/write locations are used to store state variables required for any signal processing. The read-only locations are used when it is necessary to introduce constants into the data path. Using the macrocell approach to designing the memory circuits supports this intermixing. The memory is configured from the design file declarations of variables and constants.

The lack of conditional branches in the control flow does not allow a conventional approach to implementing decision-making operations. Instead, a PLA-based finite state machine (FSM) may be included as an adjunct to the data path. The programming of the PLA is specified in the design file, with the FSM's state variables corresponding to logical values in the algorithm. A conditional write operation is supported, which conditions the assignment of a value to a variable in data memory on the state of the FSM. This fairly simple decision-making capability satisfies the requirements of signal processors.

Each processor executes its microprogram once per sample interval. Since primitive operations such as multiplies must be microcoded, it follows that the sample rate must be substantially slower than the processor's clock rate. Otherwise, there would not be time for a significant amount of processing during the sample interval. On the other hand, if the ratio of clock rate to sample rate becomes very large, the architecture exhibits an imbalance wherein the control ROM consumes nearly all of the silicon area, and the data path only a small fraction. Based on the relative sizes of the library cells, the processor architecture is most efficient if the ratio of clock rate to sample rate is between 50 and 1000. This corresponds to a sample rate range of 5 kHz to 100 kHz, since the cell library is designed to operate at a maximum 5 MHz clock rate.

4. Layout Generation

The first step in layout generation is the assembly of macrocells from the intermediate file description. This is done essentially by tiling of library cells. In the process, information on terminal locations is preserved for use as input to subsequent routing programs.

An important step in any macrocell-based design is the proper placement and interconnection of the macrocells. The silicon compiler does not require a general-purpose placement and routing program. Instead, it is only necessary to assemble those IC's the system is capable of producing. This more constrained problem is much more easily solved.

The problem is subdivided into two phases. In the first phase, the individual processors are assembled. In the second phase, the processors, the host interface (if included) and the bonding pads are assembled into the final IC.

A fixed-floorplan approach is used to assemble the processors from the separate macrocells. The floorplan consists of a left side (containing AUIO, RAM, AAU and FSM), a central wiring channel, and a right side (containing PC, ROM and SPC). The relative placement of the macrocells is defined by the floorplan, with absolute placement dependent on the sizes of the macrocells and the wiring channels. Wiring internal to either the left or right side is accomplished by repeatedly calling a simple river router.

The central channel is wired by a general purpose channel router, based on Kuh's channel routing algorithm [4]. The algorithm as originally stated in [4] required the terminals along either side of the channel to be aligned with a coarse grid, the dimensions of the grid being the minimum contact-to-contact spacing. The algorithm was modified to allow terminals to be of arbitrary position along the edges of the channel.

The single, fixed-floorplan approach does not always produce an efficient placement. Additional floorplans are being developed, and the software has the ability to select the most compact floorplan from a small number of alternatives.

Once the individual processors are assembled, the IC itself is assembled from the processors, the

VLSI Signal Processing

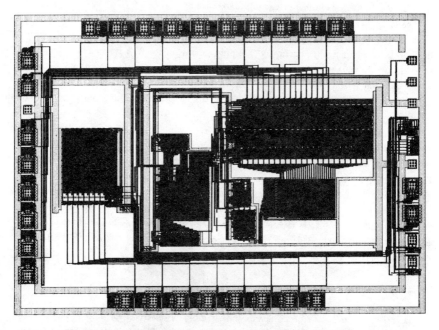

Fig.5 Circuit plot of the audio equalizer (3.6 x 5.1 mm)

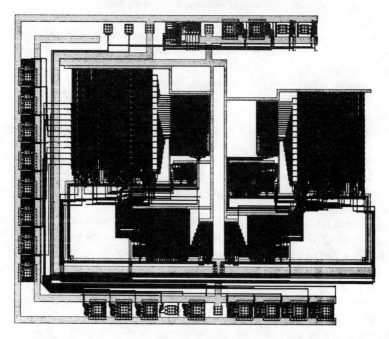

Fig.6 Circuit plot of the decision feedback equalizer (3.7 x 4.5 mm)

VLSI Signal Processing

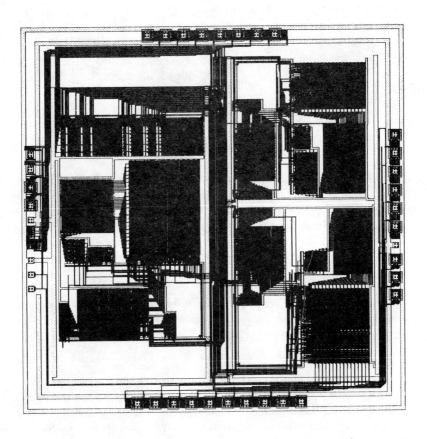

Fig.7 Circuit plot of the LPC vocoder (7.0 x 7.0 mm)

VLSI Signal Processing

host interface and the bonding pad arrays. The current approach to this global placement and routing is to partition the blocks into two columns, determining an optimal partitioning, ordering and rotational orientation for the various blocks. The channel routing routine is then called up to five times to complete the wiring of the IC.

More sophisticated global placement and routing routines are being studied.

5. Examples

Design file inputs have been prepared for a number of applications. Figs. 5-7 shows circuit layout plots of three of these, with the following characteristics:

function	processors	transistors	sample rate
audio equalizer	1	6,900	50 KHz
LPC vocoder	3	27,400	8 KHz
decision feedback equalizer	2	8,300	116 KHz

The macrocell-based silicon compiler is still actively being researched. However, it is possible at this point to draw a few general conclusions.

The macrocell approach can generate complex signal-processing circuits with far less effort than manual circuit design and layout. A few successful applications could easily amortize the development costs for the compiler and cell library. Finally, and most importantly, the design files can be prepared by signal processing researchers who have no prior knowledge of integrated circuit design. This one fact significantly expands the range of individuals who may now involve themselves in custom digital signal processing IC design, a field previously reserved for a select few.

Acknowledgements

The authors would like to thank Peter Ruetz, Jeremy Tzeng and Mats Torkelson for their contributions to this project. Research supported in part by the Defense Advance Research Projects Agency, Contract No. MDA903-79-C-0429.

References

1. S.S.Magar, E.R.Caudel, A.W.Leigh, "A Microcomputer with Digital Signal Processing Capability", *International Solid State Circuits Conference Digest*, San Francisco, 1982, pp. 32-33.

2. S. Pope, *Macrocell Design for Signal Processing*, Ph.D. Dissertation, University of California, Berkeley, CA, December 1984.

3. P.B.Denyer, D.Renshaw, "Case Studies in VLSI Signal Processing Using a Silicon Compiler", *Proc., International Conference on Acoustics, Speech and Signal Processing*, 1983.

4. T.Yoshimura, E.S.Kuh, "Efficient Algorithms for Channel Routing", *IEEE Transactions on Computer Aided Design*, V. CAD-1, Jan. 1982, pp. 25-35.

Stephen P. Pope received the Bachelor of Science degree from the California Institute of Technology, Pasadena, California, in 1978, and completed his doctoral work at the University of California, Berkeley, California in 1984. He is currently on the engineering staff of Cyclotomics Corp., Berkeley, California, where he is studying IC implementations of error detection and correction algorithms.

Jan Rabaey was born in Veurne, Belgium on August 15, 1955. He received the E.E. and the Ph.D. degrees in applied sciences in 1978 and 1983, respectively, from the Katholieke Universiteit Leuven, Belgium, where he worked on computer aided design tools for switched capacitor circuits. Since 1983 he has been a visiting research engineer at the University of California, Berkeley. His current interests are in computer aided analysis and automated design of digital signal processing circuits.

Robert W. Brodersen received the B.S. degree from California State Polytechnic College, Pomona, in 1966, and the E.E., M.S., and Ph.D. degrees from the Massachusetts Institute of Technology, Cambridge, in 1968, 1968, and 1972 respectively. From 1972 to 1976 he was with Texas Instruments, Inc., Dallas, studying the operation and application of charge-coupled devices. In 1976 he joined the faculty of the University of California, Berkeley, where he is a Professor. He is studying the application of IC technology to signal processing, image processing, and user interfaces.

FIRST - PROSPECT AND RETROSPECT

Peter B. Denyer(*), Alan F. Murray(#), David Renshaw(*)

*Department Electrical Engineering,
#Wolfson Microelectronics Institute,
University of Edinburgh.
Scotland.

1. INTRODUCTION

Design complexity has become a dominant cost limit in the development of VLSI systems. Without new design methods and tools, advances in manufacturing capability and algorithm development will far exceed our capacity for design. This effect is nowhere more pronounced than in the field of real-time signal processing; the continuous flow of data together with the complexity of many of the algorithms, imposes sever computational demands that often cannot be satisfied by general purpose machines or components.

We have addressed the issue of design complexity through the development of a powerful system synthesis tool - a silicon compiler, called FIRST. FIRST is more specialised than its contemporaries. This restricts its application (to real-time signal processing tasks), but at the same time permits an impressive functional integration density that is closer to hand-crafted custom than to other examples of automatic layout generation.

FIRST has changed our perception of silicon as a flexible development medium for rapid prototype system development. Large, complex algorithms are translated to detailed VLSI layout on short timescales by personnel previously inexperienced in VLSI design.

2. FIRST

FIRST synthesises signal processing chips and systems using exclusively bit-serial architectures. This architectural restriction has significant consequences in the simplicity and success of the environment. We shall later identify how restrictions and conventions such as these are key to a successful design methodology.

Bit-serial architectures [1] have major advantages as a VLSI strategy, especially in the areas of signal routing and communication (issues which are often important in signal processing applications), and in computation efficiency through bit-level pipelining. These features are exemplified in a range of case sudies given later.

We have built a version of FIRST around 5-micron nMOS technologies and are currently investigating a similar capability for 2.5-micron CMOS technology. The first part of this paper illustrates the principle of our approach using the nMOS facility, whilst the second part discusses early experiences in CMOS technology.

VLSI Signal Processing

We begin with a simple example system to give a flavour of using FIRST. For the purpose of this example we consider an implementation of the four-region approximation [2] to the magnitude, M, of a complex number, A+jB:

M = greater { (7/8G + 1/2L) , G }

where,

G = greater { |A| , |B| }
L = lesser { |A| , |B| }

We view this algorithm as a functional flow-graph, as shown in Figure 1. The oblong shapes are functional elements that perform fixed operations on the (bit-serial) data flowing through them. Now bit-serial elements exhibit a delay or latency of integer numbers of bit-times. It is important to equalise this delay and synchronise data paths entering each element that may have traversed Simple delay elements are shown as circles in Figure 1. In this system there is also a control requirement for a pulse coincident with the start of each new data word at each processing element. Delayed versions of this pulse are generated by a network of delay elements on the right of Figure 1.

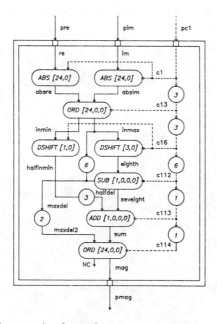

Figure 1: Flow-graph of complex-to-magnitude example.

We have deliberately reduced this algorithm to a network of functional operators (add, scale, delay, etc.) which are supported by FIRST. We call this the set of <u>primitive</u> operators. FIRST supports a finite set of primitives from which all systems are constructed. A list of FIRST primitives is shown in Table 1.

VLSI Signal Processing

TABLE 1. FIRST PRIMITIVE SET

ABSOLUTE	ADD	BITDELAY
CBITDELAY	CONSTGEN	CWORDDELAY
DPMULTIPLY	DSHIFT	FFORMAT1TO1
FFORMAT2TO1	FFORMAT3TO1	FLIMIT
FORMAT1TO2	MSHIFT	MULTIPLEX
MULTIPLY	ORDER	SUBTRACT
WORDDELAY		

Note that this is a set of relatively high level functions, well above the transistor or gate level.

Now imagine a tool capable of taking this functional flow-graph as input, of instantiating the required set of bit-serial primitives, of assembling these primitives and routing the network of flow-graph connections between them, and finally wiring input/output pads, and power and clock services to complete a custom chip design. In effect this environment acts as a compiler, delivering low-level code (in this case VLSI mask artwork) from an initial high level programme (the flow-graph). This is precisely the function performed by FIRST.

The corresponding FIRST implementation of the complex-to-magnitude algorithm in silicon is shown in Figure 2. As a VLSI device this is not

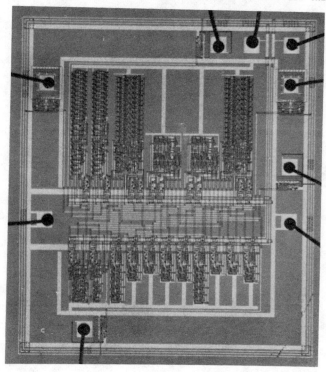

Figure 2: FIRST synthesised chip from flow-graph of Figure 1.

particularly impressive; it contains only a few thousand transistors. However as a demonstration of design methodology it is; our design cycle was very fast and required no intimate knowledge of integrated circuit design.

It is worthwhile to review several aspects of the FIRST-generated chip design of Figure 2. The overall floorplan consists of two rows of primitives, positioned above and below a central routing channel. The central channel contains all of the network wiring from Figure 1, emphasising the avantage of the bit-serial architecture. This is the standard FIRST floorplan; it may be altered (to contain several rows of function) for smaller process geometries.

The primitives themselves are irregular blocks of high-density layout, with i/o strategies configured to suit the routing channel convention. The primitives are in fact human designed and optimised. However they are not held in FIRST as single "macrocells", but rather as collections of leaf cells and procedures for their assembly. In this way the primitive set is parameterised so that users may select for example, the size of a multiplier, the length of a delay element, etc.

This simple floorplan style has proved effective in efficiently implemeting many of the systems we have attempted. Later we show a selection of such designs which exemplify this feature. We find a typical overhead of around 25 - 30% of unused silicon area.

3. SECRETS OF SUCCESS

The nMOS version of FIRST has been in use at Edinburgh University for some 18 months. Several designers have completed complex system designs in remarkably short timescales, vindicating our aims and approach. In retrospect the key to FIRST as a successful system design tool lies in two simple concepts; achitectural restriction (as already mentioned) and rigorous protocols on communication. We have already lauded bit-serial architecture. The concept of setting and maintaining system-wide communication protocols is expanded below. This is a powerful adjunct that might usefully benefit all forms of hardware achitecture.

We begin by setting rigorous conventions covering all aspects of the timing, electrical and numerical formats of signals throughout the system. If we now implement a base set of primitives that uniformly respect these conventions, then any connection of these primitives will communicate successfully, without requiring any further design attention to be devoted to the individual interfaces. Thus we can wire up a flow-graph of elements, as in Figure 1, and automatically achieve correct hardware function. This consistency in data format between operations and routines has long been taken for granted by designers of software, but has never been enjoyed to this extent by their hardware equivalents.

These advantages extend to elegant hierarchical design, because the communication conventions are automatically inherited by any network or sub-network built exclusively from such components. Thus we are able to construct higher level operators as general hardware "routines".

These features make for a very speedy system design environment.

VLSI Signal Processing

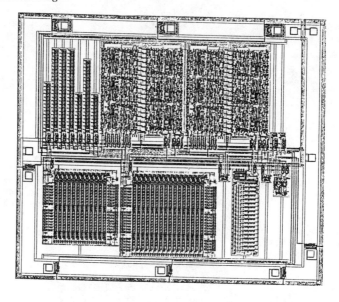

Figure 3a: Device from speech echo canceller system.

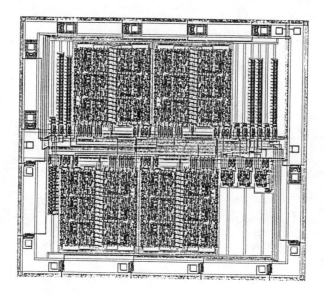

Figure 3b: Device from adaptive lattice filter system.

VLSI Signal Processing

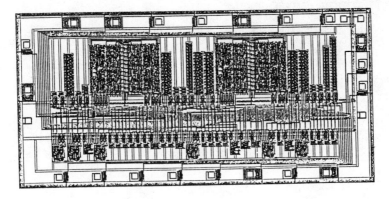

Figure 3c: LDI filter device.

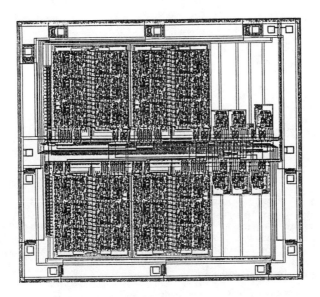

Figure 3d: Butterfly device from FFT system.

4. A PORTFOLIO OF CASE STUDIES

Table 2 summarises the scope of applications addressed by FIRST as a selection of case system design studies. The information on transistor and chip counts relates to 5-micron nMOS technology. The advent of a CMOS primitive library (as discussed below) will dramatically improve packing density and chip counts.

TABLE 2. FIRST SYSTEM CASE STUDIES

System	Computation MegaOps/sec	Chip Designs	Total Chips	Transistors (K)	FIRST Code (lines)
Speech Echo Canceller	55	3	18	252	250
Adaptive Lattice Filter	10	5	5	30	500
LDI Filter 5th Order	15	1	1	7.5	160
FFT 16-point 8 MHz	150	6	43	160	300

Most significantly this Table shows that impressive total computation rates can be realised by many slow bit-serial elements operating in parallel. The final two columns provide a comparison of the size of the system description code (FIRST input), and the transistor count of the resulting system.

Figure 3 shows a selection of FIRST-generated chip designs from these systems.

5. A REIMPLEMENTATION IN CMOS TECHNOLOGY

The technical implementation of the nMOS version of FIRST is detailed elsewhere [3,4]. We describe here the development of a bit-serial CMOS primitive library currently being undertaken at the Wolfson Microelectronics Institute.

Our design style uses a mixture of dynamic and static CMOS logic, using dynamic techniques where the benefits of the increased speed and reduction in transistor count (and therefore silicon area) can be realised and are significant. The dynamic logic structure is based on that known variously as "NP" or "NORA" CMOS, modified to render it less sensitive to clock edge times [5,6].

Testing complex digital CMOS circuits is a problem, however, due to the need for deterministically derived test vector sequences to sensitise and detect "stuck-open" faults [7]. As circuit complexities increase, random pattern testing and self-test represent elegant and efficient methods of testing VLSI devices and systems. However, the intrinsically random nature of the test vectors conflicts with the sequencing required for high coverage of stuck-open faults in CMOS. Our CMOS circuit design style circumvents the stuck-open problem, such that the ultimate complexity of VLSI systems is not limited by constraints of testability.

5.1. Design and Layout Style

Dynaic CMOS gates are either N or P type. The desired function is evaluated by "discharging" a precharged capacitance, conditional on either a positive logic (N-gate) or negative logic (P-gate) function of the gate's input variables. The technique is related to the "Domino CMOS" style, in which only N-gates are involved [8]. The mixed-gate approach is superior in that N and P gates may be cascaded without the need for an intervening inverter, and greater versatility is brought about by the positive and negative logic functions. NP, Domino and static gates may be mixed, provided the rules for inclusion are obeyed.

Figure 4a shows an exclusive-NOR (XNOR) gate taken from part of the library. The form of the dynamic latch used can be seen from Figure 4b to consist of a "clocked inverter". The logical function is executed by precharging the output nodes of the dynamic gates, and then permitting an optional discharge. Input values are sampled during ϕi, during which time the nodes C and D are precharged (or, more correctly, preset) low and high respectively. During ϕi, these nodes are conditionally "discharged", depending on the voltages on the gates of the transistors in the evaluation logic tree.

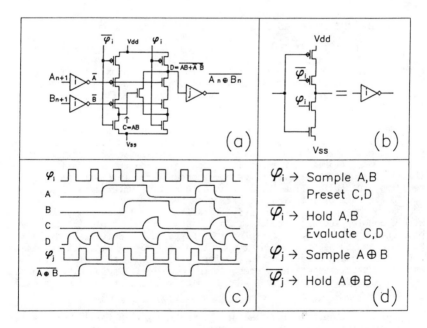

Figure 4: Dynamic CMOS XNOR stage.

Node C is pulled high if $\bar{A}$ and $\bar{B}$ are both low, forming the minterm AB. Simultaneously, node D is discharged, conditional on either C or both A and B being high, forming the exclusive OR function. During Φj, this value is sampled by the clocked inverter to give a "held" XNOR during Φj. The effect of this is shown in Figures 1c and 1d. In common with all synchronous design systems, the output of a complete gate is presented one full clock cycle after its inputs. In addition to the Φi -> [NP gates] -> Φj ordering illustrated, gates may also be constructed which sample inputs on Φj, and change outputs on Φi. Furthermore, N and P sections may be cascaded internal to a gate as far as ripple through times allow. With these capabilities, extremely tight pipelining of logic may be achieved.

A structured layout style is used for the design of the cell library, to ease checking and to impose some consistency in the layout of low-level gates. The layout style is a relaxed version of that known as "gate-matrix" layout, within which the P-channel devices of a logic gate are grouped together and the N-channel devices similarly grouped to minimise the effect of the large minimum P-channel to N-channel device separation [9]. Furthermore, the devices are arranged in rows with the source-drain direction horizontal (say), such that polysilicon gate "wires" may run vertically between P- and N-channel rows. The first metal layer is used for general interconnect (mostly vertical but some horizontal), and the second layer for (almost exclusively horizontal) distribution of power and clock lines. The resultant layout is dense, neat and amenable to pitch-matching between gates, particularly with regard to power and clock distribution.

5.2. A CMOS Bit-Serial Complex Multiplier

We illustrate this CMOS style with the design of a bit-serial complex multiplier primitive. The inputs to a multiplier are the multiplicand (n bits), and the coefficient (m bits). As with nMOS FIRST, fixed point, two's complement arithmetic is used. Any bit-serial multiplier is constructed as a set of stages through which the data flow sequentially, the number of stages being determined by the coefficient length [10].

We have developed a set of subcells to enable the design of a family of multipliers. The computational latency of a bit-serial multiplier is either 2n (for a "straightforward" serial-parallel-serial multiplier) or 1+3n/2 (for a Modified Booth's "recoding" multiplier). For the library of CMOS primitives we have concentrated initially on a non-recoding multiplier family, as this allows for tighter data pipelining due to the lack of the sign bit extension required by a Booth operator. [10]. As an example we present a full-precision complex multiply operator with a double-precision product. This represents a typical, tightly-pipelined cell. It is capable of multiplying n-bit multiplicands by m-bit data and of being clocked at a maximum of 20MHz. An incidental benefit of using a non-recoding multiplier is the enhancement of the random pattern testability of the processors in which it is included, although this is inherently high even when a Booth's algorithm operator is used [11].

The algorithm is that of a serial pipeline multiplier [10], with some extra circuitry to obviate the need for sign bit extension (two XOR gates for each of the first m-1 multiplier stages). There is also some extra multiplexing to capture the n-m-1 lowest significant bits of the output, which are normally rounded-off in a standard pipeline multiplier, to form the n+m-1 full precision product. The latency of the

VLSI Signal Processing

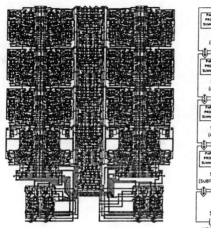
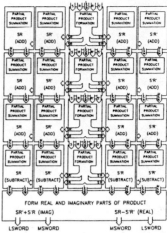

Figure 5: Layout and tesselation pattern for complex bit-serial multiplier.

multiplier, to presentation of the least significant bit of the product, is m+3 bits , of which one bit is attributable to the addition of the four raw product components to form the complex product.

In Figure 5, we show the multiplier architecture and tesselation scheme, and corresponding layout for m=4. The complex multiplicand (S+iS') and coefficient (R+iR') words flow through a central shift register network, within which the partial products are formed. The partial products are subsequently summed in four seperate, parallel data streams, to form Least and Most significant words (Lsword and Msword) of SR, SR', S'R, S'R'. These are added and subtracted together to form the components of the complex product [SR-S'R'+i(S'R+SR')] in a final stage. There are n-bit Lswords and Mswords, and only n+m-1 bits of product. When m=<n, the upper n-m+1 bits of the product Msword are sign extensions, which allow for arithmetic growth in subsequent calculations. An m-bit multiplier is formed by m-1 ADD sections, and one SUBTRACT section, followed by the final real/imaginary generator. The individual partial product summation sections, which form the bulk of the multiplier, are identical to those which would be used for a non-complex multiplier, illustrating the modularity of the approach. A complete 16-bit complex bit-serial multiply operator comprises some 10000 transistors and occupies an area of approximately 1x4 mm in a 2.5 micron, double metal p-well process.

5.3. Other Library Elements

The other functional elements necessary to provide a comprehensive primitive set (add, subtract, multiplex etc.) are realised in a similar manner to the multiplier family. For example, a major requirement is for bit- and word- oriented FIFO storage blocks (signal and control delay elements). To this end, a 3-transistor cell dynamic FIFO has been developed. This occupies an area per bit of 25*25 microns.

6. CONCLUSIONS

We have described the development and use of the FIRST silicon compiler. By setting rigorous architectural restrictions and conventions, FIRST greatly reduces the problems of system synthesis. In this environment custom VLSI hardware design becomes as productive as modern software design, encouraging innovative experimentation at higher architectural levels in the system.

Whilst the existing version of FIRST is implemented for nMOS technologies, the design of an advanced CMOS primitive library is underway. This improves computation rates by 150% and increases the integration density by greater than 1000%.

Major applicaions are apparent in:

(a) SONAR and RADAR pulse-compression filtering.

(b) Real-time super-MegaHertz FFT systems.

(c) Frequency selective filters and adaptive equalisers for telecommunications.

(d) Enhancement filters for real-time image processing.

References

1. R. F. Lyon, "A Bit-Serial VLSI Architectural Methodology for Signal Processing," in VLSI 81, ed. J. P. Gray, pp. 131 - 140, Academic Press, 1981.

2. A. E. Filip, "A Baker's Dozen Magnitude Approximation and Their Detection Statistics," IEEE Trans. AES, vol. AES-12, no. 1, pp. 87 - 89, January 1976.

3. P. B. Denyer, D. Renshaw, and N. Bergmann, "A Silicon Compiler for VLSI Signal Processors," Proc. ESSCIRC'82, pp. 215 - 218, Brussel, September 1982.

4. N. Bergmann, "A Case Study of the FIRST Silicon Compiler ," Proc. 3rd Caltech Conference on VLSI, pp. 473-430, 1983.

5. N. F. Goncalves and H. G. De Man, "NP-CMOS : A Racefree Dynamic CMOS Technique for Pipelined Logic Structures.," Proc. 8th European Solid-State Circuits Conference, Brussels., pp. 141-144, 1982.

6. N. F. Goncalves and H. G. De Man, "NORA : A Racefree Dynamic CMOS Technique for Pipelined Logic Structures.," IEEE Journal Solid State Circuits, vol. 18, no. 3, pp. 261-266, 1983.

7. V. V. Nickel, "VLSI - The Inadequacy of the Stuck Fault Model.," Proc. International Test Conference, pp. 378-381, 1980.

8. B. Murphy, R. Edwards, L. Thomas, and J. Molinelli, "A CMOS 32b Single Chip Microprocessor.," *ISSCC Digest of Technical Papers*, pp. 230-231, 1981.

9. S. M. Kang, R. H. Krambeck, H-F. S. Law, and A. D. Lopez, "Gate Matrix Layout of Random Control Logic in a 32-Bit CMOS CPU Chip.," *IEEE Trans. Computer Aided Design*, vol. CAD-2, no. 1, pp. 18-29, 1983.

10. R. F. Lyon, "Two's Complement Multipliers," *IEEE Trans. Communications*, vol. 24, pp. 418-425, 1983.

11. A. F. Murray, P. B. Denyer, and D. Renshaw, "Self-Testing in Bit-Serial VLSI Parts : High Coverage at Low Cost," *Proc International Test Conference*, Philadelphia, pp. 260-268, 1983.

Part VIII
LANGUAGES AND SOFTWARE

Edited by:
Prof. S. Y. Kung
University of Southern California

VLSI Signal Processing

28

LOGIC PROGRAMMING OF SYSTOLIC ARRAYS

R. S. FREEDMAN
Research Laboratories
Hazeltine Corporation
Greenlawn, N.Y. 11740

ABSTRACT

We show how logic can be a convenient notation for specifying VLSI signal processing algorithms that are to be implemented on systolic arrays. When implemented in Prolog, this logic specification is a very powerful tool for verifying and validating systolic algorithms. We also show how logic can be used to implement many sophisticated signal processing algorithms fairly easily, despite its lack of suitable "data structures".

INTRODUCTION: SYSTOLIC ARRAY ARCHITECTURES

A systolic array architecture is a subset of a data-flow architecture. It is made up of a number of identical cells, each cell locally connected to its nearest neighbor. The cells are usually arranged in a definite geometric pattern, corresponding to the tessellations of the Euclidean plane. The cells that are on the boundary of the pattern are able to interact with the outside world. At a given clock pulse, data enters and exits each cell; entering data is processed and stored so it can be output at the next pulse. Computational power is thus identified with the speed of input and output: a wavefront of computation is propagated in the array with a throughput proportional to the input/output bandwidth. This pulsing behavior is what gives this architecture its name. One might say that systolic arrays exhibit parallelism in time (because of this high degree of concurrency) and space (because of the high degree of geometric symmetry).

There are several good surveys of algorithms that have been implemented on systolic arrays (see Ref. 1). Unfortunately, these surveys are hampered by the lack of a convenient notation for specifying geometry and data-flow. Another problem is the lack of simulation facilities. Most programming languages cannot easily simulate parallel computation.

VLSI Signal Processing

LOGIC PROGRAMMING OF SYSTOLIC ARRAYS

The difficulty of understanding systolic algorithms is due to the fact that procedural mechanisms are used to explain them. However, these algorithms are much more easily explained non-procedurally, since data-flow is being emphasized, not procedural computation. The declarative semantics of logic can be used as a notation for these algorithms. The concurrency that is exhibited by the systolic algorithms can be elegantly described with the implicit non-deterministic AND-parallelism that is seen in Horn clause logic. When implemented in Prolog (see Ref. 2), this declarative semantics provides a powerful tool for the emulation of systolic arrays.

Conversely, the ability to describe powerful systolic algorithms in logic allows one to implement these algorithms in Prolog. Despite the fact that Prolog has no "data structures" that are convenient for numerical computation, algorithms can be implemented straightforwardly to multiply and invert matrices, solve linear systems of equations, perform Fourier transformations, and solve eigenvalue problems. Logic as implemented in Prolog can become a language for signal processing analysts.

AN EXAMPLE OF A SYSTOLIC ALGORITHM

We illustrate the semantics of a logic representation of a signal processing algorithm with an example. A very important and common numerical problem is the convolution problem:

Compute the y-sequence $(y(0), y(1), ..., y(n+1-k))$, where

$$y(i) = w(0)*x(i) + w(1)*x(i+1) + ... + w(k-1)*x(i+k)$$

given an input x-sequence $(x(0), x(1), ..., x(n))$ and a

w-sequence of weights $(w(0), w(1), ..., w(k-1))$.

Several algorithms associated with matrix algebra, pattern matching, linear least squares approximation, polynomial division, and eigenvalue determination can also be placed in this form. We examine the convolution problem for the case k=4. A systolic array that solves this problem was discussed in (Ref. 1) and is presented as Figure 1.

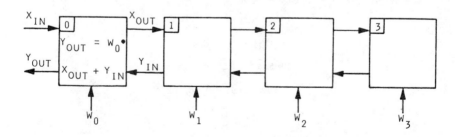

Figure 1. Systolic Array for the Convolution Problem

VLSI Signal Processing

We observe that this systolic array is a linear two-way pipeline system. The inputs are separated by two cycles to allow the successive summation to occur. At a given cycle in the computation, the different weights are broadcast to all cells. Cell(0) also receives a new input message that denotes a new element of the x-sequence. The value of the previously calculated sum from cell(1) is also passed in a message from cell(1) to cell(0). The contents of this message, the input message, and the weight message are used to compute the new stored sum in cell(0). Cell(0) in turn passes its previously calculated y-value to the output and also passes to cell(1) whatever x-value it previously received. Similar actions occur simultaneously at the other cells. It is seen how computation is identified with the speed of input and output. In addition, the sequencing of input also determines a particular algorithm. For example, if all the weights w(0), w(1), w(2), w(3) are set to be unity for all time, and if the X-sequence is input with alternating zeros, then the sum y(0) = x(0) + x(2) + x(3) is output after eight cycles. The same array with a different sequencing of data can result in the computation of other algorithms. For example, if the weights are not restricted to be constant, they can be sequenced and broadcast to form an upper triangular matrix W. Again, if the x-sequence is input with alternating zeros, the corresponding y-sequence (y(0), y(1), y(2), y(3)) will accumulate the matrix-vector product W * (x(0), x(1), x(2), x(3)).

There is a certain canonical form in the logic representation of systolic arrays. This can be seen in the Prolog representation of our example. The existing cell states and their state changes are explicitly stated. Part of a Prolog specification that emulates the systolic array in Figure 1 is given by the following single clause Prolog program:

```
cycle(N) :-      cell(N,0,X0,Y0),

                 cell(N,1,X1,Y1),

                 cell(N,2,X2,Y2),

                 cell(N,3,X3,Y3),

                 M is N +1,

                 output(Y0),

                 broadcast(W0),

                 inputNewX(A0),

                 B0 is Y1 + W0*X0,

                 becomes(cell(M,0,A0,B0)),
```

VLSI Signal Processing

> A1=X0,
>
> broadcast(W1),
>
> B1 is Y2 + W1*X1,
>
> becomes(cell(M,1,A1,B1)),
>
> A2=X1,
>
> broadcast(W2),
>
> B2 is Y3 + W2*X2,
>
> becomes(cell(M,2,A2,B2)),
>
> A3=X2,
>
> Y4=0,
>
> broadcast(W3),
>
> B3 is Y4 + W3*X3,
>
> becomes(cell(M,3,A3,B3)).

Each cell has four parameters. The first parameter represents the current cycle (or pulse) of computation. The second parameter is the cell identifier. The third and fourth parameters are the values of X and Y, respectively, of the cell at the current cycle.

We sketch a trace of the execution of this program. A logic program is invoked by providing a "goal" to satisfy; this initial goal may invoke other goals to be satisfied. In logic, the mechanism of "parameter passing" is replaced by the powerful mechanism of "unification", an operation very similar to pattern matching.

Initially, the contents of the X-values and Y-values of all cells are zero. After the first pulse (a goal to cycle(0)), the Y-value of cell(0) is output and various inputs are requested. It should be understood that even though the input and output functions are implemented in (sequential) Prolog, they should be regarded as executing concurrently; this is the intended meaning of the "broadcast" goals. We also note that time is global, and that all cells locally change their states. This is represented (following Hewitt in Ref. 3) with the "becomes" goal (which is implemented separately with Prolog assert goals). Given a goal for the next cycle (cycle(1)) we input 0 for the new X-value, reflecting this systolic algorithm in that input values are separated by one clock pulse. The values of x(0), x(1), x(2), x(3) are input at cycles 0, 2, 4, 6; zero is input at cycles 1, 3, 5, 7. For a simple convolution discussed above, the weights w(0), w(1), w(2), w(3) are constant. After listing the cells at the end of cycle(7), the sum y(0) = w(0)*x(0) + w(1)*x(1) + w(2)*x(2) + w(3)*x(3) is the Y0-value of cell(0). At the next pulse, this Y-value is output to some object that is not part of the systolic array.

VLSI Signal Processing

The space and time parallelism of the algorithm is also seen in the Prolog program. Terms corresponding to each cell in the systolic array (except those cells that interact with the outside world) are identical in form and function (reflecting space parallelism). Time parallelism can be seen with an implicit AND-parallelism of the conjunctives. This representation is also very amenable to implementation in Concurrent Prolog (see Ref 4). Concurrent Prolog implementation of a systolic algorithm would be more efficient, especially if the "becomes" goals can be implemented with some parallel assert goals, and if the input and output goals can be implemented with concurrent operations.

CONCLUSION

We have shown that logic is a convenient notation for specifying some signal processing algorithms that are to be implemented on systolic arrays. When implemented in Prolog, this specification can be used as a powerful tool to verify and validate the systolic array algorithms. Conversely, we have also shown that Prolog can be used to implement signal processing algorithms, and can thus be used in place of the more "traditional" procedural languages for many numerical applications.

REFERENCES

1. Kung, H. T., "Why Systolic Architectures," IEEE COMPUTER, Vol. 15, No. 1, January, 1982.

2. Clocksin, W. F., Mellish, C.S., PROGRAMMING IN PROLOG, Springer-Verlag, New York, 1981.

3. Hewitt, C., de Jong, P., "Analyzing the Roles of Descriptions and Actions in Open Systems," PROCEEDINGS OF THE NATIONAL CONFERENCE ON ARTIFICIAL INTELLIGENCE, AAAI, Washington, D.C., August, 1983, pp. 162-167.

4. Shapiro, E., Takeuchi, A., "Object Oriented Programming in Concurrent Prolog," NEW GENERATION COMPUTING, Vol.1, No.1, 1983, pp. 25-48.

The Specification and Verification of Systolic Wave Algorithms

C. J. Kuo   Bernard C. Levy   Bruce R. Musicus

*Department of Electrical Engineering and Computer Science
Massachusetts Institute of Technology, Cambridge, MA 02139*

Abstract

This paper proposes a rigorous and systematic approach to analyze the systolic wave algorithms described by data flows and local interactions in a regular multiprocessor array. We formulate basic equations called space-time-data (STD) equations to describe the motion of a single data element in particular, and of a whole wave in general. Using this approach, we prove the correctness of the matrix multiplication algorithm on a hexagonal array. Then, the general specification and verification procedure is presented. Finally, the computational wavefront method is explained from this new point of view, and its limitations are discussed.

1. Introduction

Systolic arrays were first proposed by Kung and Leiserson in 1978 [1]. In their paper, they used a graphical approach to show how their ideas work. The examples they considered included matrix-vector multiplication, matrix multiplication on a hexagonal array, LU decomposition of a matrix, convolution, etc.. Although these algorithms are correct, no formal mathematical proof was given. It was only in 1982 that a proof of the correctness of the matrix multiplication algorithm was given by Chen and Mead [2]. They expressed the algorithm as a recursive space-time program, and then applied inductive techniques to verify this program. This approach seems tedious and complicated. In addition, it is difficult to understand its relation with the graphical approach. It is surprising that intuitively obvious results such as systolic algorithms cannot be proved in a straightforward way. One important reason is that we lack suitable tools to describe data flow phenomena in multiprocessor arrays. Several researchers have made efforts in this direction. One popular approach is known as the computational wavefront method [3] [4]. However, to our knowledge, this approach does not provide a verification method either.

One important class of systolic algorithms can be characterized by regular data flow through a multiprocessor array of regular geometry. We call this type of algorithm a *systolic wave algorithm*. In this paper, we propose a simple approach to analyze the matrix multiplication algorithm in particular, and other systolic wave algorithms in general. Specification and verification procedures can be stated in a systematic way. In addition, we formulate the concept of "wavefront" in a mathematically rigorous way. In Section 2, we consider the algorithm of matrix multiplication on a hexagonal array as an illustrative example; then the general specification and verification approach is described in Section 3. Finally, we use

---
*This work was supported in part by the Army Research Office under Grant No. DAAG29-84-K-0005 and in part by the Advanced Research Projects Agency monitored by ONR under contract N00014-81-K-0742.

VLSI Signal Processing

our model to interpret the wavefront concept in Section 4.

2. Matrix multiplication on a hexagonal array - an example

Kung and Leiserson's algorithm is described in Fig 1. The first observation is that a hexagonal array can be described as the superposition of two rectangular grids with the same spacing (h_1, h_2), and with origins at (0,0) and at ($\frac{1}{2} h_1$, $\frac{1}{2} h_2$). Therefore, the coordinates of the processors are either ($n_1 h_1$, $n_2 h_2$) or ($n_1 h_1 + \frac{1}{2} h_1$, $n_2 h_2 + \frac{1}{2} h_2$), where n_1 and n_2 are integers. In fact, h_1 and h_2 are related to the physical space only and are not essential in our discussion, so that we denote the coordinates of the processors as the ordered pair (n_1, n_2). We may redraw the picture in the (n_1, n_2) coordinates as shown in Fig 2.

For convenience, we use $a(i,j)$, $b(i,j)$, and $c(i,j)$ to denote the entries of matrices A, B, and C, and the systolic algorithm shown in Figs. 1 and 2 is used to perform the matrix multiplication A B = C. The origin of the coordinates (n_1, n_2) is chosen to be the point at which $a(1,1)$, $b(1,1)$, and $c(1,1)$ coincide with each other. We use k to represent the global clock, which takes nonnegative integer numbers. Initially, for $k = 0$, we assume that $a(1,1)$, $b(1,1)$, and $c(1,1)$ are located at ($-\frac{1}{2}d$, $\frac{1}{2}d$), ($\frac{1}{2}d$, $\frac{1}{2}d$), and ($-d$, 0) respectively; where d is an arbitrary positive integer number. At each time step, all $a(i,j)$'s move to the right one half unit and move down one half unit, so that we may describe this motion with a velocity vector ($\frac{1}{2}$, $-\frac{1}{2}$). Correspondingly, the velocity vectors of $b(i,j)$ and $c(i,j)$ are ($-\frac{1}{2}$, $-\frac{1}{2}$) and (0, 1). The components of the same matrix retain their relative position when they are shifted because they have the same velocity. We may call the motion of the entire group a *wave*. For the matrix multiplication case, there are three waves, called wave A, wave B, and wave C. The *position function* of some component, say $a(i,j)$ at time k, expressed as $P_a^k(i,j)$, is defined to be the coordinates of the processor where $a(i,j)$ arrives at time k. Therefore, $P_a^0(i,j)$, $P_b^0(i,j)$, and $P_c^0(i,j)$ represent the initial configurations of waves A, B, and C respectively. In order to specify the motion of waves A, B, and C, we must specify $P_a^k(i,j)$, $P_b^k(i,j)$, and $P_c^k(i,j)$. These can be expressed as follows:

$$P_a^k(i,j) = P_a^0(i,j) + (\frac{1}{2}, -\frac{1}{2})k \ , \qquad (2.1.a)$$

$$P_b^k(i,j) = P_b^0(i,j) + (-\frac{1}{2}, -\frac{1}{2})k \ , \qquad (2.1.b)$$

$$P_c^k(i,j) = P_c^0(i,j) + (0,1)k \ . \qquad (2.1.c)$$

Let n_1, n_2 represent the processor coordinates of data $a(i,j)$ at time k. Usually, the processor array has only finite size. However, for convenience, we may assume it extends in a regular manner to infinity in all directions.

From Fig 3, it is not hard to see that the relation between the data indices (i, j) and the initial processor coordinates $P_a^0(i, j) = (n_1, n_2)$ for wave A is:

$$n_1 + n_2 = -i + j \qquad k = 0, \qquad (2.2.a)$$

$$3n_1 - n_2 = -2d - 3(i + j - 2) \qquad k = 0. \qquad (2.2.b)$$

Solving equation (2.2), we get

$$n_1 = -\frac{(2i + j + d - 3)}{2} \qquad k = 0, \qquad (2.3.a)$$

$$n_2 = \frac{(3j + d - 3)}{2} \qquad k = 0. \qquad (2.3.b)$$

Therefore, we have

$$P_a^0(i, j) = \left(-\frac{2i + j + d - 3}{2}, \frac{3j + d - 3}{2}\right), \qquad (2.4)$$

and from (2.1.a) and (2.4), we get

$$P_a^k(i, j) = (n_1, n_2) = \left(-\frac{2i + j + d - 3 - k}{2}, \frac{3j + d - 3 - k}{2}\right) \qquad (2.5)$$

From above, we can derive two basic equations relating the processor coordinates n_1, n_2, the time index k, and the data indices i, j of matrix A, which is called, therefore, the *Space-Time-Data* (STD) equations of wave A. They are

$$2n_1 + 2i + j + d - 3 - k = 0, \qquad (2.6.a)$$

$$2n_2 - 3j - d + 3 + k = 0. \qquad (2.6.b)$$

The motion of wave A is totally specified by its STD equations. The variable d is determined by the initial relative positions of these three waves, so we may regard it as a constant in the following discussion.

There are two equations and five variables to describe the motion of wave A. If i, j are fixed, then n_1 and n_2 can be represented as functions of k, which represent the locus of the motion of the component $a(i, j)$ in a two dimensional plane. Furthermore, if k is chosen to be some specific value, the position of $a(i, j)$ is uniquely determined. This is defined as the position function $P_a^k(i, j)$ before. On the other hand, we may fix n_1 and n_2 and view i, j as functions of k. We define the wave A component of the *memory function* of the processor (n_1, n_2) at time k as

$$M_a^k(n_1, n_2) = (i, j), \qquad (2.7)$$

i.e., (i, j) is the index of the element of wave A contained in processor (n_1, n_2) at time k. The memory function tells us what component is stored within a specific processor at a specified time. Rearranging equation (2.6), we obtain

VLSI Signal Processing

$$M_a^k(n_1, n_2) = (-\frac{3n_1 + n_2 + d - 3 - k}{3}, \frac{2n_2 - d + 3 + k}{3}) \quad (2.8)$$

For fixed n_1, n_2 and k, the computed data indices i and j of wave A are not necessarily integers. If they are not integers, where $a(i, j)$ cannot be defined, we may assign $M_a^k(n_1, n_2)$ the value "nil". In fact, it is not hard to see that each processor contains "nil" values for two succesive time steps in k, then gets a new value of $a(i, j)$ on the third time step.

Wave B and C can be treated similarly. Looking at Fig 4 and using the same approach, we get the STD equations for wave B

$$2n_1 - i - 2j - d + 3 + k = 0 \quad , \quad (2.9.a)$$

$$2n_2 - 3i - d + 3 + k = 0 \quad . \quad (2.9.b)$$

The corresponding position function and memory function are

$$P_b^k(i, j) = (\frac{i + 2j + d - 3 - k}{2}, \frac{3i + d - 3 - k}{2}) \quad . \quad (2.10)$$

and

$$M_b^k(n_1, n_2) = (\frac{2n_2 - d + 3 + k}{3}, \frac{3n_1 - n_2 - d + 3 + k}{3}) \quad (2.11)$$

Similarly, from Fig 5 and using the previous procedure, we find the STD equations of wave C

$$2n_1 + i - j = 0 \quad , \quad (2.12.a)$$

$$2n_2 + 3i + 3j - 6 + 2d - 2k = 0 \quad . \quad (2.12.b)$$

The position function and memory function are

$$P_c^k(i, j) = (\frac{-i + j}{2}, -\frac{3i + 3j - 6 + 2d - 2k}{2}) \quad (2.13)$$

and

$$M_c^k(n_1, n_2) = (-\frac{3n_1 + n_2 - k + d - 3}{3}, \frac{3n_1 - n_2 + k - d + 3}{3}) \quad (2.14)$$

respectively.

In order to prove the components relationship for the matrix multiplication, i.e.

$$c(i, j) = \sum_{p=1}^{N} a(i, p) b(p, j) \quad , \quad (2.15)$$

All we have to do here is to guarantee that the suitable data will come to the same processor at the right time. So let us keep track of the locus of motion of some arbitrary component of the matrix C, say $c(i, j)$, and see what components, which belong to other waves, are within the same processor at the same time. This statement can be described precisely by two compound functions: $M_a^k(P_c^k(i, j))$ and $M_b^k(P_c^k(i, j))$. Using (2.8), (2.11), and (2.13), we can evaluate the values of these two compound functions and get

$$M_a^k(P_c^k(i,j)) = (i, p(k)) \quad , \quad (2.16.a)$$

$$M_b^k(P_c^k(i,j)) = (p(k), j) \quad . \quad (2.16.b)$$

where

$$p(k) = -i - j + 3 + k - d \quad . \quad (2.17)$$

If $k \leq i + j + d - 3$, then $p(k) \leq 0$. However, $a(i,j)$ and $b(i,j)$ are defined only when i and j are positive integers. That means the component $c(i,j)$ has not yet encountered the other two waves. If $k \geq i + j + d - 2$, then $p(k)$ is monotonically increasing from 1 with the time clock until the upper bound, which is determined by the size of the input matrix A (or B), say N here. Let us take a simple case, say, $i = 1$ and $j = 1$. When $k = d$, these three components, i.e., $a(1,1)$, $b(1,1)$, and $c(1,1)$ meet one another at the processor $(0,0)$, which equals $P_c^d(1,1)$. This satisfies our original assumption. When $k = d + 1$, $a(1,2)$, $b(2,1)$ and $c(1,1)$ coincide at the processor $(0,1)$, which is $P_c^{d+1}(1,1)$, and so on. Therefore, we prove the correctness of the systolic matrix multiplication algorithm.

3. The Specification and Verification of General Systolic Wave Algorithms

In general, to specify a systolic wave algorithm precisely requires knowledge of processor coordinates, the action of each processor, the data wave values and their physical locations as a function of time. First, assume that there is a fixed network of processors at "coordinates" $\underline{n}$. These coordinates may be the physical location of each processor in space, or they may just be a convenient index. Let us also assume that the data flowing through the network may be partitioned into N "waves" of data $a_1(\underline{i}), \ldots, a_N(\underline{i})$, where each $a_j(\underline{i})$ is the jth set of data indexed by coordinate $\underline{i}$. As the computation proceeds, the value of each element in the wave will change, so let us call $a_j^k(\underline{i})$ the value of the ith element of wave j at time k.

As the data flows through the network of processors, differing elements of each wave will be located in different processors at different times. We assume that this data flow is independent of the values calculated by the processors. Let $\underline{i} = M_j^k(\underline{n})$ be the element of the jth wave which is located in processor $\underline{n}$ at time k. We assume that at most one element from each wave may be located in a single processor at any one time. If no element of the jth wave is at $\underline{n}$ at time k, we will say that the value of the "memory function" $M_j^k(\underline{n})$ is nil, and the values $a_j(nil)$ and $M_j^k(nil)$ we define to be nil. It is convenient to define the "position function" $P_j^k(\underline{i})$ as the inverse of $M_j^k(\underline{n})$; i.e. $\underline{n} = P_j^k(\underline{i})$ is the processor containing element $\underline{i}$ of the jth wave at time k. This function will have value nil if the element $\underline{i}$ is not assigned to any processor at time k. (The equations $\underline{n} = P_j^k(\underline{i})$ form the STD equations of section 2).

Finally, suppose that each processor at time k takes the N values $a_1, \ldots, a_N$ located in that processor from each data wave, and computes new values for the corresponding elements of each data wave. Let $a_j^{k+1}(\underline{i}) = G_{j,\underline{n}}^k(a_1, \ldots, a_N)$ be the new value of the ith element of the jth wave computed in processor $\underline{n}$ at time k given data values $a_1, \ldots, a_N$. If no element of wave j is located at processor $\underline{n}$ at time k, i.e. if $M_j^k(\underline{n}) = nil$, then we assume that $G_{j,\underline{n}}^k(a_1, \ldots, a_N)$ is also nil. If at time k an element of a wave $a_j^k(\underline{i})$ is not located in any processor, $P_j^k(\underline{i}) = nil$, then we treat the element as if it were in a temporary storage cell during this clock period, and set $a_j^{k+1}(\underline{i}) = a_j^k(\underline{i})$.

Now to verify a systolic algorithm, we start with some assumed initial values for all the waves of data, $a_1^0(\underline{i}), \ldots, a_N^0(\underline{i})$, compute the values for all data waves at all times k as the data flows through the processors, and verify that the final values $a_1^\infty(\underline{i}), \ldots, a_N^\infty(\underline{i})$ are the desired results. This verification needs to iterate over all processor nodes $\underline{n}$, all waves j, all wave elements $\underline{i}$, and all time k, and can be organized in a variety of ways. The simplest approach is to track the activity of every processor as a function of time:

For all time $k = 0, 1, 2, \ldots$

For all processor nodes $\underline{n}$:

$$\underline{i}_j = M_j^k(\underline{n}) \qquad \text{for } j = 1, \ldots, N$$

$$a_j^{k+1}(\underline{i}_j) = G_{j,\underline{n}}^k \left(a_1^k(\underline{i}_1), \ldots, a_N^k(\underline{i}_N) \right) \qquad \text{for } j = 1, \ldots, N$$

This procedure simply duplicates the calculation of every processor node over time, and thus precisely simulates the network. An alternative verification method which is more similar to that used in the matrix multiplication example above, is to track each wave of data through the array:

For all time $k = 0, 1, 2, \ldots$

For each wave $j = 1, \ldots, N$

For all elements $\underline{i}$:

$$\underline{n} = P_j^k(\underline{i})$$

$$\underline{i}_m = M_m^k(\underline{n}) \qquad \text{for } m = 1, \ldots, N$$

$$a_j^{k+1}(\underline{i}_j) = G_{j,\underline{n}}^k \left(a_1^k(\underline{i}_1), \ldots, a_N^k(\underline{i}_N) \right)$$

This approach may be more efficient in cases such as matrix multiplication where elements are only sparsely distributed through the processor array, so that $M_j^k(\underline{n})$ has many nil values.

Note that this approach also covers networks in which each processor is a finite state machine with state $\underline{x}^k(\underline{n})$ at time k. Simply add a new "data wave" $a_{N+1}^k(\underline{i})$ whose value is the state and whose position does not change with time.

4. Computational Wavefronts

In special cases, the waves of data flowing through the processor array can be treated as if they were waves propagating through a homogeneous medium. This will be the case if each wave $a_j^k(i)$ retains its "geometric shape" as it flows through the network. More rigorously, we will say that the processor array has "wavefront" behavior if the position and memory functions $P_j^k(i)$ and $M_j^k(n)$ are linear functions of the time and wave indices:

$$P_j^k(i) = \Phi_j \underline{i} + k \underline{\gamma}_j \quad \text{whenever} \quad P_j^k(i) \neq nil$$
$$M_j^k(\underline{n}) = \Psi_j \underline{n} + k \underline{\xi}_j \quad \text{whenever} \quad M_j^k(\underline{n}) \neq nil \qquad (4.1)$$

where Φ_j and Ψ_j are matrices, and where $\underline{\gamma}_j$ can be interpreted as the "physical wavefront velocity vector", and $\underline{\xi}_j$ can be interpreted as the "data wavefront velocity vector". In the matrix multiplication case, for example, the physical velocity of wave A was $\underline{\gamma}_a = (½, -½)$, and the data velocity vector of wave A was $\underline{\xi}_a = (\frac{1}{3}, \frac{1}{3})$.

Note that wave-like behavior can only occur on a processor network with the appropriate "shift-invariant" geometric regularity. Namely, there must be processors $\underline{n} = \Phi_j \underline{i} + k \underline{\gamma}_j$ for every i, j, k for which $P_j^k(i) \neq nil$.

Although this geometric regularity does not necessarily imply that all processors compute the same function $G_{j,\underline{n}}^k(a_1, \ldots, a_N)$, the wavefront behavior is particularly useful if the processors do all compute the same function at each time k, so that $G_{j,\underline{n}}^k$ is not a function of $\underline{n}$. Assume that $P_j^k(i)$ has no nil values, and is linear in k and i. In this case, it is easy to show that the mapping from the waves $a_j^k(i)$ to $a_j^{k+1}(i)$ is spatially invariant, and thus that the final data wave values $a_j^\infty(i)$ are a spatially invariant mapping from the initial values $a_j^0(i)$. This is the case, for example, in the matrix multiplication example of section 2. For such systems, our second verification procedure is particularly easy since we need only track a single element i of each wave through the processor network as time evolves, and verify that $a_j^\infty(i)$ obeys the correct mapping from the initial data wave values. Spatial invariance then guarantees that all data values will be correct.

By analogy with wave propagation through a medium, we might define a "computational wavefront" for each wave of data as the set of processors $\underline{n}$ located in a hyperplane "orthogonal" to the velocity vector, $\{\underline{n} \mid \underline{\gamma}_j^T \underline{n} = constant\}$. This implies a corresponding wavefront of data elements i which flow through these processors at time k, whose coordinates can be derived from the position function:

$$constant = \underline{\gamma}_m^T \underline{n} = \left(\underline{\gamma}_j^T \Phi_j\right) \underline{i} + k \left\|\underline{\gamma}_j\right\|^2 \qquad (4.2)$$

Since the wave moves as a unit through the processor array, these wavefronts can never cross.

While this "wavefront" idea is rather elegant, unfortunately, it has some technical problems. Most importantly, if data elements are kept in temporary storage and are not used on every cycle, then the position and memory functions $P_j^k(\underline{n})$

and $M_n^k(\underline{i})$ have numerous *nil* values. As a result, the data flow cannot be spatially invariant, and a simple "velocity" vector will not properly describe the behavior of this data flow. Another problem is that defining the "angle" between processors or data paths is a rather arbitrary concept. Thus the choice of which processors belong on a "wavefront" is rather arbitrary. In short, for systolic networks which are not spatially invariant, the existence of computational wavefronts may be intuitively pleasing and may simplify the synthesis of systolic algorithms, but it does not significantly simplify our procedure for proving correctness of a systolic algorithm.

5. Conclusion

In this paper we have used a new approach to describe data flow phenomena in systolic arrays. It is easy, rigorous and systematic. We have also clarified the concept of a "computational wavefront". Kung mentioned two open problems in [5]: one is the specification and verification of systolic algorithms, and the other is automatic algorithm design. We have solved the first problem for an important class of systolic algorithms. However, if the geometry is not very regular or the flow pattern is extremely complicated, it will not be easy to analyze the systolic algorithm. In this case, it may be preferable to start with a simple implementation of the desired algorithm which can be easily proved correct, and then systematically transform it into a parallel and pipelined form. Thus, we come to the second problem mentioned in [5] - automatic systolic algorithm design. This is still an open area of research.

6. Acknowledgement

The first author wishes to thank Wei K. Tsai for helpful discussions.

References

[1] H. T. Kung and C. E. Leiserson, "Systolic Array (for VLSI)," in *Sparse Matrix Proc. 1978*, SIAM, 1979, pp. 256-282.

[2] M. C. Chen and C. A. Mead, "Concurrent Algorithms as Space-Time Recursion Equations," in *USC Workshop on VLSI and Modern Signal Processing*, 1982.

[3] S. Y. Kung, K. S. Arun, R. J. Gal-ezer, and D. V. Bhaskar Rao, "Wavefront Array Processor: Language, Architecture, and Applications," *IEEE Trans. on Computer*, vol. 31, no. 11, pp. 1054-1066, Nov. 1982.

[4] U. Weiser and A. Davis, "A Wavefront Notational Tool for VLSI Array Design," in *VLSI Systems and Computations*, Rockville, MD: Computer Science Press, 1981, pp. 226-234.

[5] H. T. Kung, "Why Systolic Architectures?," *Computer*, vol. 15, no. 1, pp. 37-46, Jan. 1982.

VLSI Signal Processing

FIGURES

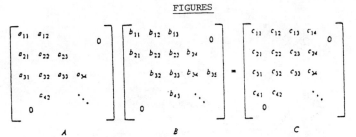

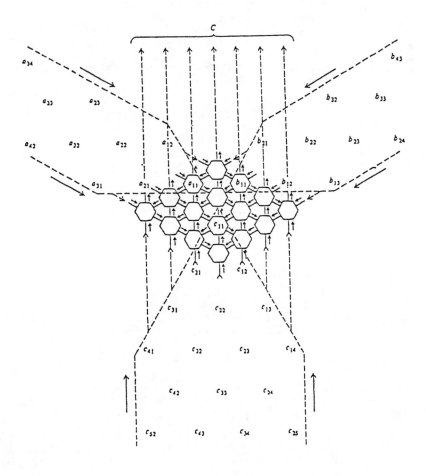

Fig. 1 Systolic matrix multiplication on a hexagonal array

VLSI Signal Processing

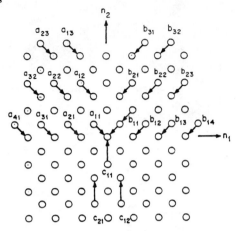

Fig. 2 Systolic matrix multiplication rearranged in (n_1, n_2) coordinates

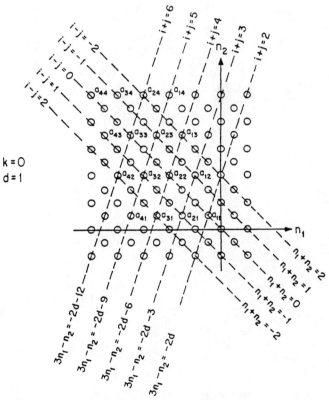

Fig. 3 Initial configuration of wave A

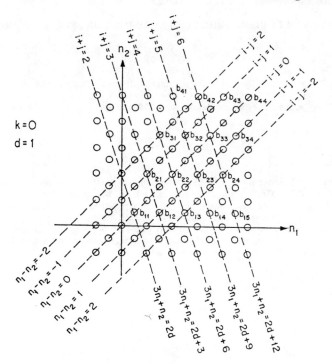

Fig. 4 Initial configuration of wave B

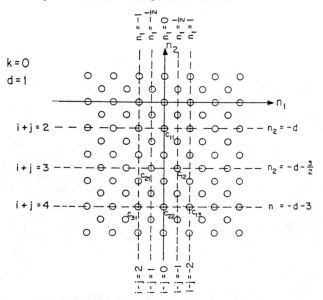

Fig. 5 Initial configuration of wave C

SIGNAL: A DATA FLOW ORIENTED LANGUAGE FOR SIGNAL PROCESSING.

P.Le Guernic, A.Benveniste, P.Bournai, T.Gautier

IRISA/INRIA, Campus de Beaulieu, F 35042 RENNES CEDEX, FRANCE

We present the language SIGNAL, which is a data flow oriented real time, synchronous, side effect free language suited to the expression and recovery of the parallelism in signal or image processing algorithms. The language is intended to be at the same time an executable simulation language, and a specification of a virtual machine implementing the algorithm. The language is semantically sound, and is suitable to program transforms, a major requirement when the ultimate goal is an aid to the architecture design.

The need for a fast implementation of signal or image processing algorithms is growing up very rapidly. The main concerned areas are: real-time control of industrial processes, telecommunications, videocommunications, speech and image coding, speech processing and real time computer vision, biomedical applications, and mostly signal and image processing encountered in radar and sonar systems.

On the other hand, the availability of
* new specialized microprocessors, such as the audio-bandwith range signal processing oriented microprocessors,
* faster technologies such as VHSIC,
* new silicon compilers
is increasing even more rapidly.

But a too long delay still remains for the fielding of these new tools and processors when the signal processing task definition and specification is considered as the entry point of the chain. This delay is mostly due to a relative lack of CAD tools from signal processing task specification to architecture specification.

Hence there is a need for a coherent set of tools with
* as input: formal specifications of signal processing tasks
* as output formal descriptions of multiprocessor architectures, or special purpose architectures (such as systolic ones),
and further providing a (partial) validation of the correctness of the implementation with respect to the specification.

For such a purpose, a basic tool is a suitable language for signal processing task specification and execution: the purpose of the present paper is to present briefly such a language, the language SIGNAL.

Complex signal processing tasks such as encountered in speech and image processing typically involve: 1/ the transform of signals by highly regular elementary algorithms (filters, LPC, FFT,...), 2/ some decision rules based on the monitoring of the signals (tests,...), thus eventually resulting in the running of signals with different clocks. Consequently, our main objectives in designing our language were the following:

* SIGNAL should be able to describe any real time signal processing task, involving possibly the running of signals with different clocks;

* SIGNAL should allow the programmer to specify , and the CAD system to recognize, in an easy way the parallelism present in a given task.

By the way, the SIGNAL specification of a task should be considered as the description of a virtual machine implementing this task.

Finally, SIGNAL should be more than a specification language, namely an executable language: this would allow the programmer to use SIGNAL as a simulation language for the validation of the algorithms at the highest level.

The paper is organized as follows. In the first section, the basic principles supporting the language are presented. The section II is devoted to the presentation of the construction of the block diagram describing the dependences between the various elementary subtasks. In the section III, the synchronization mechanisms will be presented. Finally, the section IV will be devoted to the presentation of some examples.

I. BASIC PRINCIPLES SUPPORTING THE LANGUAGE SIGNAL.

Recall the main objectives we have in mind in designing the language SIGNAL.

(i) SIGNAL should be able to describe any real time task (relevant to signal or image processing), that is to say, according to Young (1982), "any information processing activity or system which has to respond to externally generated input stimuli within a finite and **specifiable** delay"; by "specifiable", we have in mind that it is possible to know this delay prior to the running of the task.

(ii) the timing mechanisms should be described in an entirely synchronous way. The main reason for this request is that our primary interest is in the specification of algorithms independently of their eventual particular implementation, so that we can consider that every action is instantaneous, i.e. has a zero duration. As a matter of fact, this is the most easy way to prevent from any nondeterminism in the language , due to the unknown duration of the actions (as it has been pointed out by Berry & al. (1983)).

(iii) The specification or execution of a task should not require any prior knowledge of a universal time reference. The time has rather to be nothing but the ordering of the input stimuli or data of a given task; as a consequence, the notion of time will be multiform, since the cooperation between several subtasks (with their own local timing) through the exchange of signals can modify the time reference of these subtasks, as it will be shown later.

(iv) SIGNAL should allow the programmer to specify in an easy way the parallelism present in a given task. In other words, the expression of the task in SIGNAL should reflect as much as possible the dependencies between data and operations. This feature will be of primary importance for the recovery of the parallelism and pipelining capabilities.

Finally, as any modern language, SIGNAL should provide all the state of the art characteristics to ensure modularity, strong typing, readability, security. Moreover, the temporal instructions of SIGNAL should be built from a small set of primitive statements, for which a formal semantics must be given.

Some existing languages (Young (1982); ESTEREL, Berry & al. (1983)) almost satisfy these requirements. Related to the classical **real time languages**, these languages concentrate on a microcospic and explicit description of the timing of the actions. As a consequence, the recovery of the data dependencies becomes difficult for complex tasks (point (iv)).

On the other hand, **applicative languages** are tailored to express the parallelism existing in a task (LISP, Mc. Carthy (1966); LUCID, Ashcroft (1977); FP, Backus (1978); DSM, Cremers & Hibbard (1982)). But these languages are not well suited to the timing problems.

VLSI Signal Processing

Related to the the family of the applicative language is the family of **data flow languages** (Dennis (1974); Kahn (1974); VAL, Mc. Graw (1982); Ackerman (1979); CAJOLE, Hankin (1981)). In the data flow languages, the instruction firing rule can be based simply on the availability of data. These languages are suited to the processing of infinite files of data; they do not require any explicit specification of the timing of each action, because of this automatic firing rule. Howewer, purely data flow languages are asynchronous, so that they do not fulfill the requirement (ii).

The language SIGNAL exhibits several of the interesting features of both the real time and data flow languages. These are obtained in the following way:

a SIGNAL program is

(i) the specification of a **static oriented network** (or block-diagram, in the framework of signal processing) **expressing exactly the data dependencies. Data flow along the paths of the network; at the nodes, actions are performed by processes** (or black-boxes in the signal processing framework).

(ii) the description of elementary processes, referred to as **generators**, acting at each node. Among the set of all generators, we shall distinguish the **temporal generators**, which are specific to the language SIGNAL. The purpose of these temporal generators is to express the relative timing of the various data flows (or **signals**) in an entirely synchronous way. Thus the availability of a signal is completely characterized by a **clock**, so that we shall be able to refer precisely to the time at which an action is performed.

(iii) Among the set of the actions which are performed at the same instant, the firing rule is determined according to the principle of **data flow** (i.e. automatic firing by the availability of data at the input ports). Thanks to this last principle, the timing is completely specified by the static relationships between the various clocks involved in a task; the internal timing of a clock (i.e. the detailed specification of the events subject to a given clock) can be completely implicit, contrary to the classical real time languages.

To summarize, SIGNAL is a real time, synchronous, side effect free language suited to the expression and recovery of parallelism.

II. CONSTRUCTION OF THE STATIC NETWORKS.

In order to improve the understanding, the reader who is familiar with signal processing or automatic control can refer to the construction of block-diagrams through the interconnection of black-boxes. The formal semantics of this part of the language SIGNAL is given in Le Guernic (1982).

II.1 PROCESSES AND GENERATORS.

(i) **generators**: a generator is specified by
. its name
. a list of labelled input ports
. a list of labelled ouptut ports.
The labelling of the input ports is free, but the labelling of the output ports is subject to the following constraint:

(C.1) **two different output ports must have different names.**

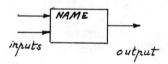

(ii) **processes**: a process is a network of interconnected generators. These connections are subject to the following constraints:

(C.2) **an input port can be connected to at most one output port,**

whereas an output port can be connected to one or several input ports. The condition (C.2) is intended to prevent from any nondeterminism in the language.

A process is characterized by
. a name P (optional)
. a labelled list of input ports, denoted by, say, ?a,b,c,...
.a labelled list of output ports satisfying (C.1), denoted by, say, !x,y,z,...

. a body, describing how the network was constructed.

The set of the labelled ports defines the interfaces of the process. These labelled ports are subject to the following constraints. Labelled output ports are or are not connected to input ports, but

(C.3) **a nonlabelled input port must be connected to some output port; conversely, a labelled input port cannot be connected to any output port of the network.**

Labelled ports will be used for the interconnection of processes; hence, the condition (C.3) prevents from any possible future violation of (C.1) or (C.2).

(iii) **connecting processes: the basic principles.**

These principles are the following:
. an existing connection cannot be removed,
. connections of processes are obtained through the identity of the names of the corresponding input and output ports.

The interconnections of ports of several processes result in a new process; this is exactly the way to build new processes from generators or already defined processes. The next paragraphs will be devoted to the description of operators for connecting processes.

II.2 PRIMITIVE OPERATORS.

These operators are the only ones for which a formal semantics has been given in Le Guernic (1982). We shall here present these operators in an informal way. In order to know the result of an operator, it is sufficient to know the **sort** of the involved process, i.e. the list of the names of their labelled ports. There are three primitive opetators.

(i) **binary operator PARALLEL:** given two processes P_1 and P_2, the operator PARALLEL replaces any pair of input ports of P_1 and P_2 with the same name by a single one, thus broadcasting the same values on these two different ports. The result is a new process, denoted by

$P = P_1, P_2$

iff it satisfies (C.1). PARALLEL is **associative** and **commutative**.

(ii) **1-ary operator LOOP:** given a process P and a name x belonging to both sets of labelled input and output ports of P, LOOP connects the output port labelled x to the input port(s) labelled x. If it satisfies (C.1), the result is a new process, denoted by

$Q = P \& x$

In any other case, LOOP has no effect.

(iii) **1-ary operators RELABELLING, HOLD:** the operator RELABELLING modifies the names of a process; the RELABELLING of an output port by " $\overline{}$ " (no name) results in a masking of this port. HOLD specifies the labelled output ports which are not masked. In any case, the result must satisfy the condition (C.1). The notation for these operators is, for RELABELLING

$Q = P <?a\!:\!b \ !c\!:\!d,e\!:\!\overline{}>$

where the old names are before : and the new corresponding ones after : ; and , for HOLD

$Q = P\{a\}$

which results in the masking of every output port except a.

These are the only primitive operators of this part of SIGNAL. The next paragrphs are devoted to the presentation of two other binary operators, and of cascades of these; all these forthcoming operators are defined as macros in terms of the primitive ones.

II.3 OTHER BINARY OPERATORS.

We give here only an informal definition, the formal one can be found in Le Guernic & al. (1984).

(i) **COMPOSITION:** given two processes P and Q, the COMPOSITION connects the output ports of P (resp. Q) to the input ports of Q (resp. P) with the same name. If the result satisfies (C.1), it is a new process, denoted by

$R = P|Q$

COMPOSITION is **commutative** and **associative**.

(ii) **SEQUENCE:** given two processes P_1 and P_2, their SEQUENCE, denoted by

$Q = P_1; P_2$

VLSI Signal Processing

is **always defined**, and is obtained as follows: every labelled output port of P_1 is connected to the input port(s) of P_2 with the same name; then, (C.1) is satisfied through the automatic masking of every labelled output port of P_1 with a name occurring in the set of the output ports of P_2. The SEQUENCE is **associative**, but not commutative. Inside a sequence, names of labelled ports can be used almost like variables in sequential languages. For example,

a := b+c ; c := b+a

can be interpreted as the corresponding sequence of PASCAL instructions, the only difference being that the values carried by the input port c are saved.

II.4 REGULAR ARRAYS OF PROCESSES.

We give here facilities for constructing regular arrays of processes indexed by a parameter. The general (informal) syntax is the following:

do... i **until** n **of** P(i) **od**

where

... = **par** for PARALLEL
    seq for SEQUENCE
    com for COMPOSITION

n is a parameter of integer type, $0<i<n+1$ is an integer, and P(i) is an indexed process with parameter i. For example,

dopar i **until** n **of** p(i) **od** = P(1),...,P(n).

II.5 EXAMPLES.

Let P(?x,y !x,y) be a process.
1/ Set
$P(i,j) = P<?x:x_{i,j}, y:y_{i,j} \; !x:x_{i,j+1}, y:y_{i+1,j}>$
Then

(**doseq** i **until** n **of**
  (**doseq** j **until** m **of** P(i,j) **od**)
od)
$\{$for i until n: $x_{i,m+1}$ for j until m: $y_{n+1,j}\}$

is the following rectangular array of copies of P:

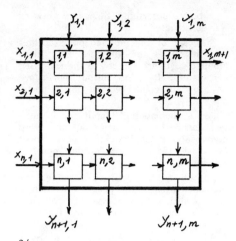

2/
doseq i **until** n **of** P **od**
is the linear array

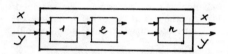

3/ Set
$P(i) = P<?x:x_i, y:y_{i+1} \; !x:x_{i+1}, y:y_i>$
then

(**docom** i **until** n **of** P(i))$\{y_1, x_{n+1}\}$

is the cascade

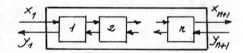

These operators are sufficient ofr constructing any static network. We have here described the first part of the language SIGNAL.

III THE SYNCHRONIZATION MECHANISMS OF THE LANGUAGE SIGNAL: THE TEMPORAL GENERATORS.

This is the most important and original part of the language SIGNAL. Among the whole instruction set of SIGNAL devoted to the description of the generators, we shall distinguish two subclasses:

* the subclass of **functions**: this subclass contains the elementary instantaneous transformations on the data (+,∗, and, or,...); functions are not specific to SIGNAL, but are chosen for a given version of the language, according to the desired class of applications. For example, the software simulation of signal processing algorithms could require the classical PASCAL- or FORTRAN-like arithmetic and boolean operators. But, for the study of rounding effects, it can be desirable to provide the programmer with "mult" or "add" with a specifiable accuracy.

* the subclass of **temporal generators**: temporal generators are specific to SIGNAL, whatever the version of the language is of interest. These generators are the basic tools for the description of the synchronization mechanisms between the various signals. The present section will emphasize on the description of the temporal generators of SIGNAL.

III.1 SIGNALS AND CLOCKS.

By a **flow**, we have in mind an ordered file of typed data of an unspecified length. Given a process (i.e. a static network) defined according to the section II, flows run along the paths of the network, and are transformed by the generators according to the following principles of synchronism:

1/ to each flow is associated a **clock**, i.e. an increasing function from $\overline{N}$ into a denumerable totally ordered set **H**; **this clock specifies at which date the i-th data of the flow is available for a given processing**. By a **signal**, we shall mean a pair (flow, clock); data never are persistent (unlike variables in classical sequential languages), but are rather lost after the date at which they are available.

2/ **relationships between clocks can be specified in a static way**: clocks are modified through the transform of the associated signals by temporal generators only.

3/ As we have mentioned in the introduction, we do not assume the existence of a universal clock. Hence, we are only interested in the relationships between clocks; in other words, the set of the clocks of a given process is defined up to a global change of time. This is of major importance, since the connection of processes generally results in such a change of time, as we shall see later.

The algebra of clocks.

The set of the clocks of a given process is endowed with a structure of partially ordered set. The relationship

$H \leq G$

(where H and G are clocks) means that **the set of the values of H is contained in the set of the values of G**, i.e. every instant of the clock H is also an instant of the clock G (this order is different from the order introduced by Caspi & Halbwachs (1982) for recognizing the causality between events). The clock $\perp$ defined by

$\perp(1) = 1$, $\perp(n) = \infty$ for $n > 1$

(i.e. the only instant of $\perp$ is the initial one) is the infimum of all the clocks. As we shall see, the set of the clocks is generally not a lattice, but we shall say that two clocks H_1 and H_2 are **compatible** if there exists a clock G such that

$H_1 \leq G$ and $H_2 \leq G$.

Then, according to Zorn's lemma, we can define the supremum and the infimum of these clocks, respectively denoted by

$H_1 \vee H_2$ and $H_1 \wedge H_2$

The relationship "H_1 and H_2 compatible" is denoted by

$H_1 -- H_2$

This structure is sufficient to specify the undersampling of a clock, but, in order to be able to specify oversampling of clocks, we shall need another structure on the set of the clocks, which we shall define now. Given an indexed set $H_1,...,H_n$ of compatible clocks, we define their **multiplexing** as follows: the resulting clock H we shall denote by

$$H = H_1 \, / \, ... \, / \, H_n$$

is the time multiplexing of the clocks $H_1,...,H_n$; the set of the values of H is then a subset of the set of the pairs $(G(t),i)$ for $t=1,...$ and $i=1,...,n$, where G is the sup of the clocks H_i, this set being totally ordered according to

$(G(t),i) \leq (G(t'),i')$ $\iff$
$G(t) < G(t')$, or $G(t)=G(t')$ and $i \leq i'$

A formal model presented in Benveniste & al. (1984) supports these notions; it is there proven that this structure on the set of the clocks is sufficient for specifying any new clock from existing ones.

III.2 THE TEMPORAL GENERATORS OF SIGNAL.

We shall present here informally the set of the temporal generators available to the programmer. Some of these generators are macros defined over primitive generators. The primitive generators are described in Le Guernic & al. (1984), where a formal semantics is given for these generators in terms of the Conditional Rewriting Rules of Plotkin (1981). These generators will be presented according to the following scheme:
. syntax
. list and type of interfaces
. result
. conditions on the clocks of the input signals
. effect on the clock(s) of the output signal(s)

(i) DELAY

. x is y[n] $m init ...

. ?x of type $\underline{T}$!y of type $\underline{T}^n$ (i.e. a vector of size n over $\underline{T}$), where n and m are parameters of integer type

. the specification ... after **init** is an element of $\underline{T}^{n+m}$ defining the initial conditions; then, writing y as follows

(init) $y_{-m-n+1},...,y_0$, (signal) $y_1,...$
x is defined by
$x_t = (y_{t-m},...,y_{t-m-n+1})$
. none
. $H(x) = H(y)$

COMMENT: the DELAY is the basic tool for organizing a SIGNAL program as an automaton, following the ideas of Cremers & Hibbard (1976) and their notion of "data space".

(ii) IF

. if C then P else P' fi

where C is a condition, P and P' are processes

. ? input of P or P', boolean C
! output of P or P'

. given an output port of P (resp. P'), the value carried by this port at a given instant is delivered at the output of the resulting process iff C is available and true (resp. false); in order to satisfy (C.1), if two output ports of P and P' have the same name, their filtered signals are merged to give a single output port of the resulting process; there is no problem of nondeterminism since C selects at a given instant at most one of the ≤ 2 available values.

. The clocks of !P, !P', and C have to be compatible

. If a is an output port of both P and P', the resulting clock of the corresponding port a_{if} of the **if...fi** process is given by

$H(a_{if}) = H(a) \wedge H(C)$

on the other hand, if b (resp. b') is an output port of P (resp. P') only, we have

$H(b_{if}) = H(b) \wedge C$
$H(b'_{if}) = H(b') \wedge \lceil C$

where the clock denoted by C (resp. $\lceil C$) is defined as the occurrences "true" (resp. "false") of the boolean signal C.

COMMENT: the primitive IF statements act on signals, not on processes, and thus result in simpler effects on the clocks, but this macro is much more powerful and secure, since the underlying "merge" of signals is always correct.

(iii) MUX

. y **is mux** x

. ?x = $|/x_1,...,x_n/|$ is a **signal collection** of type $\underline{T}$

!y signal of type $\underline{T}$

By a signal collection, we have in mind an indexed set of signals $x_1,...,x_n$ of the same type, with compatible clocks.

. the signals $x_1,...,x_n$ are time multiplexed according to the ordering defined by the index i=1,...,n.

. the clocks of the signals x_i must be compatible

. $H(y) = H(x_1)/.../H(x_n)$

COMMENT: the MUX generator is somewhat difficult to handle with, as far as the timing aspects are concerned, for the following reason. Given two compatible signal collections x and x', the reader should convince himself that there is generally no reasonable definition of the relative timing of the outputs y and y' of

y **is mux** x, y' **is mux** x'

For this reason, we have to consider that **the signals y and y' are no more compatible,** which results in the fact that **the set of all the clocks of a process is generally not a lattice, provided that the MUX generator is used.**

COMMENT: the two generators IF and MUX effectively realize the clock transforms mentioned in the preceding paragraph; the language SIGNAL thus provide the programmer with a complete set of synchronization mechanisms. However, the forthcoming generators are also useful for the following reasons: 1/ in SIGNAL, the time index is implicit, and generally not available to the programmer, a feature which is sometimes a drawback, especially when the task to be specified is modified at a given date, 2/ in SIGNAL, **the values of a signal are not persistent,** which is in agreement with our synchronization mechanisms, but is a drawback when the programmer need to save a value for an unspecified duration (like a variable).

(iv) COUNT and related generators.

. n **counts** y

. ? signal y
 ! integer n

. the integer signal n counts the past occurrences of the signal y.

. none

. $H(n) = H(y)$

The COUNT generator gives access to the date with respect to the mentioned signal. Two related macros are also available, and are presented in Le GUernic & al. (1984), which allow to specify event-based restarting from zero of the counter.

(v) EXTENDS, RESTRICTS.

. y **extends** a **at** b
 y **restricts** a **at** b

. ? signals a and b
 ! y of same type as a

. EXTENDS: the last available value of the signal a is regenerated at the output port y when a value is available at some of the input ports;

RESTRICTS: the current value of the signal a is delivered at the output port when both a and b are available.

. the inputs have to be compatible

. EXTENDS: $H(y) = H(a) \vee H(b)$
 RESTRICTS: $H(y) = H(a) \wedge H(b)$

(vi) FUNCTIONS.

As we have mentioned before, other generators will be referred to under the generic name of "functions": these generators transform only the instantaneous data, and have no effect on the clocks. A typical example is

a := b+c

We only indicate here how the clocks are concerned by functions, namely the conditions, then the result:

. all the input signals have the same clock

. the output ports have the same clock as the input ports.

This finishes the description of the generators of SIGNAL.

III.3 HOW THE CLOCKS ARE TRANSFERRED THROUGH THE INTERCONNECTION OF PROCESSES: THE CLOCK CALCULUS.

VLSI Signal Processing

Given a process P (resp. two processes P and P'), we shall here indicate the effect of the primitive interconnection operators RELABELLING, LOOP (resp. PARALLEL) on the clocks of the signals carried by the labelled ports of this (these) process(es). This effect will appear as a transfer of clocks which we shall specify by a formal calculus, we shall refer to as the **clock calculus**. This clock calculus is performed by the SIGNAL compiler according to the following principles.

Given a process P specified as follows

$P(?x_1,...,x_n,C_1,...,C_m \ !y_1,...,y_p)$
P = "body"

where the inputs C_i are boolean, the calculus begins with a set of assumptions specifying the synchronization of the input signals with equations like

$H(!x_3) = H(!x_1) \wedge !C_2$
$H(!x_5) = H(!x_1)/H(!x_2)/H(!x_3)$

When no assumptions are explicitly stated, the compiler assumes implicitly that all the input signals have the same clock. Then the body of the process provides us with 1/ a set of generators resulting in the corresponding set of conditions to be proved and of statements on the clocks, according to the section III.2, 2/ interconnection operators transferring the clocks from a generator to the other. This set of assumptions, statements, and conditions has to be solved (or an error to be detected if this set is not consistent), the final result being

* a finite (and as small as possible) nested set of clocks together with the specification of their relationships

* for each clock, the set of the signals it governs.

Note that, when the MUX generator is not used, this calculus is nothing but the classical boolean calculus, for which automatic solvers are known. We shall now indicate the resulting effect on the clocks of the primitive interconnection operators.

RELABELLING: the operator
P<?a:x !b:y>
modifies as follows the set of clock-equations of P:

"substitute ?x to ?a and !y to !b"
LOOP: the operator
P&x
results in
"substitute !x to ?x"
PARALLEL: the operator
P, Q
results in the union of the set of equations associated to P and to Q.

Given a set of clock-equations and conditions, three cases can arise:
(i) some of the conditions are violated: an error is then detected;
(ii) There is >1 free clock (by free, we have in mind that this clock was not obtained from another one); in this case, the synchronization between the input signals is not completely specified, so that it is not possible to run the corresponding process as a main program; nevertheless, this process can be used for future connections with other processes;
(iii) there is a unique free clock, we shall refer to as the **master clock**: then it is possible to run the process as a main program, the process is said to be executable.

III.4 CANONICAL FORM OF A PROCESS AS A "DATA SPACE" (Cremers & Hibbard (1976)).

A first step in the compilation of a SIGNAL process is to transform it into a "data space" form. This is obtained by setting the network in the form depicted below

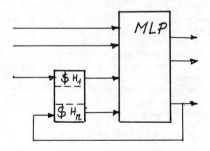

where the notation $\$(H_i)$ refers to the FIFO memory (created by a DELAY generator) ran by the clock H_i. The block labelled with the $'s is then the "state" of the automaton, whereas the main block labelled MLP (for "memoryless process") operates both the transforms on the states of the automaton, and its outputs. This form is suitable to the recovery of the ordering of the computations. Another advantage of the clock calculus is that it allow the compiler to reduce (prior to the execution of the program) the set of the possible transitions in the state space of the automaton, since the clock calculus specifies in a static way a hierarchy among the possible co-occurrences of the various signals.

IV EXAMPLES.

(i) Matrix multiplication with the WAP of S.Y.Kung (1984).

We refer the reader to Kung (1984) for a description of the WAP (Wavefront Array Processor) and the description and specification of the matrix multiplication, using a language suited to the WAP. We shall describe the multiplication of two n by n matrices A and B with result C as implemented on a virtual WAP. In the sequel,
$a_i = (a_{ik})$ is the i-th row of A
$b_j = (b_{kj})$ is the j-th column of B
c_{ij} is the (i,j)-th entry of C.
The SIGNAL program is as follows

specif MULT $\{?a_{i:i=1,...,n}, b_{j:j=1,...,n}$
                !$c_{i,j:i,j=1,...,n}\}$
parameter n

MULT =
(**doseq** i **until** n **of**
  (**doseq** j **until** n **of** C(i,j) **od**)
od)
$\{c_{i,j:i,j=1,...,n}\}$
where
C(i,j) = C<?a:a_i,b:b_j !c:$c_{i,j}$>
  where
  specif C ?a,b !a,b,c
  C=
  ((a **is** a$1,b **is** b$1);
  n **counts** a;
  (c:=zc + a∗b|zc **is** c$1))
  <zc: >;
  c **restricts** c **at** (k=n):true

end C
end C(i,j)
end MULT

COMMENTS: the body of the process C specifies the processing in a single WAP element. The signal "(k=n):true" emits a 1 when the mentioned condition (k=n) is true; hence, the instruction "c restricts c at (k=n):true" delivers the result of the accumulation at the end only (undersampling). The interconnection of the WAP elements is described using the cascade operators, according to the first example of (II.5). Finally, the pipelined structure of the WAP is simply expressed by delaying the input signals a and b of the WAP element C; the data flow firing rule of the instructions result automatically in the parallel running of the various fronts of the WAP.

(ii) On-line detection of jumps in the mean of a signal, using Page's stopping rule (Basseville & Deshayes (1984)).

This simple example will illustrate the synchronization mechanisms. An informal description of the algorithm is as follows: let y_t be a signal:
Step 1: from t>0 do
$m_t = m_{t-1} + g*(y_t - m_{t-1})$
Step 2: for $t \geq t_o$ do in parallel
$S^+(t) = S^+(t-1) + (y_t - m_t - j_{min})$
$M^+(t) = \max(S^+(s):s \leq t)$
$T^+(t) = \arg\max(S^+(s):s \leq t)$
test: $M^+(t) - S^+(t) >$ mu
and
$S^-(t) = S^-(t-1) + (y_t - m_t - j_{min})$
$M^-(t) = \min(S^-(s):s \leq t)$
$T^-(t) = \arg\min(S^-(s):s \leq t)$
test: $S^-(t) - M^-(t) >$ mu
when test:true, deliver T^+ or T^- according to which test was positive, and go to step 1.

The two thresholds are j_{min} (minimum size of the jump to be detected) and mu (threshold of the test); m_t is the current estimate of the mean; $M^+ - S^+$ (and the other with $^-$) are the loglikelihood ratios of the alternative hypothesis (a jump occurs) against the null hypothesis (no jump).

VLSI Signal Processing

The corresponding SIGNAL program is as follows.

specif PAGE'S TEST
$\{?y\ !m,\ \text{integer}\ T\}$
parameter j_{min}, mu, g integer t_o

PAGE'S TEST=
(n **counts** y;
(((MEAN, TT **extends** T **at** y);
if n-TT>t_o **then** TEST+,TEST- **fi**)|
if $(M^+-S^+ > mu)$ **or** $(S^- -M^- > mu)$
then
INIT,
   if $M^+-S^+\geq mu$ **then** T:=T^+ **else** T:=T^-
   fi
else STATE **fi**))
$\{m,T\}$

where
MEAN $\{?y,zm!m\}$
MEAN =
m := zm + g*(y-zm)
end MEAN
TEST+ $\{?y,n,zS^+,zM^+,zT^+\ !S^+,M^+,T^+\}$
TEST+ =
$S^+ := zS^+ + (y-m+j_{min})$;
if $S^+ \geq zM^+$
  then $M^+:=S^+,T^+:=n$
  else $M^+:=zM^+,T^+:=zT^+$
fi
end TEST+
TEST- $\{?y,n,zS^-,zM^-,zT^-\ !S^-,M^-,T^-\}$
TEST- =
$S^- := zS^- + (y-m-j_{min})$;
if $S^- \leq zM^-$
  then $M^-:=S^-,T^-:=n$
  else $M^-:=zM^-,T^-:=zT^-$
fi
end TEST-
INIT $\{?n\ \ !zm,zS^+,zM^+,zT^+,zS^-,zM^-,zT^-\}$
INIT =
zm:=0,zS^+:=0,zS^-:=0,
zM^+:=0,zM^-:=0,zT^+:=n,zT^-:=n
end INIT
STATE $\{?m,S^+,M^+,T^+,S^-,M^-,T^-$
    $!zm,zM^+,zS^+,zT^+,zS^-,zM^-,zT^-\}$
STATE =
zm **is** m$1,
zS^+ **is** $S^+$$1,zS^- **is** $S^-$$1,
zM^+ **is** $M^+$$1,zM^- **is** $M^-$$1,
zT^+ **is** $T^+$$1,zT^- **is** $T^-$$1
end STATE
end PAGE'S TEST

COMMENTS: 1/ As in the previous example, the statement "where" is used for a hierarchical description of the process. For example, the main program describes the block-diagram depicted in the figure of the appendix. The clock calculus cannot be described here, since it has to be applied to the primitive generators and connection operators only.

2/ We should point out the following point: if the programmer is interested in restarting the monitoring procedure, not at the instant of detection, but at the estimated instant of change (which is past from an unbounded quantity), then the corresponding algorithm cannot be described in SIGNAL. The reason is that, for this algorithm to be implemented, a buffer of unspecified size is necessary, so that **this algorithm is no more of real time type**, according to the definition we have given. This is due to the fact that **it is proven in Le Guernic & al. (1984) that every task specified in SIGNAL is necessarily a real time task**. This guaranty is at the same time an advantage (automatic guaranty of real time property) and a drawback, especially for some pattern recognition algorithms.

A much more complex virtual machine has been specified in SIGNAL, namely the machine realizing the Adaptive DPCM TV coder described in Richard & al.(1984). In this machine, around 100 different clocks are to be specified, which results in a tremendeous amount of interleaved DO loops if the time has to be processed explicitely by the programmer, like in the classical sequential languages, but also in applicative languages such as DSM (Crmers & Hibbard (1982)).

CONCLUSION

We have presented the language SIGNAL, which is a data flow oriented real time, synchronous, side effect free language suited to the expression and recovery of parallelism in signal or image processing tasks.

REFERENCES

W.B.Ackerman (1979): Data flow languages; Proc. AFIPS conf. (E.Mervin Ed.) New York.

E.A.Ashcroft,W.W.Wadge (1977): LUCID, a non procedural language with iteration; CACM, Vol 20 n°7

J.Backus: Can programming be liberated from the Von Neumann style? A functional style and its algebra of programs; CACM Vol 2 n°8.

M.Basseville, J.Deshayes (1984): Detection of abrupt changes in signals and systems, CNRS conf. Paris 21-22 March 1984.

A.Benveniste, P.Le Guernic (1984): Synchronous real time systems: a formal model; IRISA Rep.

G.Berry, S.Moisan, J.P.Rigault (1983): Towards a synchronous and semantically sound high level language for Real time applications; rep. Centre de Math. Appl. Sophia Antipolis.

J.Mc Carthy & al.(1966): LISP 1.5 Programmer's manual; MIT Press.

P.Caspi, N.Halbwachs (1982): An algebra of events: a model for parallel and real time systems; Int. Conf. on parallel processing, Bellaire, August 1982.

A.B.Cremers, T.Hibbard (1978): Formal model of virtual machines;IEEE Soft. Eng. 4, 426-436.

A.B.Cremers,T.N.Hibbard (1982): Executable specification of concurrent algorithms in terms of applicative data space notation; USC workshop on VLSI signal proc., 1982.

J.B.Dennis (1974): First version of a data flow procedure language; LNCS Vol 9 Springer Verlag.

C.L.Hankin, H.W.Glaser (1981): The data flow programming language CAJOLE, an informal introduction; SIGPLAN Notices, Vol 16 n°16.

J.R.Mc Graw (1982): The VAL language: description and analysis; ACM trans. on Prog. Lang. and sys., Vol 4 n°1.

G.Kahn (1974): The semantics of a simple language for parallel programming; Proc. IFIP congress 1974.

P.Le Guernic (1982): SIGNAL, an algebraic description of signal flows; Proc. Congrès AFCET "Architecture des machines et systèmes informatiques, Editions Hommes et Techniques.

P.Le Guernic, A.Benveniste, P.Bournai, T.Gautier (1984): SIGNAL, a data flow oriented language for signal processing; IRISA Tech. Rep.

G.D.Plotkin (1981): A structural approach to operational semantics; Daimi FN 19, Aarhus univ., Comp. Sc. Dept.

C.Richard, A.Benveniste, F.Kretz (1984): recursive estimation of edges in TV pictures as applied to ADPCM coding; IEEE COM-32 n°6.

S.J.Young (1982): Real time languages: design and development. Elis Horwood publishers

APPENDIX: Block-diagram of the second example.

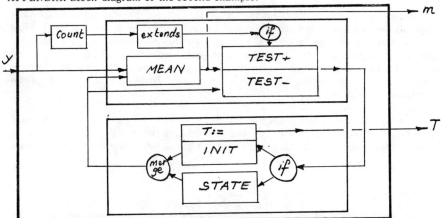

Hierarchical Iterative Flowgraph Integration for VLSI Array Processors[*]

S.Y. Kung
University of Southern California

J. Annevelink and P. Dewilde
Delft University of Technology

ABSTRACT

The structural properties of parallel recursive algorithms point to the feasibility of a Hierarchical Iterative Flow-graph Integration (**HIFI**) design method for VLSI array processor design. This HIFI method includes: the recursive algorithm decomposition, abstract notations, flow-graph (structural) and functional(behavior) description, transformation procedures, bi-directional mapping between graphic and textual codes, simulation and verification tools, and silicon compilation.

1. INTRODUCTION

In this paper we propose a new CAD methodology for highly parallel signal processors. One major issue is how to express parallel algorithms in a notation which is easy to understand by humans and possible to compile into efficient VLSI circuits and/or array processor machine codes. More specifically, we advocate a methodology based on a hierarchical and recursive mapping from algorithms to VLSI array hardwares. At a top level, a powerful (although abstract) notation is adopted to express the space-time activities in its full parallelism, which will eventually be compiled into primitive modules.

Based on a thorough taxonomy analysis on the parallel algorithms most commonly encountered in the signal processing applications, we propose a Hierarchical Iterative Flow-graph Integration (HIFI) design/description method. This includes recursive algorithm decomposition, abstract notations, flow-graph (structural) and functional(behavior) descriptions, transformation procedures, bi-directional mapping between graphic and textual codes, simulation and verification tools, and silicon compilation.

The HIFI design system will offer the designer with the <u>definitional</u> mechanisms for (1) <u>temporal</u> decomposition and (2) <u>structural</u> decomposition. Using these two basic decomposition techniques, the designer can do stepwise refinements in the hierarchical

---

[*] This research was supported in part by the Commission of the EEC under the 3744/81 Program (ICD Contract), by ZWO, the Dutch Foundation for pure scientific research, by the Office of Naval Research under contracts N00014-81-K0191 and N00014-83-C-0377 and by the National Science Foundation under Grant ECS-82-12479.

design approach. The final HIFI design/description system will offer the following desirable features:

1. Expressiveness: The HIFI system must support a notation for describing the parallel space-time activities occurring in signal processing algorithms in a natural way.

2. Freedom from detail: Information hiding, i.e. hierarchical specifications and a separation between virtual and real machine design, will allow the designer to focus his attention at the appropriate level of detail.

3. Simplicity: The definitional kernel of the design system, i.e. the design objects and the operations allowed on them, must be simple. However, care must be taken to ensure that the operations are sufficiently powerful.

As an example, let us look in the description of the space-time activities associated with a systolic array. There are already quite a number of research efforts attempting to describe the space-time activities of systolic arrays [1,2]. These attempts have in common that they try to describe the actual space-time activities, including pipelining, for example in terms of snap-shots[*], or as a recursive space-time program. As a result these descriptions are often rather complex, and therefore do not support the three desired features mentioned above. On the other hand, tracing recursion or wavefronts, as e.g. done in the wavefront array language MDFL [4], instead of snapshots, offers a contrasting (and perhaps more preferable) alternative. So a natural question to ask now is: Is there a powerful description language/notation for array processing?

The prime abstraction proposed in our approach is that of the Signal Flow Graph [3]. The Signal Flow Graph (SFG) representation derives its power from the fact that the computations are assumed to be delay free, i.e. they take no time at all. Consequently, the burden of tracing the detailed time-space activities associated with pipelining are eliminated. Moreover, any delay present in the system has to be explicitly introduced in the form of so-called Delay branches. These Delay branches allow history sensitive systems to be described in a clear and unambiguous way. Such a model is consistent with the concern Backus expressed over an "extended" functional programming, e.g. the AST systems introduced in [5,6]. Although the abstraction provided by the SFG is very powerful, transformation of a SFG description to a wavefront or systolic array description, including the pipelining, can be made rather straightforwardly. As shown in [3] there are some existing theorems which facilitate these transformations.

Therefore, our approach offers an effective starting point for the design automation and software/hardware techniques. This is because (1) SFG provides a powerful (although mathematical) abstraction to express parallelism, and yet (2) transforming from SFG to (the more realistic) systolic/wavefront arrays is straightforward.

2. HIFI design methodology

The HIFI design method is based on the concept of <u>node refinement</u>. Starting from a single node specification the system is specified in more and more detail by applying decomposition functions to the nodes.

---

[*] A snap-shot is a description of the activities at a particular time instant.

VLSI Signal Processing

By naming nodes or functions performed by the nodes, the design is (hierarchically) decomposed, allowing the designer to focus his attention on the specific node or function selected. A node refinement is used to specify both <u>structural</u> and <u>behavioral</u> refinement. Based on algorithm analysis, e.g. the recursive decomposition scheme, we will be able to identify certain useful and often occurring structures, e.g. arrays, trees etc. For an example, the concept of local recursiveness suggests locally interconnected structural primitives, applicable to most signal processing algorithms such as convolution, correlation, LU and QR decompositions.

The behavior of a node is specified in terms of the actions, i.e. offering and accepting data, that take place at the input and output ports of a node. This type of behavior definition will allow us to concentrate on the 'external' behavior, thereby deferring the details of the internal operation of the node to a later refinement.

2.1 Node - Specification

In a mathematical model, a node is given by a triple $<S, F, P>$, where S is a set of states, F is a set of functions, and P is a set of ports. The states associated with a node are called the node-control states, a specific state being denoted with a lowercase indexed s. The set of ports P consists of a set of input ports, Pin, and a set of output ports, Pout. Each of these ports carries a value of a specific data-type, together with related control information.

The mathematical model of a node is then given by the following equation:

$$F_{\tilde{s}_j, \tilde{p}_{in}}(\bar{P}_{in}) \rightarrow (\tilde{s}_{j+1}, P_{out})$$

This equation can be read as follows: There is a function, as determined by the current state and the control-part of the current input values, which maps the current input values (data-part) into the new state and output values (containing both data and control-part).

The functions and states of a node are specified with a 'behavior' construct, using so-called Bmaps. A Bmap is an applicative, i.e. side-effect free and history insensitive, mapping from a set of (input) ports to a set of (output) ports. By specifying the behavior of a node as a sequence of Bmaps we implicitly define a state.

If a Bmap is a simple mapping from input to output ports, it is specified by naming the input and output ports that are active, i.e. where an action takes place. If necessary this specification can be made more detailed by attaching names to the values accepted at the input ports. These names can then be used in expressions specifying the value made available at an output port.

2.2 Node Refinement

The purpose of refining a node specification is to give an implementation of the behavior of a node in terms of simpler behaviors with some added control. The control specifies the composition of the simpler behaviors. Composition of behavior can be specified in two way's, composition in space and composition in time. Composition in space is specified by defining a structure of nodes. Composition in time is specified with a refined behavior description of the node. In practice, a node refinement often consists of both, composition in time and composition in space of (simpler) behaviors.

Refinement steps are repeated with the simpler behaviors, until we arrive at a level where the behavior, i.e. the Bmaps that make up the behavior, are directly implementable in hardware, e.g. adders, multipliers etc. In order to give a complete hardware implementation of the behavior, we need only define a systematic way of implementing the control associated with the behavior. However, since we have only simple control conditions, that can be evaluated using data offered at the node-input ports, this does not seem to be a complex task. Note also that, due to the hierarchical refinement of node behaviors, we have at the same time given an hierarchical specification of the control. This is another advantage that will further ease the implementation of the control circuitry.

In what follows, we consider two basic refinement steps, resp. (1) Temporal decomposition and (2) Spatial decomposition. The Temporal decomposition scheme can be applied to a node whose behavior is given by a Bmap. It is used when a map can be recursively formulated. By decomposing a functional map in a recursively defined sequence of maps, the concept of state is introduced in a natural way.

Spatial decomposition is used to model concurrency. It can be applied to any node. Due to the powerful abstractions that we make when specifying the system, we need not consider pipelining here. Since we assume computations to happen with zero time delay, data-dependencies associated with pipelining will not exist explicitly.

A node can be refined using either Temporal or Spatial decomposition steps. The next two sections will discuss both these steps. Note that the hierarchical decomposition is implicit in the selection of simpler behaviors for subsequent refinement.

2.2.1 Temporal Decomposition

The temporal decomposition function, denoted by <u>Frec(ursive)</u>, can be applied to a simple functional map, a so-called Bmap. (Note that a Bmap is isomorfic to a (simple) node, whose behavior is that Bmap). The application of Frec, decomposes the Bmap in a set of simpler Bmaps. The sequence in which these Bmaps are applied is determined by the order in which they are given, as discussed in the previous section. Together with the application of Frec to the node, we have to specify the decomposition functions applied to the node ports. From the previous section we know that these functions play a major role in constructing the conditions to be included in the node behaviors. For instance, when as a result of applying TD to a Bmap (node) one of the inputs, i.e. ports of the node associated with the Bmap, is sequentialized then we are allowed to test for begin and end of this sequence. Likewise we will define typical tests for all port decompositions. As can be inferred from the above, in order to fully specify a TD, we need more information than just: apply TD. The additional information that the designer has to supply includes:

1. states: for each state the designer has to supply a name. Usually there will be only one state. This state can then be split further in subsequent refinements.

2. ports: for each port of the node that is refined the designer has to specify the decomposition function to be applied to the port.

3. map: the behavior of the node, i.e. a sequence of Bmap or cycle constructs.

VLSI Signal Processing

2.2.2 Spatial Decomposition

Different from the case with TD, for SD we will define several decomposition functions. The Spatial Decomposition functions must be applied to a node, a refinement of a node, or a named behavior created as part of a node-refinement. For now, we define only two decomposition functions, respectively Farray and Fcomp.

The function Farray will decompose a simple node into a complex node that has an array like structure. The size of the array need not be fixed at the time of decomposition. Instead we can define a parameter, e.g. the size of an input vector, that will define the actual array size.

The function Fcomp replaces a node by a graph-like structure of interconnected nodes.

2.2.2.1 Farray

The function Farray decomposes a node into an array of nodes. The number of nodes in this array is either determined by the size of a port, i.e. the size of the datatype associated with a port, e.g. a vector, or by an explicitly given parameter. The nodes of the array can be given a name in the behavioral part of the refinement specification (see also examples).

2.2.2.2 Fcomp

The function Fcomp is used to replace a node by a composition of lower level nodes. In the specification of the refinement the lower level nodes must be specified, either by giving their name (if they are already defined), or by giving their definition.

2.3 Transformations to VLSI Arrays with Pipelining, Partitioning, and Fault-Tolerance

In an earlier paper [3] a theoretical basis for automatic transformation procedures that systematically convert SFG's into pipelined structures, such as systolic or wavefront arrays, and eventually their VLSI hardware implementations, was discussed. These include an automatic (cut-set) systolization procedure, optimal scheduling, and wavefront handshaking, etc. More importantly, the hierarchical and applicative approach offers the simplicity and flexibility necessary to deal with the issue of partitioning a larger size problem to fit into smaller arrays and/or imposing fault tolerance into arrays. These additional flexibilities will be critical to the effectiveness of the arrays.

The cut-set rules are also potentially very useful for designing fault-tolerant arrays. For systolic arrays without feedback, it has been shown in the literatures that a retiming along cut-sets allows a great degree of fault tolerance. The theoretical treatment in [3] (Section III-B) offers a theoretical basis for improving fault-tolerance of arrays with feedback via the cut-set retiming procedure. More interestingly, with a slight modification, the self-timed feature of wavefront arrays offers a way of achieving the same fault-tolerance

2.4 CAD Tools and Graphics-Based Design Systems

The main reason for having a graphics based approach lies in that it provides a simple communication language for humans (one picture is worth a thousand words).

The human information processing capability makes it possible to identify at once the basic structure of an image. Together with this goes the fact that many typical constructs, like for instance 'array', have a natural graphical denotation. Especially, for the HIFI design, many array transformations are originally graphic based. More importantly, graphics workstations are undergoing a rapid evolution, making them more available and more user-friendly.

The definition of the HIFI Database, containing all the design information, including design history, will be the starting point for the development of all software tools, including a powerful graphic-editor as the main interface between the designer and the HIFI Database. The editor will use state of the art graphic techniques, like multi-window displays and many selection of editor and design functions, including the HIFI design functions. By using a menu based approach the editor will guide the designer through all the different design steps.

Another major set of tools will be related with the simulation of the SFG based designs. These will allow the user to simulate the design in different levels. This will be essential in order to timely detect major design gaps and to ensure the correctness of the design.

The definition of a Hardware Description Language (HDL) [7,8], to capture the design after all transformations (pipelining, fault-tolerance and partitioning) have been applied to it, will be useful for the documentation of the design steps as well as for interfacing the HIFI system with lower-level silicon compiler like systems. The mapping from and to this language should be one of the functions provided by the HIFI system.

3. From Algorithms to Array's

From an applicational point of view, several popular algorithms are often repeatedly utilized and very often in conjunction with a large volume of data. Typical such examples are FFT, correlation, and matrix multiplication, matrix inversion, and least-square solver. For example, the least-square solver and the 1-D and 2-D convolution find important applications in many image and signal processing algorithms. We shall therefore use them as illustrative examples for the HIFI design/description approach.

The examples given below show how a designer goes about designing a system to implement a specific algorithm. The algorithm need not be fully specified in the beginning. In fact we need to specify only a so-called B(ehavior)-function. This Bmap gives the external behavior of the system at the problem definition level. The designer then starts of by selecting one of the various decomposition functions that can be applied to the structure (SFG) implicitly associated with the Bmap. By successively applying the decomposition functions the designer gradually refines the specification of the system.

It is important to realize that in the examples below we have concentrated on textual descriptions of the node refinements. In the HIFI system as we are developing it, the emphasis will be on graphics interfaces. To get an idea of how the graphs corresponding to the textual descriptions might look like we have also included some graph pictures.

3.1 Least Square Solution of a System of Equations

In this section we will indicate how to design a system to compute the Least Squares Solution to a system of Linear Equations. As in the previous example we will first design a virtual system. After that we will show how to transform this virtual

VLSI Signal Processing

system so that it can be implemented in the form of a systolic or wavefront array.

3.1.1 Hierarchical (Temporal and Spatial) Decomposition

Problem Definition: Given a matrix A and a vector b, find the least square error solution y, such that:

$$|| Ay - b ||$$

is minimized. Determine the residual r.

At this level the system is defined with a very simple SFG (cf. Fig. 1) consisting of a single node only. The node has two input and two output ports, resp. the matrix A, the vector b, the solution y, and the residual r.

```
SFG LSS (IN A, b;  OUT y, r;)
{
type Node:    LSS;
type Branch:  B.A, B.b, B.y, B.r;

LSS: = Fnode: < > WHERE
      ports
         A = Fi;
         b = Fi;
         y = Fo;
         r = Fo;
behavior
         (A || b - fLSS → y || r);

[* Branches *]
B.A: = A <=> LSS.A;
B.b: = b <=> LSS.b;
B.y: = y <=> LSS.y;
B.r: = r <=> LSS.r;
}
```

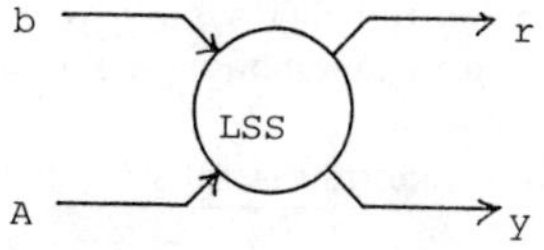

Figure 1: Problem Definition Level

At the next lower level we adopt a common method to solve the LSS problem. The problem is divided in three (smaller) subproblems, resp.:

1. Reduce A to an upper triangular matrix R by an orthogonal transformation Q, such that Q A = R.

2. Apply the same orthogonal transformation Q to b, so that b is transformed to Q b = [b^ r], where r is the residual vector.

3. Solve for y by the back-substitution procedure, R y = b^ .

We see that what we have to do is decompose the LSS node in the previous description into three separate nodes, resp. a QR node, a bb^ node and a Back-Substitution (BS) node. To specify this we have to give an implementation of fLSS, i.e. we have to use the decomposition function Fcomp to implement the mapping as the composition of the nodes given in the function specification. This results in the

VLSI Signal Processing

following description and SFG (cf. Fig. 2):

At the next and lower levels we are to repeatedly apply this same process of refinement, until we arrive at an implementable level. Here implementable simply means that the nodes model hardware primitives, and that the values communicated via the ports are scalars. How to derive a real implementation from this specification will be considered when we discuss, in a moment, the virtual to real machine mapping, including: pipelining, fault-tolerance and partitioning.

Continue on with our example, we now look more closely at node QR. As said before, node QR transforms the initial matrix A in an upper triangular matrix R by means

```
NODE_REF LSS.1 <- fLSS
{
  Fcomp  WHERE
  {
    ports
      A : Fconnect(A) o Node(QR) ;
      b : Fconnect(b) o Node(bb^) ;
      r : Fconnect(r) o Node(bb^) ;
      y : Fconnect(y) o Node(BS) ;
    internal
      QR: = Fnode: <> where
      ports
        A: Fi ,
        Q, R: Fo ;
      map
        (A - fQR → Q || R) ;

      bb^: = Fnode: <> where
      ports
        b, Q: Fi ,
        b^, r: Fo ,
      map
        (b || Q - fbb^ → b^ || r) ;

      BS: = Fnode: <> where
      ports
        R, b^ : Fi ,
        y: Fo ;
      map
        R || b^ - fBS → y) ;

    [* Internal Branches *]

      QR.Q    <=> bb^.Q ;
      QR.R    <=> BS.R ;
      bb^.b^  <=> BS.b^ ;

  }
}
```

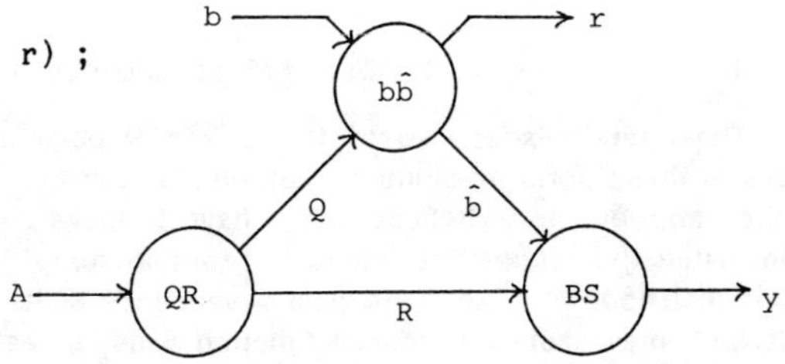

Figure 2: Basic Algorithm Decomposition

VLSI Signal Processing

of an orthogonal transformation Q, $Q\ A = R$. The first step involves the following decomposition,

$$Q_1\ A\ =\ \begin{matrix} x & x & x & \ldots & x \\ \hline 0 & & & & \\ 0 & & A^* & & \\ 0 & & & & \\ \bullet & & & & \\ 0 & & & & \end{matrix}$$

This can be modelled recursively, and leads us to the introduction of a state, modelled by a delay branch which feeds back to the same node. The initial state is given by the matrix A, the node computes the new state and the outputs. Note that the new state is different from the other outputs only in that it is delayed and fed back to the input. The node goes on computing new outputs until the state becomes undefined. In that case it inputs a new value A, and repeats the same process.

A more detailed description of QR is given below.

```
NODE_REF QR.1 → fQR
{
    Frec(state)   WHERE
    ports
        A: Fid ;
        Q, R: Fseq ;
    behavior
        (A - f1 → Q || R || new.state);
        while (state is Valid) do
            (state - f2 → Q || R || new.state);

} → QR.2
```

Figure 3: Temporal Decomposition of QR node.

The operations applied to the Q and R ports defined with QR decompose the values at these ports in sequences of simpler values. This has consequences for higher level descriptions as well, because we have to make sure that refinements made for one node match the refinements made for another node. In the example for instance, this means that node bb^ should expect a sequence at its input port Q. It also means that node bb^ may apply the control function Feos to test whether the end of a sequence of values has been reached. This interaction between node refinements can be much more complex. For instance, when we would refine the Back-Substitution part we would find that it is actually necessary to reverse the sequence of R-rows at port R of the back-substitution node. This could be easily modeled by defining a port decomposition function Frev, to reverse the sequence offered at the port. At the hardware level the application of Frev to a port would result in the inclusion of a LIFO buffer between the two nodes. The size of the buffer would be determined by the length of the sequence. This length can again be determined from the size of the original A matrix.

By a further algorithmic analysis, the QR problem is commonly decomposed into two different subproblems: One is computing the transformation matrices from the columns of A, the other is applying these transformation matrices to the columns to the right of the columns from which it was computed. A natural way of applying this decomposition to the QR node is by decomposing it into an array of nodes. Since this decomposition occurs often we have a special function Farray to do so. The result of applying Farray to a node in a SFG is not simply another SFG, but rather an entire class of SFG's, the members of the class differing in the number of nodes in the array. The nodes in this array can be addressed with the operators, 'Fall', 'Ffirst', 'Flast' 'Fnext', and 'Frest'. Applying Farray to node QR in QR.1 results in the following description:

```
NODE_REF QR.2 <- QR.1
{
   Farray WHERE
   ports
      A: Fconnect(state) o Flast ;
      Q: Fconnect(Q) o Flast;
      R: Decompose   ;
   internal
      Q: Fic_lr ;
      state: Fsfl ;
   behavior
      first node → RG
      [* RG is the name of the behavior that
      can be used for subsequent refinements *]
      while (state isnot Valid) do
         (→ )  ! NIL action
      do
         (state → Q || R)
      while (state is Valid)

      rest of nodes → RT
         while (Q isnot Valid) do
         (state → new.state )
      do
         (state || Q → new.state || new.Q || R) ;
         while (state is Valid)
      do
         (Q → new.Q ) ;
      while (Q isnot EOSeq)
}
```

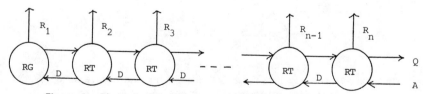

Figure 4: First Level Spatial Decomposition of QR node.

The node is split in an array of nodes, where we can indicate the type of the nodes in the behavioral specification part. This type name allows an hierarchical decomposition of subsequent refinements.

VLSI Signal Processing

The final refinements we consider give an implementation of the behaviors Rg, resp. RT, defined in the previous refinement. Both Rg and Rt are spatially decomposed using Farray as the decomposition function. These final refinements transform the simple QR node that we started with to a two-dimensional array of nodes. (See also Fig. 5 for a localized version of the complete fully refined SFG of the LSS system.)

3.1.2 Transformation to Systolic Array

As noted in [3] a regular SFG is almost equivalent to a systolic array and can be easily systolized. There exist a number of systolization procedures [9,10]. A (theoretically) straightforward scheme is essentially based on the cut-set theorem and the localization rules derived thereof. Applying these rules to the LSS example leads to the systolic array given in Fig. 5.

3.1.3 Transformation to Wavefront Array

The efficiency of pipelining as well as the potential difficulty of implementing global synchronization prompt us to consider a self-timed alternative, namely the Wavefront Array [4]. In a self-timed system the exact timing reference is ignored; instead, the central issue is sequencing. Getting a data-token in a self-timed system is equivalent to incrementing the clock by one time-unit in a synchronous system. Therefore the delay-operators D will be replaced by handshaked 'separator'[*] registers. In other words the conversion of an SFG into a data-driven system involves substituting

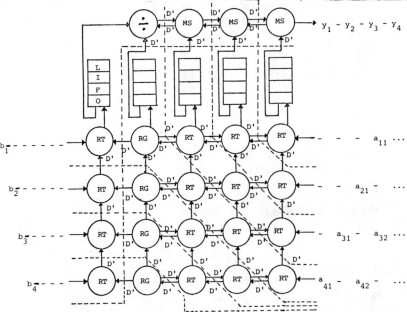

Figure 5: Temporally Localized SFG describing the complete LSS system, including backsubstitution and the LIFO buffers.

---

[*] A handshaked separator is a device which prevents any incoming data from directly passing through until the handshaking flags signal a pass

the delay D with implicit or explicit separators, and replacing the global clock by data handshaking. This process incorporates the data-flow principle into SFG's, and achieves the optimal pipelining efficiency without the need of applying an (optimal) time rescaling procedure.

4. Conclusions

The fast growing development of VLSI array processors requires a powerful array design environment. The combination of powerful abstraction mechanisms, graphic tools and (automatic) graph-based transformation procedures between array's, offers the theoretical basis for a set of CAD tools (termed HIFI) that deal with the specification and verification of the rather complex space-time relationships that occur in array processing.

The HIFI design system is tailored towards a specific class of applications. However, it is open ended and in the future we will look in the possibilities of integrating the HIFI system into a more general design environment, such as DESADE [11]. Since the ideas used in DESADE and HIFI are very consistent, such a combination will prove to be very promising.

REFERENCES

[1] H.T. Kung, Charles E. Leiserson, "Systolic Arrays (for VLSI)", Sparse Matrix Proceedings 1978, I.S. Duff and G.W. Stewart, ed. Society for Industrial and Applied Mathematics, 1979, pp. 256 - 282.

[2] Marina C. Chen, "Space-Time Algorithms: Semantics and Methodology", Ph.D. Thesis, California Institute of Technology, May 1983.

[3] S.Y. Kung, "On Supercomputing with Systolic/Wavefront Array Processors", Proc. IEEE, Vol. 72, No. 7, July 1984.

[4] S.Y. Kung, K.S. Arun, R.J. Gal-Ezer and D.V. Bhaskar Rao, "Wavefront Array Processor: Language, Architecture and Applications" IEEE Trans. on Computers (Special Issue on Parallel and Distributed Computing), Vol C-31, no. 11, pp. 1054-1066, Nov. 1982.

[5] Backus, J., "Can Programming Be Liberated from the Von Neumann Style? A Function Style and Its Algebra of Programs" Comm. ACM 21, 613 - 641, Aug. 1978

[6] Cremers, A. and Hibbard, T., "Executable Specification of Concurrent Algorithms in terms of Applicative Dataspace Notation" In: Proc. USC Workshop on VLSI and Modern Signal Processing, Los Angeles, Ca. 1 - 3 Nov. 1982

[7] J.D. Nash, "Bibliography of Hardware Description Languages" Sigda Newsletter Vol. 14 (1984) No. 1, pp. 18 - 37

[8] W. Lim, "HISDL - A Structure Description Language", Comm of the ACM, Nov. 1982, Vol. 25, No. 11

[9] Charles Leiserson, Flavio Rose, James Saxe, "Optimizing Synchronous Circuitry by Retiming" Proc. Caltech Conf. on VLSI, March 1983

[10] S.Y. Kung, S. C. Lo, J. Annevelink, "Temporal Localization and Systolization of SFG Computing Networks", Proc. SPIE 1984, San Diego.

[11] B. Kuhn, Personal Communication

Part IX
REAL TIME SPEECH PROCESSING

Edited by:
Dr. Chi-Foon Chan
Intel Corporation

DESIGNING A SINGLE-CHIP DYNAMIC-PROGRAMMING PROCESSOR FOR CONNECTED AND ISOLATED WORD SPEECH RECOGNITION

Robert E. Owen, Consultant
Saratoga, CA 95070

for
Votan Corporation
Fremont, CA 94538

ABSTRACT

The design of a 3 micron NMOS VLSI circuit for speech recognition is described. It calculates the distance between a trial utterance and reference templates stored in an attached dynamic memory. Time warping based on a linear-programming minimum distance is employed. Connected words are processed on a continuous real-time basis while isolated word recognition proceeds after the utterance is complete. Multiple stages of pipelining are used in a unique microprogrammed architecture so that memory bandwidth is fully utilized in the data-intensive distance calculation. Two seperate arithmetic logic units (ALUs) are used, one for numerical data and one for memory address generation. The processor was custom designed in a single layer metal silicon gate process. The basic instruction cycle time is 200 nanoseconds. Total chip size is 178 by 214 mils including approximately 4K bits of table look-up and 11.8K bits of microprogram read-only-memory.

INTRODUCTION

Speech recognizers generally consist of four functional blocks: a feature extractor, a pattern matcher, a reference vocabulary memory and a decision maker. The feature extractor reduces an utterance to a more compact pattern representing the speech. The pattern matcher compares previously recorded reference patterns in the vocabulatory memory with the new utterance and gives a score or distance measure on the closeness of fit for each word. The distances are compared by the decision maker and a recognition judgement made.

Contemporary VLSI digital signal processors can be programmed to provide a low cost and relatively small sized feature extractor. This connected to a general purpose microprocessor and memory as in Figure 1 supply the feature extracting and control functions. The pattern matching function is less easily implemented with available processors. As developed in a companion paper (Ref. 1) and noted by others (Ref. 2-4), the dynamic-programming time warp algorithms most commonly used for pattern matching imposes unusual processing demands. Neither digital signal processors nor general purpose microprocessors provide the combination of rapid access to two data memories, one very large, with fast non-linear arithmetic and the multiple-way data-dependent-branching which is required. The only commercially available pattern matching chip (Ref. 3) would not provide the high recognition accuracy and robustness in low

VLSI Signal Processing

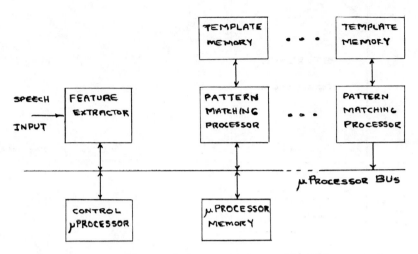

Figure 1. A complete speech recognition system with multiple pattern matching processors.

signal-to-noise environments due to its low precision arithmetic and lack of program flexibility in the algorithm.

The design of a proprietary integrated circuit for pattern matching was undertaken to equal the small size and low cost of the other speech recongition functional processors in the system and provide improved performance. The emphasis in this article will be on the design process and architectural and implementation choices.

ARCHITECTURAL CHOICES

The pattern matching or dynamic-programming processor is located on the control microprocessor bus in the speech recognition system as shown in Figure 1. The large template memory interfaces directly to this processor itself rather than being located on the microprocessor bus. This permits the processor uncontested access to its memory once the reference templates and trial utterance have been loaded. An additional feature is that more processors and memory can be paralleled to increase the vocabulary while maintaining nearly the same recognition response time.

A memory map of the attached template memory is in Figure 2. The largest area is for reference templates which represent the complete recognition vocabulary. The active vocabulary is determined by the start and end addresses for the individual word templates stored in the parameter block area of memory. Thus the recongizer can quickly change from one active subset vocabulary to another due to speech context differences with a simple parameter block change. The complete trial utterance is stored in an area of the template memory for isolated word recognition while only the latest sample is needed for connected words.

309

VLSI Signal Processing

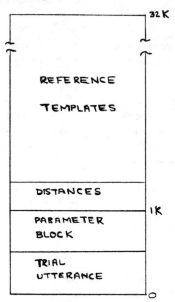

Figure 2. A template memory map.

Conversely only one word's set of path distance values are stored for isolated word recognition, while sets for all words are stored for connected word recognition in the distance area of the memory.

The dynamic-programming time warping process conceptually consists of two major steps. The first is computing the distance or difference between the features of the trial utterance and the features of the reference word for all possible time warpings of the trial utterance into the reference word's time period. The second is to sum the distances on a path from the beginning to the end of the utterance. The path is the one on which each point is the minimum total distance from the beginning up to that point. In practice the two steps can be combined with considerable time and memory savings if the individual distances are calculated and the minimum distance is accumulated for all possible paths at the same time. This proceedure was adopted here. Even with this savings each feature vector of the trial utterance must be fetched repeatedly from template memory and differenced with each feature vector of the reference vocabulary words. It is this computation and data flow which dominates the architectural choices.

Each of the steps in the distance computation: finding the vector differences, accumulating the differences, and selecting the minimum can be done in sequence so they may be pipelined using seperate circuitry for each step. This reduces the speed limitation to one of fetching data from and returning data to the template memory. Fast address generation can make the complete process limited only by the template memory bandwidth itself. This was the approach used in this design.

IMPLEMENTATION

Figure 3 is a block diagram of the implemented dynamic-programming processor. The internal numeric data processor and memory address processor share a common 8 bit data bus which interfaces to the external template memory and control microprocessor. Both internal processors are controlled by an on-chip program sequencer and program read-only-memory (ROM).

The data processor has an input register file to store a single trial utterance feature vector. Individual elements from it and the reference template feature vector stored in the reference register feed a ROM look-up table. The look-up table output is a difference function of the two feature elements. This 3840 bit ROM is an area-efficient means of computing a highly nonlinear function. A register at the ROM output completes the first pipeline stage. Some vector elements are shifted in a data dependent way in the A shifter at the beginning of the second stage before being added by the ALU. The accumulator stores the sum at the end of the second pipeline stage. The total individual distance at a point is scaled by shifter B and added to the minimum of J, K or L, the nearest neighbor total distances. The comparator addresses two registers at a time to determine the minimum. Additional minima are stored for comparison to terminate computation for large distance individual paths or whole words. The data processor precision is twelve bits for high accuracy with vocabularies of connected words. Each pipeline stage takes one instruction cycle of 200 nanoseconds. Also included in the data processor is a token register file. For connected speech identifying tokens are associated with each path distance.

The pipelining of the data processor increases processing speed only if a steady flow of feature elements and distances can be maintained with the template memory. Economy dictated that low-speed dynamic RAM be used for the large template memory. The lower cost of the memory chips themselves contributed but also the lower processor pin count for their time-multiplexed address lines without need for external address latches. Keeping the pin count low for a lower cost package also limited the memory data path, and therefore the internal data bus, to 8 bits. To assure full utilization of the memory a separate address processor with its own register file and ALU were included. The memory cycle time of 400 nanoseconds allows two ALU cycles per memory read or write operation. Thus an increment and end-of-count comparison can be made each memory cycle without slowing data flow. In addition to an address register for the current reference template, parameter block, distance and trial utterance, a memory refresh counter is included for the dynamic RAMs. This programmed refresh during data processor initialization and overhead instructions makes the refresh operation invisible rather than directly reducing the useful memory bandwidth. Addresses are 15 bits allowing up to 32 Kbytes of template memory to be attached. The address processor connection to the internal data bus is for loading reference template starting and end addresses from the parameter block in the template memory.

VLSI Signal Processing

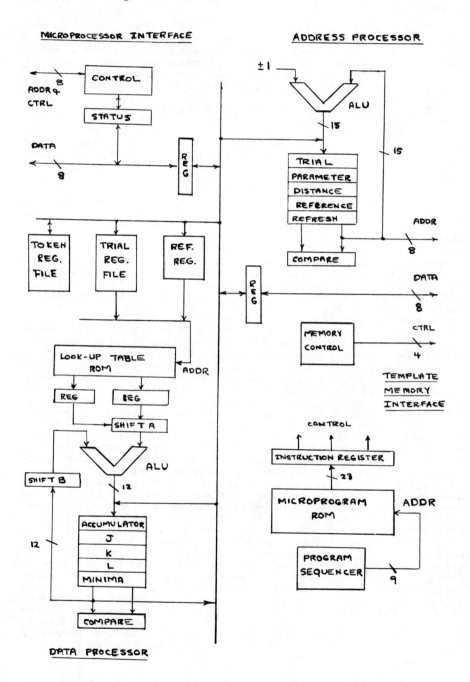

Figure 3. Block diagram of the single-chip processor.

VLSI Signal Processing

A full 8 bit microprocessor interface to the internal data bus is used for loading reference templates, parameter blocks and the trial utterance, and reading out computed distances. Two command and status registers in the interface select the various initializing, operational and test modes. Data transfers can be initiated by monitoring the status registers or by a control microprocessor interrupt at the end of the proceeding operation.

Operation of both internal address and data processors and the microprocessor and memory interfaces is controlled by a microprogram sequencer and microprogram ROM. A primitive sequencer with only increment, conditional branch and branch to origin on reset instructions was required since in-line code was used for maximum speed. Independent control of all functions would have required a microinstruction word of over 40 bits. Because there are two instruction cycles per template memory operation, arithmetic and data move operations tend to alternate with memory and address generation operations. This allowed splitting the microinstruction word into a DATA and MEMORY instruction class of 23 bits each. Two additional classes are subsets of DATA and MEMORY to allow conditional branching. The instruction set fields are shown in Figure 4. The 512 word microprogram ROM includes isolated and connected word recognition routines, loading and initialization, and about twenty percent for self test routines.

MEMORY OPERATION	ADDR SOURCE	ADDR OPERATION
BRANCH ADDRESS & CONDITION	ADDR SOURCE	ADDR OPERATION
DATA OPERATION	DATA SOURCE	DATA DESTINATION
BRANCH ADDRESS & CONDITION	DATA SOURCE	DATA DESTINATION

Figure 4. The 23 bit microinstruction set has four operate classes: DATA and MEMORY with and without conditional branching.

The relatively slow 200 nanosecond instruction cycle time, possible because of the slow 400 nanosecond template memory cycle time, allowed four clock phase dynamic circuitry to be used to save chip area. The four phases are derive from a 10 MHz externally supplied system clock. A single metal layer silicon gate NMOS process was used with a 3.0 micron minimum gate length. The resulting chip size was 178 by 214 mils or about 38.1 K square mils. The microphotograph in Figure 5 shows the layout. The 11.8 Kbit microprogram ROM is in the upper center with the table look-up ROM on the lower left. The nearly equal sized data processor and address processor ALUs and register files are in the lower right below the microprogram ROM and instruction set decode circuits. There are a total of forty signal, clock and power supply pins. Power dissipation is less than 750 milliwatts.

VLSI Signal Processing

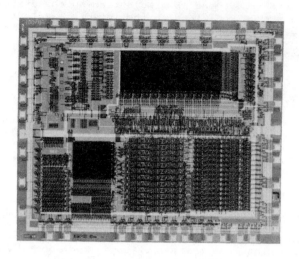

Figure 5. A microphotograph of the 178 by 214 mil chip.

ACKNOWLEDGEMENTS

It is a pleasure to recognize the contributions of Joe Nance, Roger Visser and Ray Ho of Votan and Mike Assar of International Microelectronic Products (IMP).

REFERENCES

1. R. E. Owen, "A single-chip dynamic-programming processor for connected and isolated word speech recognition," submitted for presentation at IEEE ICASSP 85.

2. M. Lowy, H. Murveit, D. M. Mintz, R. W. Brodersen, "An architecture for a speech recognition system," Digest of 1983 IEEE ISSCC, pp 118-119, Feb 1983.

3. T. Iwata, H. Ishizuka, M. Watari, T. Hoshi, Y. Kawakami, M. Mizuno, "A speech recognition processor," Digest of 1983 IEEE ISSCC, pp 120-121, Feb 1983.

4. N. Weste, D. J. Burr, B. D. Ackland, "Dynamic time warp pattern matching using an integrated multiprocessing array," IEEE Trans. Computers, Vol. C-32, pp 731-744, Aug 1983.

Brief Biography

Mr. Owen has over fifteen years of industrial experience in digital signal processing. He was involved with minicomputer signal processing systems at GenRad's Time/Data Corporation and then peripheral floating-point array processors at ESL-TRW. His work on VLSI has been first with Intel and then Fairchild in their digital signal processor design groups. Since 1982 he has been an independent engineering consultant to both manufacturers and users of standard and custom VLSI circuits for digital signal processing.

Mr. Owen has a BEE degree from Rensselaer Polytechnic Institute and a MSEE from Case Institute of Technology. He is a charter member of the IEEE ASSP Technical Committee on VLSI.

Mailing Address

Mr. Robert E. Owen
19348 Columbine Court
Saratoga, CA 95070

(408) 996-2060

VLSI Signal Processing

Single Chip Implementation of Feature Measurement for LPC-Based Speech Recognition

John G. Ackenhusen
Y. H. Oh

AT&T Bell Laboratories
Murray Hill, New Jersey 07974

ABSTRACT

A single chip implementation of LPC-based feature measurement has been developed using the AT&T Bell Laboratories Digital Signal Processor and has been verified by both numerical simulation and system use.

The feature measurement technique is identical to that used in numerical simulations of LPC-based isolated and connected word recognition that use combinations of dynamic time warping, vector quantization, and hidden Markov modeling. As a result, this feature measurement processor represents a single chip common building block for real time hardware implementation of most speech recognition techniques under investigation at AT&T Bell Laboratories.

The feature measurement circuit, called the FXDSP, performs eighth-order LPC analysis continuously in real time. It receives mu-law-encoded telephone bandwidth speech at a 6.667 kHz sampling rate from a standard CODEC and produces a feature vector consisting of the log energy, nine amplitude-normalized autocorrelation coefficients, and nine LPC-based test pattern coefficients for each analysis frame of speech. Feature vectors are output continuously at a frame period of 15 msec.

1. Introduction

Most speech recognition work at AT&T Bell Laboratories has been based on a standard form of feature measurement first proposed by Itakura.[1] Speech recognition features are computed from an eighth-order linear predictive coding (LPC) calculation on 45 msec analysis frames spaced by 15 msec. The autocorrelation method is used, and the speech is of telephone bandwidth (3.3 kHz) and is sampled at 6667 samples/sec.

With this front end, numerical simulations have demonstrated successful word recognition algorithms based on dynamic programming for both isolated words[2] and connected words.[3] Real time hardware that uses this front end for isolated word recognition has been reported.[4] More recent simulations have used the same front end in recognizers that use vector quantization for isolated[5] and connected word recognition[6] and for recognizers using hidden Markov modeling.[7]

Comparative tests of the LPC front end with a variety of filter banks have found the LPC technique to provide superior performance for complex vocabularies over telephone bandwidths.[8]

This paper describes a real time implementation of this LPC feature measurement technique that is of single-chip complexity. The implementation uses a programmable signal processor, the AT&T Bell Laboratories Digital Signal Processor (DSP).[9] In this implementation, called the FXDSP (Feature eXtracting Digital Signal Processor), the output continuously provides results of LPC analysis of whatever input signal is present with less than 1 frame (15 msec) of delay. An implementation of the same LPC feature measurement algorithm based upon a special-purpose integrated circuit has been described previously.[10]

2. LPC Feature Measurement

The requirement of LPC is to determine a unique set of predictor coefficients, a_k, $k=1, 2, ..., p$, that minimize the sum of squared differences, E_n, between actual speech samples, $s(n)$, and approximated speech samples, $\tilde{s}(n)$. The approximated speech samples $\tilde{s}(n)$ are formed from a linear combination of speech samples over a short segment of the speech waveform. Thus, the approximate speech samples are given by

$$\tilde{s}(n) = \sum_{k=1}^{p} a_k s(n-k) \qquad (1)$$

where $p = 8$ in this analysis. The task of minimizing the prediction error, E_n, is to choose a_k such that

$$E_n = \sum_m e_n^2(m). \qquad (2)$$
$$= \sum_m [s_n(m) - \tilde{s}_n(m)]^2 \qquad (3)$$
$$= \sum_m [s_n - \sum_{k=1}^{p} a_k s_n(m-k)]^2 \qquad (4)$$

is a minimum.

Techniques for calculating the linear prediction coefficients, a_k, from the speech samples, $s(n)$, are described in the literature.[11] The method used here is a block-processing technique based on the autocorrelation method and Durbin's recursion (Fig. 1).

Speech which has been band limited to 100 - 3300 Hz and sampled at 6667 samples/sec is first preemphasized with a first-order network:

$$s'(n) = s(n) - as(n-1); \quad a = 0.95. \qquad (5)$$

The preemphasized speech is then blocked into frames of 300 samples (45 msec) which are spaced by 100 samples (15 msec). Thus, the *l*th frame of speech, $\tilde{x}_l$, is given by

VLSI Signal Processing

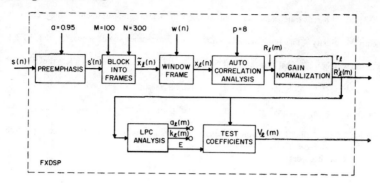

Fig. 1   Signal processing for extracting LPC features for recognition

$$\tilde{x}_l = s'(Ml+n), \quad n = 0, 1, ..., N-1; \quad l = 0, 1, ..., L-1 \qquad (6)$$

where $M = 100$ and $N = 300$ for an input sequence length of L frames. As a result of this choice of M and N, each speech sample contributes to three consecutive analysis frames.

Each frame is then smoothed by a Hamming window:

$$x_l(n) = w(n) \cdot \tilde{x}_l(n), \qquad (7)$$

$$w(n) = 0.54 - 0.46 \cos\left[\frac{2\pi n}{N-1}\right], \quad N = 300. \qquad (8)$$

The resulting windowed frames of speech data are used to perform an autocorrelation calculation, given by

$$R_l(m) = \sum_{m=0}^{N-1-m} x_l(n) x_l(n+m) \; ; \; m = 0, 1, ..., 8. \qquad (9)$$

The logarithm of the frame energy, $R_l(0)$, is then calculated:

$$r_l = \log_2 R_l(0). \qquad (10)$$

The autocorrelation coefficients are gain-normalized such that $R'_l(0) = 1$, as follows:

$$R'_l(m) = \frac{R_l(m)}{2^{r_l}} \qquad (11)$$

This normalization is required so that later computation of Durbin's recursion uses the full integer precision of the machine. The log energy, r_l, is used for endpoint detection and frame energy information during the recognition process.

Durbin's recursion is then applied to calculate a set of PARCOR coefficients, k_i, $i = 1, 2, ..., 8$, and a prediction residual from the $R'_l(m)$ for each frame as follows (the frame index l is suppressed):

VLSI Signal Processing

$$E^{(0)} = R'(0) \tag{12}$$

For $i = 1, 2, ..., 8$, do Eq. (13 - 16):

$$k_i = \frac{\left[R'(i) - \sum_{j=1}^{i-1} \alpha_j^{(i-1)} R'(i-j)\right]}{E^{(i-1)}} \tag{13}$$

$$\alpha_i^{(i)} = k_i \tag{14}$$

$$\alpha_j^{(i)} = \alpha_j^{(i-1)} - k_i \alpha_{i-j}^{(i-1)}; \quad (j = 1, 2, ..., i-1; i \neq 1) \tag{15}$$

$$E^{(i)} = (1-k_i^2) E^{(i-1)} \tag{16}$$

Extract final residual, E, and LPC coefficients, a_j:

$$E = E^{(8)} \tag{17}$$

$$a_j = a_j^{(8)} \tag{18}$$

Test pattern coefficients are then formed by computing:

$$V_l(m) = \frac{R'_l(m)}{E}, \quad m=0, 1, ..., 8. \tag{19}$$

The FXDSP output consists of r_l, $R'_l(m)$, and $V_l(m)$ for $m = 0, 1, ..., 8$. The PARCOR coefficients k_i and LPC coefficients a_j are calculated as a result of calculating E; however, since they are not used directly in real-time pattern matching, they are discarded. Reference templates are made up of autocorrelations of a_j which are produced during a non-real-time vocabulary training session. In the current robust training algorithm, a reference pattern is made up of an autocorrelation average of two tokens that correspond to two different repetitions of a word.[12] Therefore, no use can be made of the LPC coefficients in real time.

3. Hardware

The hardware for this implementation consists of a mu-law CODEC with filters which is run at a 6.667 kHz sampling rate, and the AT&T Bell Laboratories DSP which is run at 10 MHz (Fig. 2). Separate oscillators control the sampling rate of the CODEC and the clock of the DSP.

A block diagram of the DSP is shown in Fig. 3. The version used here, known as DSP-2, is an improved version of the original signal processor described in Ref. 9 in which both speed and RAM size have been doubled.

The DSP-2 has a 400 nsec instruction cycle time. The processor consists of a read/write memory of 256 20-bit words and a mask-programmable program read-only memory of 1024 16-bit words. Alternatively, the DSP can be run from 1024 words of external program memory, usually made of

VLSI Signal Processing

EPROM or downloadable RAM. An address arithmetic unit (AAU) contains registers for controlling memory access. A data arithmetic unit (DAU) contains a 16-bit x 20-bit multiplier, a 40-bit accumulator, a 40-bit adder, and a 20-bit rounding-overflow circuit. Input and output occur through two serial data pins.

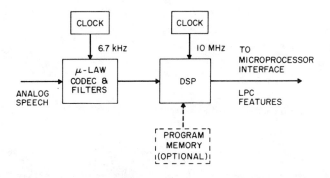

Fig. 2 LPC feature measurement hardware

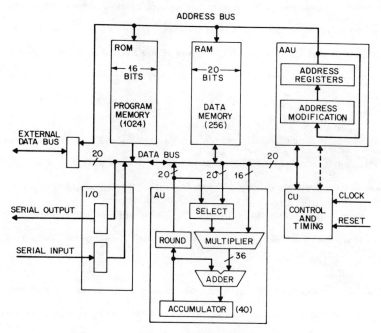

Fig. 3 Block diagram of DSP

In one 400 nsec machine cycle, the DSP can 1) decode an instruction, 2) fetch data and perform a multiplication, 3) accumulate output products from the multiplier, and 4) store data in memory.

4. Program Architecture

A conflict arises between the input time scale of the FXDSP, one sample every 150 μsec, and its output time scale, 19 coefficients of a feature vector every 15 msec. The FXDSP is required to process a new sample every 150 μsec regardless of any other operation in process, else the input sample is lost and the resulting frame feature vector is incorrect. Thus, two timescales exist, a sample timescale and a frame timescale.

As a result of the two timescales, the program architecture really consists of two separate programs, a sample update program that updates autocorrelation vectors every four samples (Eq. 5 - 9) and a frame recursion program that calculates the output feature vector from the autocorrelation vectors from the previous frame (Eq. 10 - 19). The frame recursion program is divided into smaller pieces that are interposed with repeated executions of the sample update program (Fig. 4).

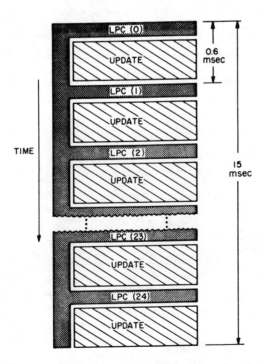

Fig. 4   Interleaving of sample update and frame inversion programs

VLSI Signal Processing

The sample update program operates on four samples each time it is executed. This four sample operation is a compromise between fully block processing, in which autocorrelation vectors are calculated on a frame of 300 samples all at once, and fully stream processing, in which the autocorrelation vectors are updated on receipt of each new sample.[13] Fully block processing, however, requires enough read/write memory to store all 300 samples, which is more memory than the single chip DSP has available. Fully stream processing, which has been used in other implementations, [4] executes too slowly for real time analysis on the DSP.[13] This is because before any autocorrelation update occurs, address pointers must be set up for accessing samples and autocorrelation vectors, and each autocorrelation coefficient must be accessed and placed in the accumulator of the arithmetic unit. These overhead operations are necessary for any number of samples used in the update, and can only be tolerated in real time if the updates occur for more than one sample at a time.

The frame period of 100 samples and the updating of autocorrelation vectors by four samples at a time require that the update program be executed 25 times per frame period. Therefore, an output operation of one frame coefficient is added to the sample update program to provide 25 output coefficients per frame, spaced at 4 sample intervals. The 19 frame coefficients (r_l, $R'_l(m)$, and $V_l(m)$, $m = 0, 1, ..., 8$) and six consecutive zeroes are output for each frame. The sequence of six zeroes provides a synchronization marker for identification of the 19 coefficients by the processor that receives the output of the FXDSP.

Fig. 5 shows a more detailed view of the timing of operations. The frame recursion is divided into

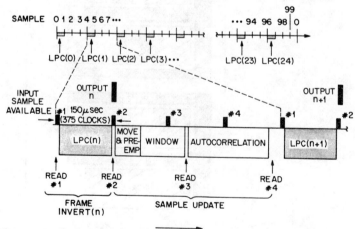

Fig. 5 Timing of input, output, and program sections

twenty-five pieces numbered LPC(0) through LPC(24). Between the first and second samples of the group of four sample inputs, one piece of the frame recursion program is executed. Each piece is completed within the sample period of 150 μsec. In Table 1, the function of each piece of the frame recursion is shown, as well as the time required for execution and the number of 16-bit words in program ROM required for that piece. As shown at the bottom of Table 1, the final eleven time slots, LPC(15) through LPC(24), are unused.

TABLE I
FRAME RECURSION
TIMING AND PROGRAM MEMORY
(BY FUNCTION)

LABEL	FUNCTION	EXECUTION TIME (μsec)	PROGRAM LOCATIONS
LPC (0)	READ $R_\ell(m)$ TO FRAME RECURSION INPUT BUFFER, SHIFT WINDOW	95	97
LPC (1)	CALCULATE r_ℓ	60	143
LPC (2)	CALCULATE $R'_\ell(m)$, m = 1, 2, 3, 4	144[1]	50
LPC (3)	CALCULATE $R'_\ell(m)$, m = 5, 6, 7, 8	144	8[2]
LPC (4)	SET UP FOR DURBIN'S RECURSION ($E_0 = R'_\ell(0)$)	50	32
LPC (5)	CALCULATE $1/E_{i-1}$ AND DURBIN'S RECURSION (i=1)	128	226
LPC (6)	CALCULATE $1/E_{i-1}$ AND DURBIN'S RECURSION (i=2)	128	6[2]
LPC (7)	CALCULATE $1/E_{i-1}$ AND DURBIN'S RECURSION (i=3)	128	6
LPC (8)	CALCULATE $1/E_{i-1}$ AND DURBIN'S RECURSION (i=4)	128	6
LPC (9)	CALCULATE $1/E_{i-1}$ AND DURBIN'S RECURSION (i=5)	128	6
LPC (10)	CALCULATE $1/E_{i-1}$ AND DURBIN'S RECURSION (i=6)	128	6
LPC (11)	CALCULATE $1/E_{i-1}$ AND DURBIN'S RECURSION (i=7)	128	6
LPC (12)	CALCULATE $1/E_{i-1}$ AND DURBIN'S RECURSION (i=8)	128	6
LPC (13)	CALCULATE $1/E$	128	6
LPC (14)	CALCULATE $V_\ell(m)$, m = 0, 1, ..., 8	12	41
LPC (15) THRU LPC (24)	IDLE	12 EACH	26
	TOTAL (% USED OF AVAILABLE)	1777 (12%)	688[3] (67%)

[1] FOR SIGNAL 51dB DOWN FROM PEAK; SHORTER EXEC. TIME FOR STRONGER SIGNALS.

[2] LOCATIONS INCLUDE ONLY THE SUBROUTINE CALL; SUBROUTINE PREVIOUSLY COUNTED.

[3] TOTAL INCLUDES (17 LOCATIONS) POWERUP INITIALIZATION ROUTINE NOT LISTED ABOVE.

During the time following the second, third, and fourth samples, the sample update operation is performed. The operations associated with the sample update operation are described in Table 2. A

VLSI Signal Processing

TABLE 2

**SAMPLE UPDATE
TIMING AND PROGRAM MEMORY
(BY FUNCTION)**

LABEL	FUNCTION	EXECUTION TIME (μsec)	PROGRAM LOCATIONS
READ #1	READ AND μ-TO-LINEAR CONVERT SAMPLE	3	11
OUTPUT	OUTPUT ONE FRAME FEATURE COEFFICIENT	6	29
READ #2	READ AND μ-TO-LINEAR CONVERT SAMPLE	60	54
MOVE & PRE-EMP	SHIFT SAMPLE BUFFER BY 4 SAMPLES AND PRE-EMPHASIZE 4 SAMPLES		
WINDOW	CALCULATE WINDOW VALUES & APPLY 3 TIMES TO 4 SAMPLES	111	123
READ #3	READ AND μ-TO-LINEAR CONVERT SAMPLE	193	118
AUTOCOR-RELATION	USE 4 SAMPLES TO UPDATE 9 AUTOCORRELATION VECTORS FOR 3 OVERLAPPED FRAMES		
READ #4	READ AND μ-TO-LINEAR CONVERT SAMPLE		
	TOTAL (% USED OF AVAILABLE)	373 (62%)	335 (33%)

sample is available every 150 μsec and is placed in the FXDSP input buffer by the CODEC. The update program reads that sample at a convenient time, but before the next sample, arriving 150 μsec later, overwrites it. Each sample is immediately converted from mu-law to linear encoding by the FXDSP and is then written into a four-sample buffer without any further processing until all four samples are obtained.

As a result, the sample update program has a pipeline delay of four samples. The frame recursion program calculates on the frame just completed and produces the output of a feature vector within one frame period after the end of the corresponding frame.

5. Program Implementation

This section describes several novel programming techniques that were required to implement the FXDSP. The most scarce resource was program memory; execution time and read/write memory were available in sufficient quantities. Therefore, most innovations were directed toward reducing the amount of program memory required at the expense of increasing execution time or read/write memory requirements. The specifics of program module size, execution time, and execution sequence are covered in Tables 1 and 2.

One major problem, the negotiation between the input sample timescale of 150 μsec and the output frame time of 15 msec, was solved by the program architecture discussed in the previous section.

A second problem was the Hamming window computation. Because of the frame size and overlap, each sample falls into the first third of one analysis frame, the second third of the previous analysis frame, and the final third of the twice previous frame. Additionally, every 100 samples, when one of the three frames is completed, the relationship of the three analysis windows rotates cyclically. As a result, the Hamming window presented both the problem of producing the cosine-based values and rearranging the segments of the window upon completing a frame.

In earlier implementations, [4] [13] the Hamming window was stored as a table in program memory. In this implementation, program memory was too scarce, so a Taylor series expansion was used instead. Each third of the Hamming window (100 samples) was computed from a third order Taylor series expansion about its midpoint (sample 50, 150, and 250).

To conserve program memory, several pieces of program modules were shared for multiple functions, sometimes with multiple exit points. For example, to perform the division required by Eq. (13), the reciprocal of the energy E was calculated. An efficient reciprocal routine developed by Daugherty[14] was used, but required that the number for which the reciprocal was being formed be between 1 and 2. To build a general-purpose reciprocal routine, the number was first normalized to fall within the desired range. The reciprocal was readjusted to its true value to compensate for the normalization. The reciprocal normalization is the same operation as the amplitude normalization and log energy calculation performed at LPC(1) (Eq. 10) and the same program performs both functions. However, for amplitude normalization, the program is exited before the reciprocal calculation is executed. Thus, the reciprocal routine, which requires 181 program locations, shares 99 of these locations with the gain normalization program, saving on overall program space.

An important way to conserve program space was the development of a means of computing Durbin's recursion with a common piece of code for all orders, $i = 1, 2, ..., 8$. Although the recursion is readily executed as a subroutine in microprocessor and FORTRAN implementations, it

VLSI Signal Processing

is difficult to perform as a subroutine in a programmable signal processor. This is because a digital signal processor does not allow enough addressing capability to handle the two dimensional array of α and the one dimensional arrays of k, E, and R. A digital signal processor typically provides only indirect addressing with the ability to increment one of two or three pointer registers by a fixed amount. An implementation of Durbin's recursion, if strung out, requires 536 program locations (not including the reciprocal calculation). With the iteration-independent form used here, that figure drops to 119 program locations.[15]

6. Summary

A single chip basic building block for LPC-based connected and isolated word recognition systems has been described. The single chip is an appropriately-programmed Digital Signal Processor of AT&T Bell Laboratories.

Because the major limitation in attaining single chip implementation was the amount of program memory available, several novel programming techniques were used to conserve program memory. These included 1) development of a program architecture that interleaved a background main frame inversion program with a foreground sample update program, 2) development of a form of Durbin's recursion suitable for implementation as an iteration-independent subroutine, 3) use of overlaid subprograms with multiple exit points, and 4) use of a Taylor series expansion, rather than a lookup table, to store and permute segments of a Hamming window.

Comparison with numerical simulations shows that the error introduced by the implementation is negligible. This good match renders the chip suitable for use in systems that use quantities calculated in floating point on general purpose computers, such as statistically clustered templates or frames for speaker independent word recognition or for recognition based on vector quantization or hidden Markov modeling.

REFERENCES

1. F. Itakura, "Minimum Prediction Residual Principle Applied to Speech Recognition," *IEEE Trans. Acoust., Speech, Signal Processing*, vol. ASSP-23, pp. 67-72, Feb. 1975.

2. B. Aldefeld, L. R. Rabiner, A. E. Rosenberg, and J. G. Wilpon, "Automated Directory Listing Retrieval System Based on Isolated Word Recognition," *Proc. IEEE*, Vol. 68, No. 11, pp. 1364-1379, November, 1980.

3. C. S. Myers and L. R. Rabiner, "A Level Building Dynamic Time Warping Algorithm for Connected Word Recognition," *IEEE Trans. Acoust., Speech, Signal Processing*, Vol. ASSP-29, pp. 284-297, April, 1981.

4. John G. Ackenhusen and L. R. Rabiner, "Microprocessor Implementation of an LPC-Based Isolated Word Recognizer," *Proc. IEEE ICASSP-81*, pp. 746-749, 1981.

5. L. R. Rabiner, M. M. Sondhi, and S. E. Levinson, "A Vector Quantizer Incorporating Both LPC Shape and Energy," *Proc. IEEE ICASSP-84*, pp. 17.1.1-17.1.4, 1984.

6. S. C. Glinski, "On the Use of Vector Quantization for Connected Digit Recognition," (submitted to *Bell System Tech. J.*)

7. L. R. Rabiner, S. E. Levinson, and M. M. Sondhi, "On the Application of Vector Quantization and Hidden Markov Models to Speaker Independent, Isolated Word Recognition," *Bell System Tech. Journal*, Vol. 62, No. 4, pp. 1075-1105, April, 1983.

8. B. A. Dautrich, L. R. Rabiner, and T. B. Martin, "On the Effect of Varying Filterbank Parameters on Isolated Word Recognition," *IEEE Trans. Acoust., Speech, Signal Processing*, Vol. ASSP-31, pp. 793-807, Aug., 1983.

9. Special Issue on the DSP, *Bell Syst. Technical J.*, Vol. 60, No. 7, Part 2, Sept. 1981.

10. Y. H. Oh, J. G. Ackenhusen, L. M. Breda, L. F. Rosa, M. K. Brown, and L. T. Niles, "Architecture for a Real-Time LPC-Based Feature Measurement Integrated Circuit," *Proc. IEEE ICASSP-84*, pp. 25B.2.1-25B.2.4, 1984.

11. L. R. Rabiner and R. W. Schafer, *Digital Processing of Speech Signals*, Prentice Hall, Inc., Englewood Cliffs, N. J., 1978.

12. L. R. Rabiner and J. G. Wilpon, "A Simplified, Robust Training Procedure for Speaker-Trained, Isolated Word Recognition Systems," *J. Acoust. Soc. Amer.*, Vol. 68, No. 5, pp. 1271-1276, November, 1980.

13. J. W. Daugherty, "Using a Digital Signal Processor in an Exploratory Speech Recognition System," AT&T Bell Laboratories Internal Memorandum, 1980 (unpublished).

14. J. W. Daugherty, "A Division Subroutine for the Digital Signal Processor Device," AT&T Bell Laboratories Internal Memorandum, 1980 (unpublished).

15. John G. Ackenhusen, "Iteration Independent Subroutine Form of Durbin's Recursion for Programmable Signal Processors," (to appear).

Applications in Speech Recognition Using Speech-specific VLSI

P. Ramesh and C-F. Chan

Intel Corporation

THIS MANUSCRIPT UNAVAILABLE FOR PUBLICATION

35

VLSI Signal Processing

A PROGRAMMABLE VLSI DIGITAL SIGNAL PROCESSOR

R.J. Dunki-Jacobs, R.M. Hardy, W.J. Premerlani, and J.E. Wheeler

General Electric Company
Corporate Research and Development
Schenectady, New York 12301

ABSTRACT

There is always the need to match new implementation technologies with applications. Architectures selected for implementation in VLSI should provide problem solutions that were economically or technically limited with earlier implementations. This document discusses a multi-application signal processor device that was custom designed on a chip (see Figure 1). A team of four VLSI System Architecture Program members at CRD conceived, designed, and implemented a custom chip during the last half of 1983 with development software and work stations provided by VLSI Technology Incorporated [1], a California company with a branch office in Boston. The following sections provide a general description of the chip, some applications, the CAD tools used, a detailed functional description of the elements on the chip, and some aspects of testability.

FOREWORD

Integrated circuit technology has evolved at a rapid and sustained rate for over 20 years. The technology now is capable of one to six hundred thousand transistors on a single monolithic chip. Millions are expected by the end of the decade. At some recent point in time, this rapidly evolving technology transitioned from a component technology to providing the capability for revolutionary system design. This is the key point. Chip technology has developed from providing the capability for sophisticated components, including microcomputers, to providing monolithic subsystem capability. Future competitive products depend on understanding this.

VLSI technology will continue to develop rapidly. System design requires revolutionary changes. It is essential that chip design be an integral part of system and product design. This requires system designers to partition conceptual product designs into conceptual chip designs and iterate alternatives.

There must be revolutionary changes in chip design methods. Design time must be reduced dramatically. Current methods cannot use VLSI technology cost-effectively except in limited "component" ways. The rapidly architected and designed chips must yield and perform satisfactorily.

Yet the large cadre of system engineers who must become expert in VLSI system design methodology cannot become expert classical IC designers nor process engineers. (Conversely, IC designers cannot become expert system or application engineers except in limited application areas such as specific telecommunications high-volume component needs. Thus the technology utilization problem is generated.) The methodology and CAD tools must provide system engineers the capability to be tall, thin designers without being naively thin.

It was in this backdrop that this project began. System engineers would architect and design a chip aimed at understanding these issues. They would use a relatively aggressive commercial technology, 3μm silicon gate HMOS. They would adopt a

VLSI Signal Processing

hierarchical macrocell design methodology commensurate with VLSI system design objectives rather than classical IC component design methods. They would use the most advanced, and therefore far from mature, design tools available. They would incorporate system features aimed at configurable data and control flow for many on-chip processors executing concurrently. They would design this chip in four months.

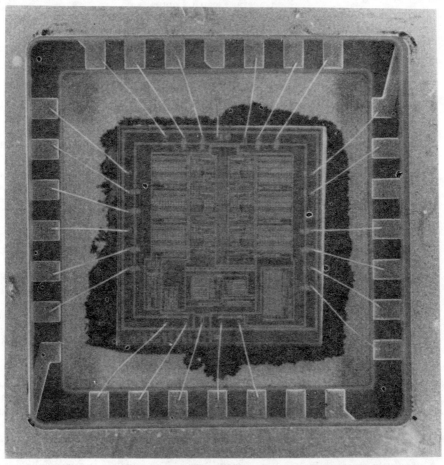

Figure 1. Multi-application signal processor chip.

GENERAL DESCRIPTION

The device described herein is a *programmable digital signal processor* implemented with HMOS three-micron technology. The *digital signal processor* consists of approximately ten thousand transistors on a chip slightly smaller than a quarter inch square, packaged in a 28-pin DIP. A picture of the integrated circuit and a floor plan are shown on Figures 2 and 3, respectively. Figure 4 shows the pin function list.

The device is made up of a set of basic elements, the relationships of which can be

VLSI Signal Processing

programmed. One is a *signed arithmetic multiplier/adder* implemented six times with input sources and output destinations programmable for a variety of applications requiring concurrent data processing. Sets of data can flow in the same or in opposite directions, or one data set can hold constant while another moves through the process.

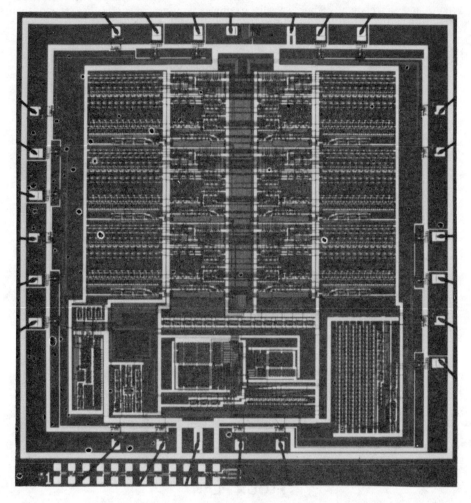

Figure 2. The picture of the finished signal processor chip.

There is also a *programmable truncator* and a *programmable delay*. Control is provided internally by a PLA controller with external features for interfacing with other devices and concatenating with like units.

In addition to the functional programming, certain elements on the chip can be programmed out of service if determined to be defective. Data can be routed to and from only those elements known to be functionally good.

VLSI Signal Processing

The block diagrams in Figures 5 and 6 are useful to understanding the operations. For a specified operation, the unique configuration for that operation is loaded into a *configuration register* to steer data and to control the operating modes (eg., truncation, delays, sign extension).

Application examples include finite and infinite impulse response filters, convolution, and correlation. Some specific examples are described in the section that follows.

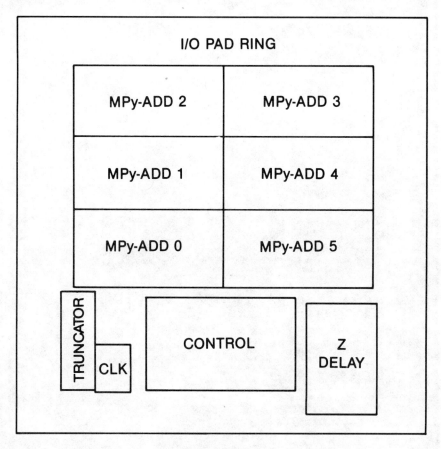

FLOOR PLAN

Figure 3. A diagram of the signal processor's floor plan.

APPLICATIONS

Application examples include finite and infinite impulse response filters, convolution, and correlation. Other signal processing applications in which multiplication and delay steps are required can also be handled.

As described in the section on configuration, the multiplier-adder-delay cells are arranged into the proper connection via the serially loaded configuration registers.

Each cell can be connected to its left or right neighbor or to one of four data busses. A chip with six cells can be configured to be up to four separate filters with the sum of all the orders less than or equal to six. For example, a single sixth-order filter, or two third-order filters, or a second- and a fourth-order filter can be implemented with a single chip. Larger filters can be implemented by chaining chips.

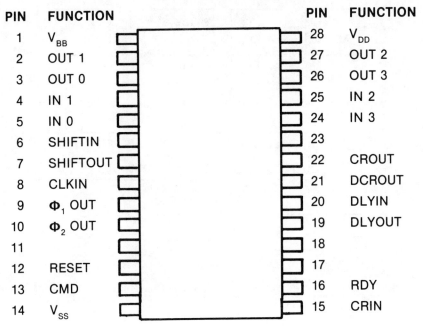

Figure 4. Pin function list.

Finite Inpulse Response Filter

A typical finite impulse response filter is shown in Figure 7. The output of the filter in Figure 7 is a weighted sum of the data points in the input sequence. The number of stages required is equal to the order of the filter. For example, a second order filter requires two stages. For an Nth order finite impulse response filter, the output is related to the input samples by

$$y_k = \sum_{n=0}^{N-1} a_n x_{k-n} \tag{1}$$

The weights a_n are initially loaded into the M1 registers. After they are loaded the filter is turned on and the data samples x_k are fed into the filter bit serially in two's complement signed format. Depending on the application, either a 12, 24, or more bit result is available as the filter output. The summation of the outputs of N filter stages, each producing 24-bit results, leads to a result with 24 plus $\log_2(N)$ bits. A provision has been made to program the chip to extend the result to the number of bits required. The truncator can be programmed to select any 12-bit portion of the result.

VLSI Signal Processing

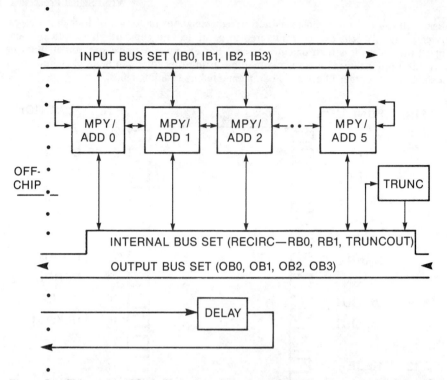

Figure 5. Chip system block diagram—through configuration programming data flow and concurrent processing can serve a variety of applications.

Infinite Impulse Response Filter

An infinite impulse response filter is shown in Figure 8. It is basically made of two strings of the form shown in Figure 7. One of the strings forms a feed forward branch; the other forms a feedback path. The equation relating the output to the input is

$$y_k = \sum_{n=0}^{N-1} a_n x_{k-n} + \sum_{n=1}^{M-1} b_n y_{k-n} \tag{2}$$

Note that the truncator is required in the feedback path to convert the output of the summer to the 12-bit input required by the multiplier. The truncator is also used to produce a 12-bit output. As is detailed in the section describing it, the truncator selects any 12 input bits, with the output delayed one time step from the input. This was done to give application flexibility, to place the implied binary points in the data and the coefficients anywhere. When the truncator is used for feedback purposes, it is programmed to match the binary point at the node receiving the feedback.

Inner Product

Another application, forming an inner product, is shown in Figure 9. In this case two input streams are multiplied together according to the following equation:

$$z_k = \sum_{n=0}^{N-1} x_{k-n} y_{k-n} \qquad (3)$$

In this example the chip is programmed to shift the contents of both M1 and M2 registers from cell to cell.

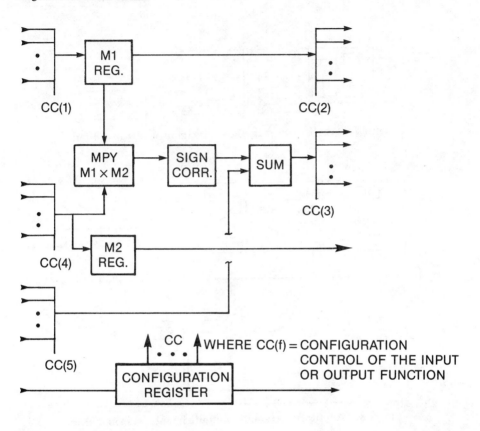

Figure 6. Multiplier/adder—configuration control of inputs and outputs provide application flexibility.

Truncated Convolution

Another example is shown in Figure 10, in which truncated convolution is performed. It is similar to the previous example, except the x and y data streams flow in opposite directions. In this case the output is given by:

$$z_k = \sum_{n=0}^{N-1} x_{k-N+n} y_{k-n} \qquad (4)$$

VLSI Signal Processing

There is another important difference between this example and the previous one. With the data streams flowing in opposite directions the data must be fed in on every other multiplication cycle. The number of cells required is two times N. This apparent inefficiency can be overcome by interleaving two separate calculations.

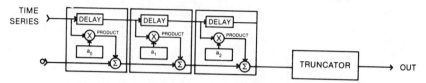

Figure 7. Application example—finite impulse response filter.

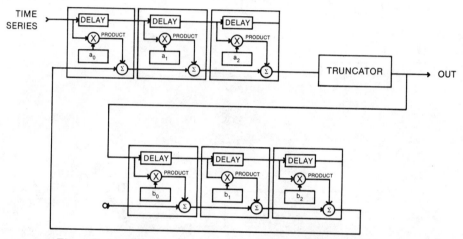

Figure 8. Application example—infinite impulse response filter.

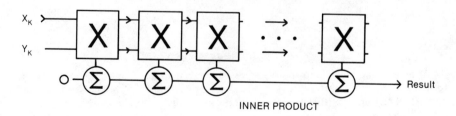

Figure 9.

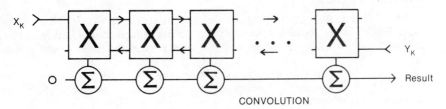

Figure 10.

COMPUTER-AIDED DESIGN TOOLS

In order to complete the detailed layout and design of an integrated circuit, the design team requires an extensive set of computer-aided engineering and design tools. This section describes a suite of integrated tools, supplied by VTI (VLSI Technology, Inc. [1]), used by the design team.

Computer Environment

The tools suite was designed to execute on either VAX11/780 equipment or Apollo Domain work stations. The design group rented computer facilities from VTI at their design center in Boston, Massachusetts. The design center was equipped with several Apollo nodes, a Versatec printer/plotter, and backup devices.

Software Environment

The software suite used is an integrated collection of VLSI design tools. These tools include the

- Hierarchical schematic editor
- Sticks editor
- Layout editor
- Composition editor
- Cell compiler library
- Logic/timing simulator
- Analog simulator
- Design rule checker
- Circuit extractor
- Netlist comparator

At any point in the physical design process a geometry file in CIF (Caltech Intermediate Form) can be produced. This file can be used by various utilities. The following sections will give a brief description of the tools and how they are used in the design process.

Hierarchical schematic editor

A schematic editor provides a familiar exchange media for the designer. The computer data base represents the design. A hierarchical schematic editor allows the designer to define a design in modules. Modularization allows the designer to deal with complex design issues and interfaces easily.

VLSI Signal Processing

The VTI schematic editor provided convenient schematic drawing aids and interfaced directly to the available simulators and the net list comparison facility. By iterating between the schematic editor and the logic simulator a valid design is quickly prepared.

It should be noted that, while the schematic editor is a very worthwhile tool, it received only limited use in the design of the chip described, because it became available late in the project.

Sticks editor

The sticks editor is a computer embodiment of the the sticks layout system presented by Mead and Conway [2]. A sticks editor allows the designer to place major circuit structures (whole transistors, contacts, lines, etc.) relative to each other. When the designer has finished placing the circuit elements, a compaction process is entered. The compactor takes the "suggested" layout and moves elements about until all design rules are obeyed with minimum overall cell area.

Layout editor

The layout editor is used to achieve totally minimized cell sizes. Using a layout editor the skilled layout artist can achieve higher circuit densities than are achievable with the sticks editor. In a layout editor the designer has complete control over the geometry of all objects on all mask levels. The price paid for this freedom is slower turn around of even simple cells. Because of this price, its use was limited in the current design.

Cell compiler library

The tools include a cell compiler library which contains parameterized building blocks such as

- I/O pads
- ROMs
- PLAs
- Buffers
- Counters
- ALUs
- ANDs, NANDs, ORs, NORs, etc.
- Text generators

Composition editor

The composition editor is used to "compose" groups of cells to form another higher level cell in the design hierarchy. The cells which can be used during composition are any cells generated by the above tools and other composition cells.

The composition editor provides an automatic interconnect facility, based on a previously entered schematic. After either automatic or manual interconnect, a compaction phase (similar to the sticks compaction) is entered. At any level a Caltech Intermediate Form (CIF) representation of the design can be produced for use by extractors, design rule checkers, and plotting utilities.

Logic/timing simulators

The logic/timing simulators provide design feedback during the schematic generation phase. After layout, real circuit capacitances are extracted and simulation can be rerun.

A transistor level, nine state logic analyzer was provided. In timing mode the simulator assumed simple RC models for transistors. The simulation could be run interactively or under a command file. The state of watched nodes is displayed as the simulation takes place in a familiar timing diagram format.

Analog simulator

The tool set included a utility to create a SPICE input deck from the logic/timing simulator input stream. Analog and device simulation was conducted using Berkeley Spice 2G.5 with process parameters supplied by VTI.

Design rule checker

The design rule checker is used to verify that a CIF file meets the design rules for the stated technology. Design rule violations are marked graphically on a plot of the cell being checked. When a cell is determined to be design rule correct it is then usable in higher level constructions.

Circuit extractor

The circuit extractor recovers electrical connectivity information from the CIF file for a cell. Node names given during leaf cell generation are retained as the root name for a unique name based on the cell hierarchy. The extraction includes primary and secondary capacitance effects. The output of the extractor is usable directly by the logic/timing simulator and through post processing by SPICE.

Netlist comparator

The netlist comparator verifies that the output of the extractor matches a corresponding hierarchical schematic drawing.

Miscellaneous utilities

There are several utilities available to perform useful functions connected with the design process. These utilities include

- CIF plotting on raster or vector plotting devices
- Schematic plotting on raster or vector plotting devices
- Layer filtering
- CIF merging
- Simulation plotting
- Programmable logic array assembling

FUNCTIONAL DESCRIPTION

The paragraphs that follow provide a description of the system on a functional block-by-block basis.

Multiplier

The object of the operation is to provide the signed product of two n-bit operands

VLSI Signal Processing

(M1 × M2) to the input of an adder (see Figures 5 and 6). For the duration of the multiplication of the two words (M1 × M2) the operand M1 remains stationary in a register while M2 is shifted in one bit at a time. The multiplier is made up of n number of bit slices (cells). Each bit slice consists of one AND gate, a FULL ADDER, and two FLIP-FLOPS. The bit slice is of the carry-save type, whereby the carry bit stays in place while the shifting and adding of partial sums takes place. The flip-flops SUMff and CYff together accumulate the running sums of a conventional bit-shift and parallel-add multiply operation.

There are 12 identical cells for a 12-bit by 12-bit multiply. The logic of Figure 11 (one cell) can be described as follows:

SUMout: = ~CYff and ~SUMff and   M1 and M2
or ~CYff and SUMff and (~M1 or ~M2)
or CYff and ~SUMff and (~M1 or ~M2)
or CYff and SUMff and   M1 and M2;

where

SUMff: =SUMout[i-1,t-1];

CYff: =SUMff[t-1] and M1[t-1] and M2[t-1]
or SUMff[t-1] and CYff[t-1]
or CYff[t-1] and M1[t-1] and M2[t-1];

t=clock time

i=position in array from LSB to MSB

After the shifting of the 12 M2 bits, the M2 operand has been loaded. An additional 12 clock times (t) are required to accumulate the 2n-bit product. During these second 12 clock times, the sign correction function is made to operate, thereby effecting no time penalty. This will be described in the next section.

Sign Correction

The previous section described a single stage in the multiply process. This stage is replicated for n-number of bits. In this case M1 and M2 each consist of 12 bits. In addition to the simple positive product of 24 bits, the following three cases can exist:

1. M1>0,M2<0 (ie., M1 positive, M2 negative)
2. M1<0,M2>0 (ie., M1 negative, M2 positive)
3. M1<0,M2<0 (ie., both negative)

Case 1:

$$P = M1*(-M2)$$
$$P' = M1*abs(2**n - abs(M2))$$
$$P' = (2**n)*abs(M1)$$
$$\quad - abs(M1)*abs(M2)$$
$$P = 2**2n - abs(M1)*abs(M2)$$
$$\quad = - abs(M1)*abs(M2)$$

where 2**2n falls off the most significant end and can be ignored

$$P = P' - (2**n)*(M1)$$

340

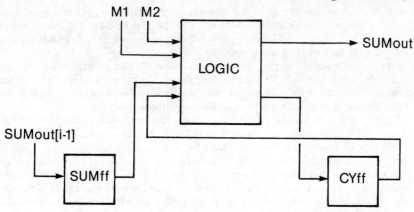

Figure 11. One of n cells in a n-by-n bit multiplier.

From the above statement, it can be seen that by subtracting M1 from the most significant 12 bits of the product, the correct signed results can be obtained. The least significant 12 bits already generated are not affected by the sign. Therefore this sign correction is accomplished during the second half of the product generation and does not affect the multiply time.

Case 2 can be analyzed by swapping M1 and M2.

Case 3 can be analyzed similarly and the product P will be as follows:

P = (−M1)*(−M2)
P' = 2**2n − (2**n)*abs(M2)
     − (2**n)*abs(M1)
     + abs(M1)*abs(M2)
P = P' − (2**2n − (2**n)*abs(M2)
     − (2**n)*abs(M1))
     and again ignore 2**2n
P = P' + (2**n)*(abs(M2)
     + abs(M1))
P = P' − (2**n)*(M2 + M1)

Therefore, signed products for all cases are produced by a simple detection of the signs of the operands and the enabling of a subtract operation for each M1 and M2 at the appropriate time.

In summary, sign correction, if required, is accomplished by subtracting the operand(s) from the most significant half of the product serially as the product is generated a bit at a time. The least significant half remains intact, the most significant half is modified, and no additional time is taken.

Sign Extension

Each signed product may be directed to an adder circuit for addition to the summation of other products. This can be seen by examination of the block diagrams of Figures 5 and 6. To prevent the occurrence of an overflow condition, provision has been made to extend the sign bit for a programmable number of bit-times.

Programmable Configuration

There are four INPUT and four OUTPUT BUSSES on the chip. Each multiplier/adder has the capability through configuration programming of connecting to these busses. In additional, there are RIGHT and LEFT NEIGHBOR connections which allow forward and reverse flow of data operands. An INTERNAL BUS SET provides for recirculation and truncation.

The Applications section shows that different configurations are required for different applications. These configurations involve a specified number of multiplier/adders, forward and reverse data flow paths, a number of sign extension bits, storage of constants, portions of data to truncate, and delays.

Figure 5 shows the data flow paths that provide the capability to reconfigure the functional elements on the chip to meet different functional requirements.

Figure 12 shows the data flow path for setting up a configuration. Configuration bits are loaded and stored in CONFIGURATION REGISTERS. Multiplexors and demultiplexors make use of these configuration bits and route input and output data accordingly. Once a configuration has been set up during a command input sequence, the configuration remains for all operations until such time as a new command input sequence is loaded.

Table 1 is a representation of programmable configurations. Fourteen bits program each of the six multiplier/adders. Two bits are for the truncator, four for the delay, and twelve for the controller. These bits together with six leading Test Bits make up the command input sequence of 108 bits divided into 9 12-bit words.

Control

Timing and control signals are provided to each of the six multiplier/adders, the truncator, and delay by a central CONTROLLER, along with the two-phase clock signals. In addition the controller drives and receives external signals.

Control sequences exist in the form of microcodes permanently stored in two PLAs operating on the clock phases. One PLA is loaded with codes relating to bit timing, and the other PLA is loaded with codes relating to states that represent a major operational function. These states can be seen in Figure 13, along with the overall control picture of the chip's operation. The physical location of the controller is shown on the diagram of the floor plan (see Figure 2). The two PLAs, TIMER and STATE, operate through the various states as shown in Figures 14 and 15, respectively. The PLAs are embedded within the controller block (shown on the system floor plan) along with input signal conditioning, control signal drivers, and the controller's own configuration register. Its own configuration register determines the following information: whether to load a new M1 operand or to hold M1 constant and move M2 only, the number of bits of sign extension,and the bits for truncation. Table 1 shows a bit description of the programmer's options.

There are 2 basic operating modes,SETUP and OPERATE. The first is the loading of the configuration register which sets up the chip's configuration in preparation for operation. Setup mode is selected externally by the activation of the line titled CMD. The Operate mode involves operating on input and stored operands as specified by the configuration. The handshake lines of WAITING and RDY are provided at the interface.

At any point in time the chip is operating in one of the states shown horizontally in Figure 13 (except at power-up time when it goes through State 0). Each state is subdivided into t-times that may range from t0 through t11. The t-times actually represent bit times. States 2, 3, 4, 5, and 7 always go through the 12 t-times. States 1 and 6 may be terminated early to maximize the through-put speed. State 6 is programmed to be exactly long enough to accommodate the extension of sign bits, depending on the expected word length resulting from concatenations of multiplies

VLSI Signal Processing

and adds. State 1 (waiting) is designed to terminate quickly upon receipt of an external signal indicating that new input data is ready.

Functions are listed on the left of the Figure 13 chart and the associated actions are embedded within the chart to show state and timing relationships.

While the drawings and text of Figures 13, 14, and 15 serve to document the controller functions, it is not necessary for one to peruse these or to read the remainder of this section on control, unless interested in the lower levels of detail.

External inputs to the controller are as follow:

RESET — Returns the controller to the reset state
RDY — Signals that external data or command is ready
CMD — When coincident with RDY indicates that the data to be entered is CONFIGURATION data rather than M1 or M2
CRIN — Configuration data input

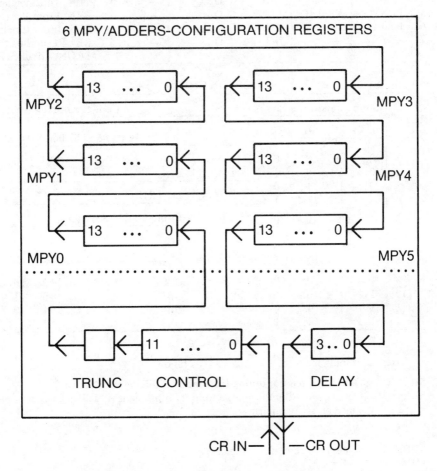

Figure 12. Configuration register data flows into the chip, and serially flows until all bits reach their destinations to set up the chip's operating configuration, with CR out providing leading test bits.

Table 1
CONFIGURATION REGISTER CONTROL DATA FORMATS

A) MULTIPLIER/ADDER (each of six)

	Configuration Register					
	M1 In 13 12 11	M2 IN 10 09 08	Sum In 07 06 05	Sum Out 04 03 02	M1 Out 01 00	
0	IB0	IB0	ZERO	OFF	TRN1	where:
1	IB1	IB1	OB0	OFF	TLN1	IB: Input Bus
2	IB2	IB2	OB1	OFF	RB0	OB: Output Bus
3	IB3	IB3	OB2	OFF	RB1	TRN: To R Neighbor
4	FRN1	FLN1	OB3	OB0	—	TLN: To L neighbor
5	FLN1	TRUNC	FLN2	OB1	—	FRN: From R neigh
6	RB0	—	IB2	OB2	—	FLN: From L neigh
7	RB1	—	IB3	OB3	—	RB : Recirc Bus
						TRUNC: Trunc Out

B) DELAY

	Configuration Register			
Delay	03	02	01	00
1	0	1	0	0
2	0	1	0	1
3	0	1	1	0
4	0	1	1	1
5	1	0	0	0
6	1	0	0	1
7	1	0	1	0
8	1	0	1	1
9	1	1	0	0
10	1	1	0	1
11	1	1	1	0
12	1	1	1	1

C) CONTROL

	Configuration Register		
	11	10 09 08	07 06 05 04 03 02 01 00
M1 Constant	0	SX1	TX1 TX2
M1 variable	1		

D) TRUNC

Configuration Register		
Trunc In	01	00
OB0	0	0
OB1	0	1
OB2	1	0
OB3	1	1

External outputs from the controller are as follow:

SHOUT — Data is shifted out during the time that this signal is active.
SHIN   — Data is shifted in during the time that this signal is active. The absence of SHIN implies waiting.

Chip control signals are as follow:

SHCR   — Shift in configuration data
SHM1   — Shift M1
SHM2   — Shift M2
S1   — Reset state

VLSI Signal Processing

SHTRUNC — Shift truncator
SHZ — Shift delay
SX1TX1 — Programmable state and time to cut off truncator
TX2 — Programmable bit-time-out for sign extension
TSTSIGN — Signal to multiplier to test sign to determine if sign correction is required
RECIRC — Recirculate M1 and M2 for sign correction during second half of multiply cycle

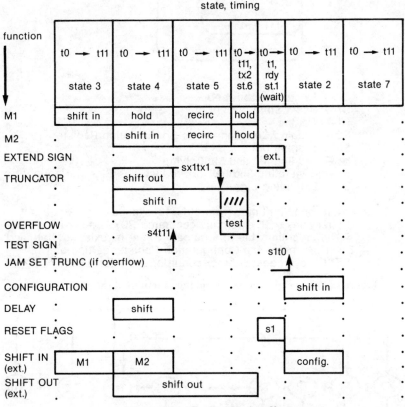

Figure 13. Control timing diagram.

Clock

The clock requires a symmetrical (or nearly so) signal at two times the intended operating clock rate of the chip. The internal CLOCK circuit generates PHI1 and PHI2 which are actually phases one and three of a four phase clock. Since the other two phases are not used, there are guaranteed spaces between the two phases that are used. See Figure 16.

Control signals are generated by PHI2 and extend from PHI2 to PHI2. The logic of the entire chip is wired such that control signals are strobed by PHI1 at the point of use. For example, if control signal SHM2 becomes active at PHI2 then the next PHI1 ANDed with SHM2 will cause data to be shifted into each M2 bit position. The next PHI2 will then latch the data into the output stage of each M2 bit position.

VLSI Signal Processing

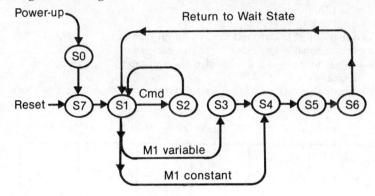

- S0— Exists after power-up
- S7— Reset State
- S1— Waiting
- S2— Loading Commands into Configuration Register
- S3— Loading M1
- S4— Loading M2 and Multiplying
- S5— Generating 2nd half of Product with Sign Correction
- S6— Extending Sign

States exist for 12 bit-times (where n = 12) except:
    S1 recycles until externally generated RDY is received.
    S6 is programmable to terminate when the existing State and Time match the Configuration Register setting of TX2.
    RESET overrides all other conditions and forces S7.

Figure 14. The eight states control the functions of the chip.

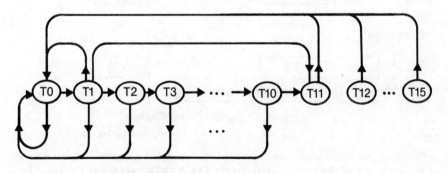

Tx → T0 if S6 and Tx = TX2
T1 → T0 if S1 ~ RDY
T1 → T11 iff S1RDY
T11,T12..T15 → T0 unconditional
T0 → T1 unconditional
Ti → Ti + 1 if 2 < = i < 11

Figure 15. The control uses a Mod 12 Bit-Timer with truncation to avoid time loss during a waiting state or an arithmetic termination state.

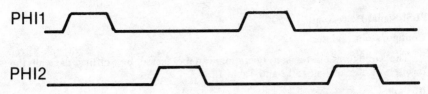

Figure 16. Guaranteed non-overlapping clocks.

Truncator

The TRUNCATOR is a shift register n-bits (12) in length that can be programmed to span any n-bits of the 2n + sign extension bits from any of the multiply/adders. In other words, each multiplier receives two 12-bit operands and produces a 24-bit product. The adder receives this 24-bit product plus sign extension bits (as programmed) and another similar input from its input programmed source. The user may want to save only the 12-bits from the adder that he considers significant to be passed on to another multiplier as a 12-bit operand. The values placed in SX1 and TX1 of the configuration register of Table 1B determine which 12 bits will be saved. These 12 bits will then be shifted out during the time that M2 is shifting in the next operand.

Delay

The DELAY circuit provides delays of z^{-1} to z^{-12} under program control. This feature is useful for certain types of filters. Input and Output are connected to input and output pins, respectively, on the chip. Programmed delays are shown in Table 1C.

TESTABILITY

Definition

The term testability informally defined as a measure of the ease of generating and applying test patterns to detect physical failures in a circuit was elaborated on by Jacob Abraham, University of Illinois [3].

No VLSI design should be started without consideration for testability at the outset. This is agreed upon universally. The degree or level of testability considered to be acceptable is not agreed upon. The question of whether testability would be a limiting factor in the progression of VLSI design was considered by a panel at the March 1983 IEEE VLSI Test Workshop [4].

The manufacturer and user are both interested in adequate testing in a short period of time. This generally precludes testing a device functionally for its intended purpose, particularly if its intended purpose is general. Therefore, there is a need for innovative circuits to rapidly test a system by other means. Much literature has been written on the subject and many techniques have been presented. Jacob Abraham [3] provides a survey. In addition, trade journals and conferences address the issue.

Testable Design

The signal processor design contains no built-in test equipment. The circuits are laid out in such a way as to make testing convenient through the functionally serialized data paths.

VLSI Signal Processing

Configuration register

The configuration register operation can be tested by shifting data all the way through. Refer to Figure 12 and Table 1.

Bus set

Selected configurations can be set up to route data in and out in such a way as to verify free flow through the input and output data busses. Refer to Figure 5 and Table 1.

M1 and M2 registers

The M1 and M2 registers are serial devices with capability for linking under configuration control. Data can be verified to flow through all the M1 and M2 registers linked serially by configuration programming of the data paths and the appropriate control operations. Refer to Figure 6 and Table 1C.

Data source and destination

Source and destination paths are set for each functional block (ie., multiplier/adder, truncator, delay) by the information in the configuration register.

Optimized test

As seen by a glance at Table 1, it would not be realistic to attempt to test all combinations of configurations. However, the system architecture is suitable for an optimum set of test vectors.

Redundancy

In cases where certain system elements (such as a multiplier/adder) may be determined to be defective, the entire chip may not be a loss. Because of the configuration programming capability, defective functions may be deleted from the operational data paths. This action may be effective when many like chips are concatenated in a larger system.

REFERENCES

[1] VLSI Technology, Inc. 1101 McKay Drive, San Jose, CA 95131.
[2] Carver Mead and Lynn Conway, *Introduction to VLSI Design*, 1980, Addison-Wesley, Reading, MA.
[3] Abraham, Jacob A., "Design for Testability," *Custom Integrated Circuits Conference,* 1983.
[4] "Impact on Test Complexity," *1983 IEEE Test Workshop,* Atlantic City, NJ.

Part X
SYSTOLIC ARCHITECTURE
Edited by:
Dr. J. G. Nash
Hughes Research Laboratories

Partitioned QR Algorithm for Systolic Arrays[1]

D.I. Moldovan

Department of Electrical Engineering-Systems
University of Southern California
Los Angeles, CA. 90089 - 0781

1. Processing QR algorithms

1.1. Statement of the problem

A widely accepted method for computing the eigenvalues of a matrix $A \in R^{n \times n}$ is to first transform A into an upper Hessenberg form and then to apply iteratively a QR algorithm to this upper Hessenberg matrix [1]. If Q_k is an orthogonal matrix, then after a similar transformation

$$A_{k+1} = Q_k^T A_k Q_k = R_k Q_k$$

the eigenvalues of A_{k+1} are the same as the eigenvalues of A_k. Q_k is chosen such that $R_k = Q_k^T A_k$ is upper triangular. QR algorithm applied directly to a full matrix A leads to $O(n^4)$ multiplications. This is the reason why we prefer to transform first A into an upper Hessenberg form by a sequence of elementary reflectors transformations which require $O(n^3)$ multiplications. Then, the application of QR algorithm to the Hessenberg form requires only $O(kn^3)$ operations, where k is the average number of iterations for one eigenvalue. If A is symmetric, then, the upper Hessenberg form reduces to a tridiagonal matrix and a repeated QR algorithm applied to such matrix diagonalizes the matrix leaving eigenvalues as diagonal elements. QR algorithms have remarkable numerical stability, and as we will see, our partitioning and mapping procedure does not degrade this property.

The purpose of this paper is not to propose a new algorithm for the QR problem, but rather to show how a selected sequential QR algorithm can be mapped methodically into a VLSI array processor with only near neighbor connections. Furthermore, we address the problem of mapping an arbitrarily large size QR algorithm into a fixed size array. This is achieved by partitioning the QR algorithm. The technique used in this paper for partitioning and mapping QR algorithm into a VLSI array processor is applicable to any algorithm with cyclic loops.

---

[1] This research was supported by NSF Grant ECS-8307258 and JSEP Contract No. F49620-81-C0070.

1.2. Previous results

Since the QR technique was first introduced by Francis [2] in 1961 many researchers have proposed sequential and parallel QR algorithms. Sameh and Kuck [3] proposed a parallel QR algorithm for symmetric tridiagonal matrices. This algorithm requires O(log n) time steps and O(n) processors, but the issue of data communication between processors was ignored. Gentleman and Kung [4] were first to propose systolic arrays for QR factorization of matrices. In this scheme the rotations are moving in the array and the elements are fixed (hold R in the array). Other systolic arrays where both rotations and matrix elements are moving were proposed by Heller and Ipsen [17]. Johnsson [5] studied Householder transformations for band matrices. Brent and Luk [6] describe a scheme for eigenvalue decomposition of a symmetric matrix based on parallel Jacobi method. In their square array architecture the row and column rotation parameters need to be broadcast in constant time and this constitutes a limitation. Other parallel Jacobi methods for computing eigenvalues are given in [7, 8]. Kung and Gal-Ezer [9] proposed an array of O(n) processors to triangularize a symmetric matrix in $O(n^2)$ time. They conclude that not much is gained in terms of speed if a square array is used instead of one-dimensional array. A lucid presentation of the state of the art in eigenvalue computations is given by Parlett [11]. He considers two distinct cases: small matrices, when the matrix is contained in the computer main memory (for many computers 100 x 100 matrix is small), and large otherwise (safely 1000 x 1000 is large). All schemes mentioned above refer to small matrices.

Partitioning schemes for QR factorization were proposed by Schreiber and Kuekes [10] and Heller [12]. Schreiber and Kuekes present an analysis for a partitioning method using a p x q trapezoidal systolic array (a truncated q x q Gentleman-Kung triangle). The main steps are:

partition $A = (A_1 A_2 ...)$, A_1 is n x p and A_i is n x (q-p), $2 \leq i$.

reduce A_1 to $\binom{R}{0}$, $R = Q_1^T A_1$, upper triangular p x p.

recycle the rotations, obtaining $Q_1^T A_i$.

omit the first p rows and columns of the updated A and repeat.

All who studied partitioning for matrix processing algorithms approached the problem by partitioning the original A matrix in columns, rows or diagonal bands. Although this division seems natural, it fails for many algorithms. For instance, we don't know yet of any previous partitioning scheme for the QR algorithm for computating eigenvalues (not to be confused with QR factorization). Perhaps the cause is that eigenvalue computations lead to some data dependencies which forbid the partitioning of matrix in its original form. As we will see, our partitioning method transforms first the algorithm into a new form where dependencies are rearranged in a convenient way, and then we partition the modified matrix. This partitioning does not necessarily correspond to any row, columns, or diagonal bands, but results directly from an index transformation function.

In spite of its practical importance, only few researchers were concerned with the general case of partitioning algorithms. Hwang and Chang [13] considered the partitioning of algorithms for LU decomposition, inversion of nonsingular triangular matrices, solution of linear systems and others. They proposed a set of primitive functional devices which interconnected properly could perform partitioned

VLSI Signal Processing

algorithms. Heller [12] proposed some clever partitioning approaches, but they are applicable only to some types of algorithms.

In the next section, we present a generalized technique for partitioning algorithms for the purpose of their VLSI implementation, and then in section 3 we apply this general technique to the specific problems of this paper, namely QR algorithm.

2. General method for mapping algorithms into smaller VLSI arrays

2.1. Introduction

A method for designing algorithmically-specialized VLSI devices was recently proposed by Moldovan [14]. Briefly, this method is based on finding a transformation matrix which transforms a sequential algorithm index set into a new index set such that the transformed data dependencies are mapped into local interprocessor communications. This mapping technique can be easily extended to the case when the size of algorithm is larger than the size of VLSI array. A natural solution to this problem is to divide the computational problem into smaller problems. The partitioning of algorithms for VLSI devices is not a simple problem for several reasons. First, partitioning may introduce undesired side effects degrading the numerical stability of algorithms. Secondly, a poor allocation of computations to processors may lower speed up factor. This may be caused by the amount of overhead operations resulting from communication between partitions. In what follows we will use symbol I to denote the set of non-negative integers and Z to refer to the set of all integers. The nth cartesian powers of I and Z are denoted as I^n and Z^n respectively.

2.2. VLSI array model

We assume that the computational resource consists of a mesh connected network of processing cells.

<u>Definition 1</u>.

A mesh connected array processor is a tuple (J^{n-1}, P) where $J^{n-1} \subset Z^{n-1}$ is the index set of the array and $P \in Z^{(n-1) \times r}$ is a matrix of interconnection primitives.

Although, we consider for the sake of generality that VLSI arrays are (n-1)-dimensional, practical arrays have a planar layout. The position of each processing cell in the array is described by its cartesian coordinates. The size of the array is expressed by the number of processors $M = m_1 \times m_2 \times \cdots \times m_{n-1}$. The interconnections between cells are described by the difference vector between the position of adjacent cells. The matrix of interconnection primitives is

$$P = [\bar{p}_1\ \bar{p}_2\ \cdots\ \bar{p}_r]$$

where $\bar{p}_j$ is a column vector indicating a unique direction of a communication link.

For example the array shown in figure 1 is described as (J^2, P) where

$$J^2 = \{(j_1, j_2):\ 0 \leq j_1 \leq 2,\ 0 \leq j_2 \leq 2\}$$

$$P = \begin{bmatrix} 0 & 1 & -1 & -1 & 1 & 0 & 0 & 1 & -1 \\ 0 & 1 & -1 & 1 & -1 & 1 & -1 & 0 & 0 \end{bmatrix} \begin{matrix} \vec{j}_1 \\ \vec{j}_2 \end{matrix} \qquad (1)$$

The structural details of the cells and the timing is derived from the algorithm which is mapped into such arrays.

2.3. Algorithm model
In this paper we consider the class of numerical algorithms with nested loops. In order to partition and to map algorithms into VLSI array processors it is convenient to define an algorithm model.

Definition 2

An algorithm A is a five tuple $A = (J^n, D, C, X, Y)$ where

J^n is a finite index set of A, $J^n \subset I^n$
D   is the set of data dependencies of A
C   is the set of computations of A
X   is the set of input variables of A
Y   is the set of output variables of A

Most algorithms for processing matrix or vector operations have an inherent regularity in the sense that computations repeat at different index points $\bar{j} \in J^n$. Also, the data dependencies which define some precedence relations in computation have a regular structure and can be simply described by a matrix $D \in Z^{n \times m}$ whose columns are dependence vectors associated to some variables. A dependence vector $\bar{d}$ is defined as the difference $\bar{d} = \bar{j}^2 - \bar{j}^1$ where $\bar{j}^1 \in J^n$ is an index point at which a variable is generated and $\bar{j}^2 \in J^n$ is an index point where the same variable is used. We write

$$D = [\bar{d}_1 \ \bar{d}_2 \ldots \bar{d}_m] \qquad (2)$$

The model presented above is particularly useful for numerical algorithms normally written in the form of nested loops in conventional programming languages. A distinct class of algorithms are those for which data dependencies are constant over the entire index set. Algorithms in this class are easier to partition and to map into VLSI arrays. The set of inputs X is mapped by the algorithm into the set of outputs Y.

2.4. Algorithm transformations
In [14] a transformation function T was introduced for mapping algorithms into systolic arrays. Necessary and sufficient conditions were given for the existence of such transformations. In this section we discuss briefly this transformation and show that partitioning of algorithms and their mapping into smaller VLSI arrays can be

VLSI Signal Processing

done by the same transformation if an additional constraint is satisfied. Consider that an algorithm $A = (J_a^n, D, C_a, X, Y)$ is input/output equivalent to another algorithm $B = (J_b^n, \Delta, C_b, X, Y)$.

Definition 3

An algorithm transformation T which transforms algorithm A into algorithm B is defined as:

$$T = \begin{bmatrix} \Pi \\ S \end{bmatrix} \quad (3)$$

where:

(i) T is a bijection function (4)

(ii) $\Pi : J_a^n \rightarrow J_b^1$ and $S : J_a^n \rightarrow J_b^{n-1}$ (5)

(iii) $\Pi(\overline{d_i}) > 0$ for $i = 1, 2, ... m$. (6)

If T is a linear transformation, i.e., $T \in Z^{n \times n}$ than condition (6) becomes $\Pi \overline{d_i} > 0$, and the new dependencies $\Delta = T D$. The purpose for separating T into two linear functions Π and S is that Π alone is related to the timing properties of the new algorithm and S is related to the space properties of the new algorithm. Condition (6) indicates that the transformed data dependencies are positive in lexicographical sense, thus preserving the precedence relations of the original algorithm and assuring a proper execution ordering. Transformation S can then be selected such that the transformed dependencies map into a VLSI array modeled as (J_b^{n-1}, P) as discussed above. This can be written as:

$$S \cdot D = P \cdot K \quad (7)$$

where matrix K indicates the utilization of primitive interconnections in matrix P. Matrix $K = [k_{ji}]$ is such that

$$k_{ji} \geq 0 \quad (8)$$

$$\sum_j k_{ji} \leq \Pi \overline{d_i} \quad (9)$$

Expression (8) requires that all elements of K matrix are nonnegative and (9) requires that communication of data associated with dependence $\overline{d_i}$ must be done using some primitives $\overline{p_j}$ exactly $\sum_j k_{ji}$ times. Transformation S is selected such that the number and the length of interprocessor connections are as small as possible. Since K is the utilization matrix, it means that K should have as few nonzero elements as possible. In section 3 we discuss the selection of transformation T for the QR algorithm.

2.5. Partitioning the index set

We consider now the case when the size of the algorithm is larger than the size of the VLSI array. A natural approach to this problem is to divide the algorithm into smaller subproblems and map these subproblems into the array. Our idea of partitioning is to divide the n-dimensional index space in bands using (n-1) hyperplanes. These hyperplanes Π_{pk} are linearly independent and actually are the rows of transformation S. We write

$$S = \begin{bmatrix} \Pi_{p1} \\ \Pi_{p2} \\ \vdots \\ \Pi_{p(n-1)} \end{bmatrix} \qquad (10)$$

Transformation S is normally selected from (7), but in the case of partitioning the following additional conditions must hold:

$$\Pi_{pk} \overline{d_i} > 0 \quad \text{for } 1 \leq k \leq n-1 \text{ and } 1 \leq i \leq m \qquad (11)$$

The meaning of (11) is that the transformed dependencies are oriented in the same direction, which in turn allows for the partitioning of the new index set into bands. For instance if a band B_1 follows another band B_0 in the direction of hyperplane Π_{pk}, then (11) assures that no computation in B_0 depends on data generated in band B_1. The mapping of indices to processors is simply done by associating a coordinate of a processor to one partitioning hyperplane Π_{pk}. Each index point $\overline{j} \in J_a^n$ is processed in a processor whose kth coordinate is

$$x_k = \Pi_{pk} \overline{j} \bmod m_k \qquad k = 1,2,...,n-1. \qquad (12)$$

Let us label the bands by their n-1 coordinates $(b_1 b_2 \cdots b_{n-1})$. Then, it is easy to see that a computation indexed by $\overline{j} \in J_a^n$ belongs to a band whose coordinates are

$$b_k = \left\lfloor \frac{\Pi_{pk} \overline{j}}{m_k} \right\rfloor \qquad \text{for } k = 1,\cdots,n-1. \qquad (13)$$

In our partitioning scheme, index points in each band are swept by the time hyperplane Π and then the bands are executed in some order such that data dependencies are not violated. In [15, 16] more details can be found regarding this partitioning scheme. In section 3 we will apply this general methodology to a QR algorithm.

2.6. Software tools

The partitioning method outlined above was implemented at USC as a software package called ADVIS (Automatic Design of VLSI Systems). This program is written in Pascal and runs on VAX under Unix operating system. The results obtained for the QR algorithms described in the next section were derived using this program.

The input information consists of: algorithm data dependencies D, algorithm index set J^n, matrix P of allowable interconnections in the VLSI array and the size of the

VLSI Signal Processing

VLSI array $m_1 \times m_2 \times \cdots \times m_{n-1}$.

The output information consists of: a valid transformation T including timing hyperplane Π and partitioning hyperplanes $\Pi_{p1}\cdots\Pi_{p(n-1)}$, allocations of computations to processors, the geometry of the VLSI array (actual interconnections), external hardware connection for recirculating data between partitioning bands and timing information. The program selects T such that all required conditions are satisfied, and moreover it optimizes the processing time and reduces the number of interconnections in the actual VLSI array. More details about finding an optimum time transformation Π_{opt} can be found in [16].

3. Partitioned QR algorithm

3.1. QR algorithm model

The QR algorithm used here is taken from Steward's book [1] and it is shown for convenience in figure 2. In this algorithm, the input A matrix is upper Hessenberg of order n and the shift scalar is x. The algorithm overwrites A with $Q^H A Q$ where the orthogonal matrix $Q = P_{12}^T \; P_{23}^T \cdots P_{n-1,n}^T$, where $P_{k,k+1}$ is a plane rotation in the (k, k+1) plane.

Before we apply to this algorithm the transformation technique outlined in the previous section we have to determine the index set J^n and data dependencies D. Following a technique discussed in [14], the data dependencies of an algorithm are derived easier from a pure sequential form (or pipelined form) where all indexes are specified at each computation, thus, eliminating any possible data broadcasts. The equivalent pipelined form for the QR algorithm is shown in figure 2.

It can be seen that the algorithm index set is $J^3 : 1 \leq k \leq n, 1 \leq i \leq k, k+1 \leq j \leq n$. A data dependence vector is easily picked as the difference between the index points where a variable is used and the index point where that variable was generated, i.e., $\bar{d} = (k^2, i^2, j^2)^T - (k^1, i^1, j^1)^T$. From figure 2, we find $\bar{d}_1 = (1,0,0)^T$ from statements (10), (11), (17) and (18); $\bar{d}_2 = (1,-1,0)^T$ from line (10); $\bar{d}_3 = (1,1,0)^T$ from line (11); $\bar{d}_4 = (0,0,1)^T$ from lines (12) and (13); $\bar{d}_5 = (1,0,-1)^T$ from line (17), $\bar{d}_6 = (1,0,1)^T$ from line (18) and $\bar{d}_7 = (0,-1,0)^T$ from line (19) and (20). It results

$$D = [\bar{d}_1 \bar{d}_2 \bar{d}_3 \bar{d}_4 \bar{d}_5 \bar{d}_6 \bar{d}_7]$$

$$= \begin{bmatrix} 1 & 1 & 1 & 0 & 1 & 1 & 0 \\ 0 & -1 & 1 & 0 & 0 & 0 & -1 \\ 0 & 0 & 0 & 1 & -1 & 1 & 0 \end{bmatrix} \quad (14)$$

We want to map this algorithm into the simplest possible array of size m x m. For beginning we start with a set of allowable interconnections as indicated by (1), but as we will see the actual final array is much simpler.

3.2. Time transformation π
We seek a transformation Π such that (6) is satisfied. For a given set of dependencies D, ADVIS found the optimum Π in the sense of shortest execution time. Π is

$$\Pi = [\,2\ \text{-}1\ \ 1\,] \tag{15}$$

It results

$$\overline{\delta} = \Pi D = [2\ 3\ 1\ 1\ 1\ 3\ 1] \tag{16}$$

Each number in $\overline{\delta}$ indicates the time taken by the variable associated to that dependence to travel between two consecutive processors. Now, we can easily obtain the parallel time $\hat{k}$ at which a computation indexed by (k, i, j) takes place. (This is the time without partitioning yet).

$$\hat{k} = \Pi\,[k\ i\ j]^T = 2k - i + j \tag{17}$$

$\hat{k} =$ constant is the equation of a plane which contains index points unrelated by any dependence, therefore, executable in parallel. This is the key to parallelism. The total processing time is $T_{proc} = \hat{k}_{max} - \hat{k}_{min} + 1 = 3n - 3$. We will see that because of the partitioning this time will change.

3.3. Utilization matrix K
The next step is to generate all possible K matrices which satisfy the conditions (8), (9) and (11). This step is the most computationally intensive step. The program found many utilization matrices K; the one leading to the "best" transformation T is

$$K = \begin{bmatrix} 0&0&0&0&0&0&0 \\ 1&1&0&0&1&1&0 \\ 0&0&0&0&0&0&0 \\ 0&0&0&0&0&0&0 \\ 0&0&0&0&0&0&0 \\ 0&0&1&0&0&0&0 \\ 0&0&0&0&0&0&0 \\ 0&1&0&0&0&0&1 \\ 0&0&0&0&0&0&0 \end{bmatrix} \tag{18}$$

Notice that rows 1,3,4,5,7, and 9 of K matrix contain zero elements; this means that the respective columns of P matrix are not essential in solving equation $SD = PK$. In other words, matrices P and K can be reduced to:

$$P = \begin{bmatrix} 1&0&1 \\ 1&1&0 \end{bmatrix} \quad (19) \qquad K = \begin{bmatrix} 1&1&0&0&1&1&0 \\ 0&0&1&0&0&0&0 \\ 0&1&0&0&0&0&1 \end{bmatrix} \tag{20}$$

Notice that each nonzero entry in the columns of K matrix indicate the usage of an interconnection link by some data dependence vectors. If there is only one nonzero in

VLSI Signal Processing

a column of K, it simply means that a data dependence $\bar{d_i}$ will use only one communication link. Obviously, this is desired because it simplifies the control logic. In our case the program did not identify any K matrix with only one nonzero element per column. Then, by relaxing this constraint we have found the next best, one column (the second one) has two ones. This means that variable a_{ij}^k leading to data dependence $\bar{d_2}$ uses two communication links.

To matrix P from [19] it corresponds a VLSI array with interconnections as shown below in figure 4.

Obviously, the actual interconnections network shown in figure 4 is much simpler than the allowable interconnections network of figure 1.

3.4. Space transformation S

After a simple utilization matrix K has been identified as in (20), we want to find all possible transformations S such that

(i) diophantine equation SD = PK can be solved for S

(ii) matrix transformation T is nonsingular

ADVIS has found several S matrices as a result of this step, of which we selected

$$S = \begin{bmatrix} \Pi_{p1} \\ \Pi_{p2} \end{bmatrix} = \begin{bmatrix} 1 & -1 & 0 \\ 1 & 0 & 0 \end{bmatrix} \quad (21)$$

The rows of S matrix are the partitioning hyperplanes which divide the three-dimensional index space J^3 of the QR algorithm into some partitioning bands.

From (15) and (21) results the final transformation T which transforms the sequential algorithm into a parallel algorithm.

$$T = \begin{bmatrix} \Pi \\ S \end{bmatrix} = \begin{bmatrix} 2 & -1 & 1 \\ 1 & -1 & 0 \\ 1 & 0 & 0 \end{bmatrix} ; \begin{bmatrix} \hat{k} \\ \hat{i} \\ \hat{j} \end{bmatrix} = T \bullet \begin{bmatrix} k \\ i \\ j \end{bmatrix} \quad (22)$$

The parallel version of the QR algorithm is now easily obtained by replacing indices (k,i,j) from figure 2 with $(\hat{k},\hat{i},\hat{j})$ as they result from $(k,i,j)^t = T^{-1} (\hat{k},\hat{i},\hat{j})^t$.

3.5. VLSI architecture for algorithm without partitioning

In figure 3 a computer printout is shown indicating the original index points (k,i,j) and the transformed index points $(\hat{k},\hat{i},\hat{j})$ n=9. $\hat{k}$ represents the processing time and $(\hat{i},\hat{j})$ is the coordinate of the processor. Based on this output, and on matrix P from (19), it is trivial now to construct the VLSI array; it is shown in figure 4. The operation of this array can be easily determined by following how each dependence vector and its respective variables are mapped into the array. The new data dependencies are given by the columns of matrix Δ where

$$\Delta = TD = \begin{bmatrix} 2 & 3 & 1 & 1 & 1 & 3 & 1 \\ 1 & 2 & 0 & 0 & 1 & 1 & 1 \\ 1 & 1 & 1 & 0 & 1 & 1 & 0 \end{bmatrix} \begin{matrix} k \\ i \\ j \end{matrix} \quad (23)$$

For instance $\bar{d}_1 = (1\ 0\ 0)^T$ corresponding to variable a_{ij}^k in statement (10) in the QR algorithm is transformed into a new dependence $\bar{\delta}_1 = (2\ 1\ 1)^T$. This means that in the transformed algorithm this variable requires two time units to travel between consecutive processors and that the data transfer is in direction $(1, 1)^T$, i.e., using interconnection primitive $\bar{p}_1$. The operation of entire array is summarized in the matrix Δ.

3.6. Allocation of computations to bands

Next, we consider the partitioning case, when $m < n$. (In our example $m = 3$ and $n = 9$). The rows of the transformation S, namely $\Pi_{p1} = (1\ -1\ 0)^T$ and $\Pi_{p2} = (1\ 0\ 0)^T$ divide the index space J^3 into some bands. It is difficult to visualize how index set J^3 is partitioned into bands. Instead, we can observe easier the projection of these bands on the $(\hat{i},\hat{j})$ plane; $\hat{i}$ = constant corresponds to partitioning hyperplane Π_{p1} and $\hat{j}$ = constant corresponds to partitioning hyperplane Π_{p2}. This is shown in figure 4. In figure 3, a printout is shown, indicating the original index points (k,i,j) partitioned the index set into bands denoted by (b_i, b_j), the allocation of computations to processor $(\hat{\hat{i}},\hat{\hat{j}})$ and the timing $\hat{k}$ with partitioning, ($\hat{i}$ and $\hat{j}$ is shown only for convenience). The band number was derived from

$$b_i = \frac{i}{3} \quad \text{and} \quad b_j = \frac{j}{3}$$

The coordinates $\hat{\hat{i}}$ and $\hat{\hat{j}}$ resulted from

$$\hat{i} = \Pi_{p1} \begin{pmatrix} k \\ i \\ j \end{pmatrix} \bmod m_i = [1\ -1\ 0] \begin{pmatrix} k \\ i \\ j \end{pmatrix} \bmod 3$$

$$\hat{j} = \Pi_{p2} \begin{pmatrix} k \\ i \\ j \end{pmatrix} \bmod m_j = [1\ 0\ 0] \begin{pmatrix} k \\ i \\ j \end{pmatrix} \bmod 3$$

The processing time $\hat{k}$ depends upon the band scheduling policy. In our case, we select a lexicographical ordering for $(b_i\ b_j)$. It can be seen from figure 4 that this band scheduling policy respects condition $\Pi_{pk}\bar{d}_i > 0$; e.g., band B_{01} is scheduled after band B_{00} because dependencies point from B_{00} to B_{01}. Next, we show that the partitioning method described above does not affect the accuracy of the QR algorithm. The theorem below indicates that the oredering resulted from partitioning is a correct execution ordering.

VLSI Signal Processing

<u>Definition 4</u>
A correct execution ordering is an ordering which does not violate the precedence ordering imposed by the data dependencies D.

<u>Theorem</u>
The mapping and the partitioning of the QR algorithm provides a correct execution ordering.

<u>Proof</u>

Consider two index points $\bar{j}_1, \bar{j}_2 \in J^3$. We have to prove that for any $\bar{d}_i = \bar{j}_2 - \bar{j}_1$ relation $\hat{\bar{k}}(\bar{j}_2) > \hat{\bar{k}}(\bar{j}_1)$ holds. Consider two cases:

a). $\bar{j}_1$ and $\bar{j}_2$ are in the same band, thus implying that

$$\left\lfloor \frac{\Pi_{pk}\, \bar{j}_1}{m} \right\rfloor = \left\lfloor \frac{\Pi_{pk}\, \bar{j}_2}{m} \right\rfloor \text{ for } k = 1, 2$$

In this case $\hat{\bar{k}}(\bar{j}) = \hat{k}(\bar{j})$. From (6) we have that $\hat{\bar{k}}(\bar{j}_2) - \hat{\bar{k}}(\bar{j}_1) = \Pi\, \bar{d}_i > 0$.

b). $\bar{j}_1$ and $\bar{j}_2$ belong to different bands, thus implying that

$$\left\lfloor \frac{\Pi_{pk}\, \bar{j}_1}{m} \right\rfloor \neq \left\lfloor \frac{\Pi_{pk}\, \bar{j}_2}{m} \right\rfloor \text{ for } k = 1, 2$$

Since according to our requirement that any band scheduling policy respects conditions $\Pi_{pk}\, \bar{d}_i > 0$, it means that $\bar{j}_2$ belongs to a band which is scheduled for execution after the band to which $\bar{j}_1$ belongs. it follows that $\hat{\bar{k}}(\bar{j}_2) > \hat{\bar{k}}(\bar{j}_1)$. This can be verified in the printout of figure 3. QED.

3.7. VLSI architecture for partitioned QR algorithm

In figure 5 it is shown the architecture for the partitioned QR algorithm. It consists of a 3 x 3 VLSI array surrounded by shift registers. The first in first out (FIFO) registers connect peripheral processors such way that they extend outside the array the internal interprocessor connections. They are needed for temporary storage of variables which cross the band boundaries. Because we selected a band scheduling policy which accesses adjacent band in a subsequent order, it results that external data communication is performed in an orderly manner and no complicated outside control or memory management is necessary. The size of FIFO registers is related to the size of the problem. The largest FIFO's are required along the diagonal interconnection. Data leaving an output processor has sufficient time to reach an input processor.

The input data is fed from outside only into the first row of processors, and it is time multiplexed with the data from the output of some FIFO's.

In the QR algorithm from figure 2, the main operations are the primultiplications (the first loop) and the postmultiplications (the second loop). Transformation T allocates computations to processors such that all premultiplications are assigned to the first row of processors while the postmultiplications are distributed to the rest of the processors in the array. The structure of the cell results from the computations required by the QR algorithm. All cells are identical.
Since the cells are simple enough the controls are also simple: load data from previous cell, perform multiply and add operations and output data to next stage. The control unit also provides control signals to FIFO registers and multiplexer.

4. Conclusions

In this paper we presented a general technique for partitioning and mapping algorithms into smaller VLSI arrays. The QR algorithm was considered as a case study. In order to develop a methodical mapping of partitioned algorithms into VLSI arrays, an algorithm model and an array model were introduced. The mapping then was reduced to relating these two models. Since VLSI arrays are characterized by regular interconnections, cartesian representation was the base for our models. The algorithm model given in definition 2 is useful especially for algorithms with regular repetitive computations. Typically, signal processing algorithms fall into this category.

QR algorithms belong to the class of algorithms for which a natural mapping (one to one) from index set to processors does not exist if VLSI restrictions are imposed. This is because of the rather complex data dependence matrix. Especially the partitioning of such algorithms is difficult to derive heuristically. The design process was carried out with the help of a program called ADVIS which implements our general partitioning and mapping technique. The power of this technique (and program) was demonstrated by the short time needed to derive the results. However, the derivation of data dependencies is not computerized yet, and this takes some time, especially for nonspecialists. Although, ADVIS needs more improvements to be utilized effectively by users with different backgrounds, we believe that it represents a major step in computerizing the design of algorithmically-specialized VLSI processor arrays. The mapping of complex algorithms can now be studied in relatively short time.

References

[1] G.W. Stewart, "Introduction to matrix computations," Academic Press, 1973.
[2] J.G.F. Francis, "The QR transformation," I and II, Computer Journal, 4, pp. 265-271 and pp. 332-345, 1961 and 1962.
[3] A.H. Sameh and D.J. Kuck, "A parallel QR algorithm for symmetric tridiagonal matrices, *IEEE Trans. on Computers*, vol. C-26, No. 2, Feb. 1977, pp. 147-153.
[4] W.M. Gentleman and H.T. Kung, "Matrix triangularization by systolic arrays," *Proc. SPIE Symp.* 1981, vol. 298, Real-Time Signal Processing IV.
[5] L. Johnsson, "A computational array for the QR method," *Proc. MIT Conf. on Advanced Research in VLSI*, Jan. 1982, pp. 123-129.
[6] R.P. Brent and F.T. Luk, "The solution of single-value and symmetric eigenvalue problems on multiprocessor arrays," Technical Report 83-562, Dept. of Computer Science, Cornell Univ., July 1983.
[7] K.W. Chen and K.B. Irani, "A Jacobi algorithm and its implementation on parallel computers," *Proc. of 18th Annual Allerton Conference on Communication, Control and Computing*, 1980, pp. 564-573.
[8] A.H. Sameh, "On Jacobi and Jacobi-like algorithms for a parallel computer," Math. Comput. 25, 1971, pp. 579-590.

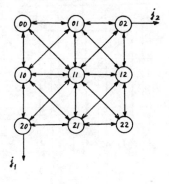

Figure 1: Near neighbor connected VLSI array

VLSI Signal Processing

[9] S.Y. Kung and R.J. Gal-Ezer, "Linear or square array for eigenvalue and singular value decomposition?," *Proc. USC Workshop on VLSI and Modern Signal Processing*, Los Angeles, Calif., Nov. 1982, pp. 89-98.

[10] R. Schreiber and Kueckes, "Systolic arrays for eigenvalue computation," *Proc. SPIE Symp.* 1982, vol. 341, Real-Time Signal Processing.

[11] B.N. Parlett, "State of the art in eigenvalue computations," *Proc. USC Workshop on VLSI and Modern Signal Processing*, Los Angeles, Calif., Nov. 1982, pp. 119-122.

[12] D. Heller, "Partitioning big matrices for small systolic arrays," Report CS-83-02, Computer Science Dept. Pennsylvania State Univ. Feb. 1983.

[13] K. Hwang and Y.H. Chung, "Partitioned algorithms and VLSI structures for large-scale matrix computations," in *Proc. 5th Symposium on Computer Arithmetic*, May 1981, pp. 222-232.

[14] D.I. Moldovan, "On the design of algorithms for VLSI systolic arrays," *Proc. of IEEE.* vol. 71, No. 1, Jan. 1983, pp. 113-120.

[15] D.I. Moldovan and J.A.B. Fortes, "Partitioning of algorithms for fixed size VLSI architectures," Technical Report PPP 83-5, Dept. of Electrical Engineering-Systems, USC, Los Angeles, 1983.

[16] J.A.B. Fortes, "Algorithm transformations for parallel processing and VLSI architecture design," Ph.D. Thesis, Dept. of Electrical Engineering-Systems, USC, Los Angeles, Dec. 1983.

[17] D.E. Heller and I.C.F. Ipsen, "Systolic networks for orthogonal eigenvalue transformations and their applications," *Proc. MIT Conf. on Advanced Research in VLSI*, Jan. 1982, pp. 113-122.

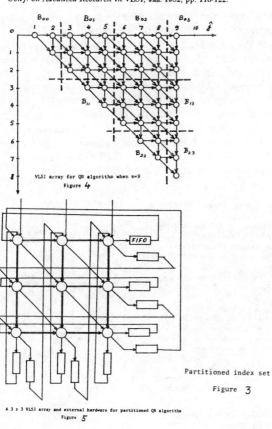

Figure 2 — QR algorithm - pipelined

Figure 3 — Partitioned index set

Figure 4 — VLSI array for QR algorithm when n=9

Figure 5 — A 3 x 3 VLSI array and external hardware for partitioned QR algorithm

37   VLSI Signal Processing

Fault-Tolerant Array Algorithms for Signal Processing [+]
(Extended Summary)

P.J. Varman
Department of Electrical Engineering
Rice University
Houston, Texas 77001

I.V. Ramakrishnan
Department of Computer Science
University of Maryland
College Park, Maryland 20742

1. Introduction

In recent years several VLSI algorithms for signal processing applications have been proposed (for example, see [5,9]). These algorithms typically execute on processor arrays comprised of simple processing elements interconnected in a regular manner. The simplest example of such an array is a linear array in which each processor is connected to neighbors on its left and right. Such array algorithms employ extensive pipelining and parallel processing to achieve the high throughput rates needed by signal processing applications.

Improvements in packing density have made it possible to build larger and more complex systems on a single silicon chip. By reducing the number of discrete components required to build a system, reliability is increased and cost is lowered. In addition, the speed of a system built as a single integrated circuit can be higher than one built from several discrete components since the need to drive many signals off-chip is reduced. *Wafer-scale integration* wherein the entire system is assembled on a single silicon wafer is being actively pursued as a means of achieving the benefits stated above. A major obstacle in wafer-scale integration, however, is that portions of the circuit fabricated on the wafer will be malfunctional due to the presence of imperfections in the silicon or caused by imperfections in the manufacturing process [12].

In this paper we present fault-tolerant linear array algorithms for some signal-processing applications that are amenable to wafer-scale integration. The algorithms are based on the concept of systolic processing [7] and execute on networks of processors that can be efficiently configured around the pockets of faults on a silicon wafer. A discussion of the various approaches to the wafer-scale integration of such networks follows.

The *structural approach* to wafer-scale integration of linear arrays studied by several researchers [1,3,4,10,11,13] is directed towards structuring ensembles of randomly distributed fault-free processors into *physical* linear arrays wherein processor i is directly connected to processor i+1. The throughput and response time of algorithms on such a restructured physical linear array is limited by the time required to communicate between adjacent processors.

In contrast, our scheme is based on a combined structural and algorithmic approach towards wafer-scale integration of linear arrays [15,16]. We structure the fault-free processors into a *logical* linear array wherein processors i and i+1 may be separated by an arbitrary number of unit-delay buffers. The idea in this approach is to divide the burden for providing fault-tolerance between the underlying hardware (structural approach) and the program that is executing on the machines (algorithmic approach). Communication between processors i and i+1 in a logical linear array is accomplished in *some variable number* δ_i clock cycles where δ_i represents the number of unit-delay buffers between them. We will see later on that the clock rate in a *unidirectional* logical linear array (wherein all data elements move from left to right) is *independent* of this separation. A recent paper by Kung and Lam [6] employs a similar notion of programmable delays to achieve fault-tolerance. We will also show that the degradation incurred by algorithms on a bidirectional logical linear array (data elements move in either directions - left to right and right to left) is less than that incurred on bidirectional linear arrays obtined by the structural approach.

The rest of this paper is organized as follows. In section 2 we develop models of unidirectional and bidirectional logical linear arrays. In section 3 we describe several signal processing algorithms on logical linear arrays and conclusions are presented in section 4.

VLSI Signal Processing

2. Logical Linear Array

We now describe models for logical linear arrays and their realization on a wafer. The concept of a logical linear array to obtain fault-tolerance was proposed in [15,16]. An algorithm on a logical linear array is comprised of several data streams. Data elements (henceforth referred to as *tokens*) traverse these streams in multiple directions at multiple speeds. Each processor in the array receives tokens from each of the streams, performs some simple operations and pumps them out (possibly updated).

We dsitinguish between unidirectional logical linear arrays that are comprised of streams in which all tokens travel from left to right and bidirectional logical linear arrays in which some tokens travel from left to right and the rest from right to left. We will denote unidirectional and bidirectional logical linear arrays as *Type-1* and *Type-2* arrays respectively.

Type-1 Arrays

One method of realizing Type-1 arrays is based on the *Diogenes* approach [13] to structuring linear arrays. The Diogenes scheme is augmented by incorporating a unit-delay buffer for each data line between every adjacent pair of processors in the fabricated linear array. Such a modification is shown in Fig. 2.1 below.

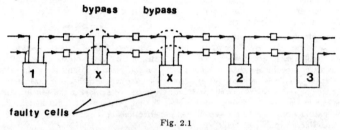

Fig. 2.1

Faulty processors are bypassed by directly connecting the input and output lines of the faulty processors. The resulting structure is a logical linear array with a delay δ_i between processors i and i+1 where δ_i is is one more than the number of bypassed faulty processors between i and i+1. Note that without the addition of the buffered delays between processors the speed of the algorithm executing on such a network would be degraded by a *factor* equal to the longest delay between adjacent processors. The throughput and response time of the algorithm would both be reduced by this factor. With our scheme, the extra delay encountered in bypassing faulty processors only adds to the total response time and does not affect the throughput, as the clock rate is independent of the separation between adjacent processors.

A second method of realizing Type-1 arrays is to configure a CHiP-like architecture [14] by "wrapping" a pipeline around the periphery of an arbitrary spanning tree that connects the fault-free processors in the CHiP-like machine.

Consider a 6-node processor tree (see Fig. 2.1) where the nodes are numbered by a depth-first traversal of the tree. The vertex numbered i will be referred to as v_i.

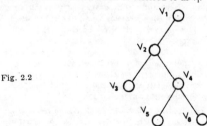

Fig. 2.2

Now replace each edge in the tree by a pair of edges between the two vertices and consider a closed path (eulerian path) in this graph from v_1 back to itself that visits all the vertices in the order $v_1, v_2, ..., v_6$ (see Fig. 2.3) Such a path is comprised of forward edges (those encountered while traversing from v_i to v_j, i<j) and reverse edges (those used to backtrack over previously visited vertices). Each reverse edge is assumed to have a unit delay associated with it whereas a forward edge has a delay which depends on the stream. In a Type-1 array, streams are edge-disoint closed paths.

These closed paths traverse the same sequence of vertices through disjoint edges.

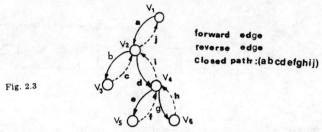

Fig. 2.3

A node in the tree must route the data on its incoming edges onto the appropriate outgoing edges. Such routing can be accomplished by the programmable switches in the CHiP-like machine. To facilitate description of the routing accomplished by the switch we introduce a conceptual model of a processing element (PE) used in the machine. The Type-1 array is assumed to be comprised of k streams.

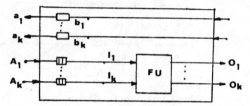

Fig. 2.4

The "▥" in the figure denote buffers that provide appropriate delays to the tokens traversing forward edges and "▭" denotes a buffer that provides unit delay to the tokens traversing the backward edges. FU denotes a functional unit that performs the same computation in every cycle using the elements at its input ports $I_1,I_2,...,I_k$ and placing them (possibly updated) at the corresponding output ports $O_1,O_2,...,O_k$. By appropriate interconnection of the buffers a Type-1 array is obtained as described below.

Consider a rooted tree wherein P_i is the father of P_j (that is, $i<j$ in the depth-first ordering and i and j are connected by an edge) and the sons of P_j are $P_{j_1}, P_{j_2},...,P_{j_r}$ with $j_1<j_2<..<j_r$ and $j_1 = j+1$. Note that $r=0$ implies that P_j is a leaf node in the tree. The interconnection of buffers at each P_j, $j=1,2,...,n$ is given in Table 2.1. (For the sake of clarity we have shown the interconnection for buffers in stream h. The interconnection for buffers in other streams is obtained by replacing h in Table 2.1. The I/O ports have been subscripted with the processor index.)

connect → ↓ To	O_{h_j}	$a_{h_{j_m}}$ $m=1,2,...,r-1$	$a_{h_{j_r}}$
$r \geq 1$	$A_{h_{j_1}}$	$A_{h_{j_{m+1}}}$	b_{h_j}
$r=0$	b_{h_j}		

Table 2.1

The tree in Fig. 2.5 realized as a Type-1 array on a two-corridor CHiP-like machine is shown in Fig. 2.6. In a double corridor CHiP-like machine there exists two rows (columns) of switches between every row (column) of processors. In Fig. 2.6 ○, the large ☐ and the small ☐ denote switches, processor and a belay buffer respectively. The ⊠'s denote faulty processors. A link between clusters (a cluster refers to the combination of a processor and the three surrounding switches) provides a buffered delay to signals passing through the link. Traversing reverse edges in the tree corresponds to traversing successive buffers in the links between clusters. The delay corresponding to a forward edge in the tree is the sum of one link delay and the extra delay required for that data stream. This extra delay is provided by buffers within the processor.

VLSI Signal Processing

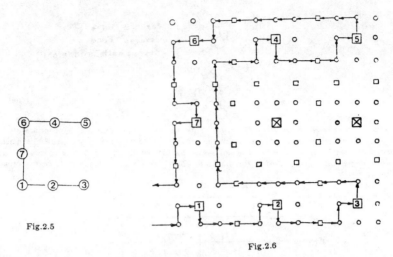

Fig.2.5

Fig.2.6

Type-2 Array

We now describe a method of realizing Type-2 arrays. Consider again the 6-node rooted tree of Fig. 2.2. Recall that in a bidirectional logical linear array some tokens traverse from left to right and the rest from right to left. The tokens traversing the array from right to left traverse a closed path (jihgfedcba) in the tree as shown in Fig. 2.7. The tokens traversing from left to right in the bidirectional array are broadcast to the processors in the tree through a series of *local broadcast* steps. On each clock cycle, a processor that has received a token from its parent uses the token for its computation and then broadcasts the token to all of its children. The ⊸ edges in the figure denote broadcast paths.

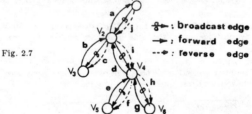

Fig. 2.7

Fig. 2.8 is the model of the processor used in a Type-2 array is comprised of w streams in which tokens travel from left to right and z streams in which tokens travel from right to left.

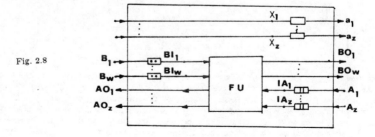

Fig. 2.8

"⊡" and "⊞" are buffers that provide delays to tokens traversing the forward edges and broadcast edges respectively (see Fig. 2.3). "◻" are buffers that provide unit delays to the reverse edges. The interconnection of buffers at a processor P_j is given in Table 2.2 below. P_j is the father of P_j and $P_{j_1}, P_{j_2},..., P_{j_r}$ are the sons of P_j with $j_1 < j_2 < .. < j_r$ and $j_1 = j+1$. (Again, for simplicity we show the interconnection for stream h which uses broadcast edges and stream q comprised of edges in the closed path.)

Table 2.2

connect → ↓ To	a_{q_j}	A_{q_j}	$Ao_{q_{j_s}}$	Bo_{q_j}
$r \geq 1$	x_{q_r}	$Ao_{q_{j_1}}$	$x_{q_{j_{s-1}}}$, s=r,r-1,..,2	$B_{q_{j_m}}$, m=1,2,...,r
r=0	A_{q_j}			

A Type-2 array realized on a two-corridor CHiP-like machine is similar to a Type-1 array realization (see Fig. 2.5) and also includes broadcast links between fathers and their sons.

3. Logical Linear Array Algorithms for Signal Processing

We now describe algorithms for some important signal processing problems that include FIR and IIR filtering, One-Dimensional (1-D) and Two-Dimensional (2-D) convolution and Discrete Fourier Transform (DFT) on logical linear arrays. These arrays are assumed to be realized on rooted spanning trees.

3.1. FIR Filtering

A Finite Impulse Response filter (FIR) is defined as $y_n = \sum_{l=0}^{k-1} a_l x_{n-l}$ where the a_l's, x_{n-l}'s and y_n's are the filter coefficients, input and result tokens respectively.

Algorithm for Type-1 Array

The algorithm is comprised of two streams - one for x_i's and the other for results. k processors are used and a_i is preloaded in processor indexed i. Processor indexed 0 serves as the root. The x_i's and the result tokens encounter a delay of 2 and 1 respectively while traversing the forward edges. The model of a processor (PE) used in the array is shown below. Let x_t and y_t be the value of the tokens at X_{In} and Y_{In} respectively at the beginning of a clock cycle. The functional unit (FU) computes $y_t + x_t a_i$ and places it in Y_{out} at the end of the cycle. It also transfers x_t unchanged to X_{out}.

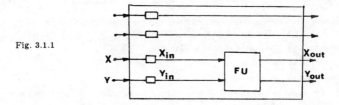

Fig. 3.1.1

The algorithm is then the following.
1. Insert x_i and y_i at ports X and Y respectively of the root at time i.
2. Extract y_i from root at time i+2k-1.

Example: We will now illustrate FIR computation using four filter coefficients on a Type-1 array realized on the tree shown below.

VLSI Signal Processing

Fig. 3.1.2

The time at which x_i and y_i are inserted into the root is i. Table 3.1.1 is the trace of y_3 as it traverses the tree. The first column denotes the time at which y_3 appears at the input port (Y) of the processor whose index appears in the second column and the last column shows the tokens at the input port (X) of that processor. For instance x_1 which was inserted at time 1 will reach input port X of processor 3 at time 5.

Table 3.1.1

Time	i	Token at X_{in}
3	1	x_3
4	2	x_2
5	3	x_1
8	4	x_0

Algorithm for Type-2 Array

Herein the x_i's traverse the broadcast edges whereas the results traverse the closed eulerian path. The delay encountered by the x_i's on any broadcast edge is 1 and the delay encountered by the result elements on any forward edge is also 1. k processors are used by the algorithm and a_i is preloaded in processor i. Fig. 3.1.3 is the model of a PE used in the array.

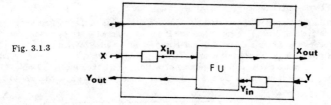

Fig. 3.1.3

The computation performed in every cycle by the PE is the same as that performed by a PE in a Type-1 machine (Fig. 3.1.1). The algorithm then, is the following.

1. Insert y_i initialized to 0 into port Y of root at time 2i.
2. Insert x_i into port X of the root at time $2i+2k-2$.
3. Extract y_i from root at time $2i+2k-1$.

Example: We trace y_3 again in FIR computation using four filter coefficients on a Type-2 array realized on the tree shown in Fig. 3.1.4.

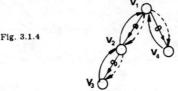

Fig. 3.1.4

Columns two and four in Table 3.1.2 indicates the times at which x_i's and y_i's respectively are inserted at the root and Table 3.1.3 is the trace of y_3 as it traverses the tree on a closed path. The

VLSI Signal Processing

x_l's however traverse the broadcast edges.

Table 3.1.2

x_0	6	y_0	0
x_1	8	y_1	2
x_2	10	y_2	4
x_3	12	y_3	6

Table 3.1.3

Time	i	Token at X_{in}
7	4	x_0
10	3	x_1
11	2	x_2
12	1	x_3

3.2. IIR Filtering

An infinite response filter is defined by $y_n = \sum_{l=0}^{k-1} a_l x_{n-l} + \sum_{l=1}^{k-1} b_l y_{n-l}$.

Let $y_n^1 = \sum_{l=0}^{k-1} a_l x_{n-l}$ and $y_n^2 = \sum_{l=1}^{k-1} b_l y_{n-l}$. Clearly y_n^1 and y_n^2 can be evaluated using the arrays for FIR computation.

Our algorithm for fault-tolerant IIR filtering uses two Type-2 arrays that are cascaded. The first array computes y_n^1 and its results are fed as inputs to the second array that computes y_n^2. Fig. 3.2.1 illustrates such an arrangement.

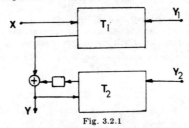

Fig. 3.2.1

T_1 and T_2 are Type-2 tree arrays that compute FIR filtering. $\oplus$ denotes an adder. The output from T_1 and the output from T_2 (delayed by a cycle) are added and fed as the input to T_2. T_1 and T_2 use k and k-1 processors respectively. a_i is preloaded in processor indexed i in T_1 and b_i is preloaded in processor indexed i in T_2. Processors indexed 0 and 1 serve as the roots for T_1 and T_2 respectively. The algorithm then, is the following.

1. Insert y_i^1 initialized to 0 at root of T_1 at time $2i$.
2. Insert x_l at root of T_1 at time $2i+2k-2$.
3. Insert y_{i+1}^2 initialized to 0 at root of T_2 at time $2i+3$.
4. Extract y_i from Y at time $2i+2k-1$.

3.3. 1-D Convolution

Given the input sequence of weights $\{w_1, w_2, ..., w_k\}$ and the input sequence $x_1, x_2, ... x_n$, the 1-D convolution problem is concerned with computing the result sequence $y_1, y_2, ... y_{n+1-k}$ where $y_i = w_1 x_i + w_2 x_{i+1} + .. + w_k x_{i+k-1}$. Type-1 and Type-2 array algorithms for 1-D convolution are identical to the algorithms for FIR filtering after relabelling the X and Y streams in the FIR filtering algorithms as the Y and X streams of 1-D convolution and replacing the filter coefficients by the weights in 1-D convolution.

3.4. 2-D Convolution

Given the weights w_{ij} for $i,j=1,2,...,k$ that forms a $k \times k$ kernel and inputs x_{ij} for $i,j=1,2,...,n$, the 2-D convolution problem is concerned with computing the output sequence $y_{ij} = \sum_{t=1}^{k} \sum_{s=1}^{k} w_{st} x_{l+s-1, j+t-1}$

Algorithm for Type-1 Array

The algorithm is comprised of two data streams for x_{ij} (one for odd j and the other for even j), 1 result stream and a control stream (1-bit wide) similar to the algorithm in [8]. k^2 processors are used and weight w_{ij} is preloaded in processor indexed $k(k-j)+(k-i+1)$, $1 \leq i,j \leq k$. Processor indexed 1

VLSI Signal Processing

serves as the root. Fig. 3.4.1 is the model of a PE used in the array.

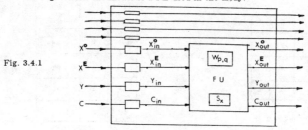

Fig. 3.4.1

S_x is a 1 bit state indicator in the processor. $S_x=1$ denotes odd state and $S_x=0$ denotes even state. Let x_{odd}, x_{even}, r and z be the value of the tokens at $X_{in}^O, X_{in}^E, Y_{in}$ and C_{in} respectively at the beginning of a clock cycle. The transformation performed by the functional unit on them in that clock cycle is described by the program below.

If $z=1$ then
    If $S_x=$odd then $S_x=$even
        else $S_x=$odd;
If $S_x=$odd (even) place $r+w_{pq}x_{odd}$ ($r+w_{pq}x_{even}$) at Y_{out}.
Place z, x_{odd}, x_{even} at C_{out}, X^O, X^E respectively.

Initially, if a processor indexed $k(k-j)+(k-i+1)$, $1 \leq i,j \leq k$ and j is odd (even), then its state is set to odd (even). The elements in the odd and even streams encounter a delay of 2 clock cycles while traversing the forward edges. The result and control elements encounter a delay of 1. The result sequence is obtained in $\lfloor \frac{n}{k} \rfloor$ passes, each pass generating a k×n matrix of result elements. The algorithm below is for the first pass.

1. $\forall i,j$, $1 \leq i \leq 2k-1$, $1 \leq j \leq n$ and j odd,
   insert x_{ij} into port X^O of root at time $(i-1)+2kw$ where $j=2w+1$.
2. $\forall i,j$ $1 \leq i \leq 2k-1$, $1 \leq j \leq n$ and j even,
   insert x_{ij} into port X^E of root at time $(i-1)+k(2w-1)$ where $j=2w$.
3. $\forall i,j$, $1 \leq i \leq k$, $1 \leq j \leq n$ do,
   insert y_{ij} initialized to 0 into port Y of root at time $(i-1)+k(j-1)+k^2-1$.
4. Insert a control token initialized to 1 into port C of root at time k^2-1+km, $m=0,1,..,n-1$.
5. Extract y_{ij} from root at ime $(i-1)+3k^2-2$.

For the rth pass the same program is applicable after replacing x_{ij} and y_{ij} in the above program by $x_{i+(r-1)k,j}$ and $y_{i+(r-1)k,j}$ respectively.

Algorithm for Type-2 Array

Herein the x_{ij}'s traverse the broadcast edges whereas the result and control tokens traverse the closed eulerian path. The delay encountered by any broadcast edge is 1 and the delay encountered by any result and control token on the forward edge is 1. The model of a PE is shown below.

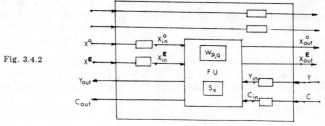

Fig. 3.4.2

The function computed by the FU is the same as the one in Type-1 algorithm. The number of processors used, the preloading scheme and the initialization scheme for the 1-bit state in every processor is identical to algorithm for Type-1 machine. The output sequence is again computed in $\lfloor \frac{n}{k} \rfloor$

VLSI Signal Processing

passes. The algorithm for the first pass then is the following.

1. $\forall i,j,\ 1 \leq i \leq 2k-1,\ 1 \leq j \leq n$ and j odd,
   insert x_{ij} into port X^O of root at time $2(i-1)+4kw$ where $j=2w+1$.
2. $\forall i,j\ 1 \leq i \leq 2k-1,\ 1 \leq j \leq n$ and j even,
   insert x_{ij} into port X^E of root at time $2(i-1)+2k(2w-1)$ where $j=2w$.
3. $\forall i,j,\ 1 \leq i \leq k,\ 1 \leq j \leq n$ do,
   insert y_{ij} initialized to 0 into port Y of root at time $2(i-1)+2k(j-1)$.
4. Insert a control token initialized to 1 into port C of root at time $2km$, $m=1,2,...,n-1$.
5. Extract y_{ij} from root at time $2(i-1)+2k(j-1)+2k^2-1$.

Example: We illustrate the algorithm using a 3×3 kernel on a Type-2 array realized on the tree in Fig. 3.4.3.

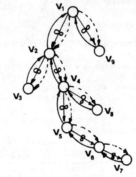

Fig. 3.4.3

Let X, W and Y denote a 5×5, 3×3 and 3×3 matriix of inputs, weights and result elements respectively. Tables 3.4.1 and 3.4.2 denote the times at which x_{ij}'s and y_{ij}'s respectively are inserted at the root. The starred entries in Table 3.4.2 (entries 1,2 and 1,3) denote that the control tokens inserted into the root at those times have their values initialized to 1. During the other times the inserted control tokens have their values initialized to 0.

j\i	1	2	3	4	5
1	0	6	12	18	24
2	2	8	14	20	26
3	4	10	16	22	28
4	6	12	18	24	30
5	8	14	20	26	32

Table 3.4.1

j\i	1	2	3
1	0	6 *	12 *
2	2	8	14
3	4	10	16

Table 3.4.2

Time	i	Weight	State	Token at X_{in}^O	Token at X_{in}^E
9	9	w_{11}	even	-	x_{22}
13	8	w_{21}	even	-	x_{32}
17	7	w_{31}	even	x_{13}	x_{42}
18	6	w_{12}	odd	x_{23}	x_{52}
19	5	w_{22}	odd	x_{33}	-
20	4	w_{32}	odd	x_{43}	x_{14}
22	3	w_{13}	even	x_{53}	x_{24}
23	2	w_{23}	even	-	x_{34}
24	1	w_{33}	even	x_{15}	x_{44}

Table 3.4.3

VLSI Signal Processing

Table 3.4.3 is the trace of y_{22} as it traverses the tree. Note that when y_{22} reaches the input port Y_{in} of a processor indexed i, all y_{ij}'s that were inserted before y_{22} would have already passed through the FU of processor i. From Table 3.4.2 it can be seen that only the control token inserted alongwith y_{12} had its value initialized to 1. Therefore every processor would have toggled its state exactly once when y_{22} reaches processor i. Thus P_4, P_5, P_6 will be in the odd state and the rest will be in the even state.

The first column denotes the time at which y_{22} appears at the input port Y_{in} of the processor whose index is given by the entry in the second column. Column three indicates the kernel weight stored in the processor and column four gives the state of the processor when y_{22} reaches its port Y_{in}. Columns 5 and 6 show the tokens at the ports X_{in}^O and X_{in}^E at the corresponding time. For instance, x_{32} which was inserted at the root at time 10 would reach I_{in}^E of processor 8 at time 13. (recall that x_{ij}'s use broadcast edges) Noting that a processor in the odd (even) state uses elements at I_{in}^O (I_{in}^E) to update y_{ij}, it can be easily verified from Table 3.4.3 that y_{22} will indeed accumulate all the partial sums it needs by the time it emerges from the root.

3.5. Discrete Fourier Transform (DFT)

The Discrete Fourier Transform (DFT) is defined by $y_k = \sum_{m=0}^{n-1} \omega^{mk} x_m$ for k=0,1,2,...,n-1 where $\omega = \dfrac{e^{-2\pi i}}{n}$. The DFT can be considered as a special case of a matrix-vector product.

We will describe a Type-1 algorithm for DFT. The algorithm uses 2n-2 processors. It is comprised of a data stream for x_m's and a result stream. Inorder to generate the powers of ω on the fly, the algorithm requires a fast stream, a slow stream and a constant stream that carries the powers of ω. A 1-bit wide control stream is used to transfer the ω's from the slow stream to the constant stream. The model of a PE is shown below.

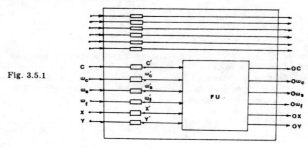

Fig. 3.5.1

ω_c, ω_s and ω_f are the constant, slow and fast streams that carry the powers of ω. X, Y and C are the input, result and control streams respectively. The delay encountered by the tokens of ω_c, ω_s and ω_f, X, Y and C while traversing the forward edges are 1,2,1,1,1 and 2 respectively.

Let γ, ω^p, ω^q, ω^r, a and b be the value of the tokens at ports C', ω_c', ω_s', ω_f', X' and Y' respectively in a clock cycle. The transformation on these values performed by the functional unit in that clock cycle is described by the program below.

If $\gamma = 1$ then
  begin
    place γ in OC_t
    place ω^q in OC_s and $O\omega_f$ /transfer token in slow belt to constant and fast belts/
    place ω^{q+1} in $O\omega_s$ /multiply token on slow belt by ω/
    place $\omega^q + b$ in Ob /compute inner product /
  end
If $\gamma \neq 1$ then
  begin
    place $b + a\omega^{r+p}$ in Ob /compute the inner product/
    place ω^{p+r} in $O\omega_f$
    transfer the other tokens unchanged to the outputs
  end

The algorithm requires 2n-2 processors. The algorithm is then the following.

VLSI Signal Processing

1. Insert x_i ($i \geq 2$) at port X of root at time n-i.
2. Insert y_i ($i \geq 1$) initialized to x_1 at port Y of root at time (i-1)+(n-2).
3. Insert a control token initialized to 1 at port C of the root at time n-2.
4. Insert a 1 at port ω_s of root at time n-2.
5. Extract y_i from root at time (i-1)+3(n-2)-2.

Example: We illustrate the algorithm to compute a four point DFT on a Type-1 array realized on the tree in Fig. 3.5.2.

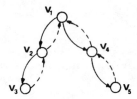

Fig. 3.5.2

The times at which y_0, y_1, y_2, y_3 are inserted at the roots are 2,3,4 and 5 respectively. The times at which x_4, x_3, and x_2 are inserted at the roots are 0,1 and 2 respectively. Table 3.5.1 is a trace of y_2 and its entries have the same interpretation as Table 3.4.3.

Table 3.5.1

Time	i	ω_s	ω_f	C_s	C_t	X
4	1			0	0	
5	2			0	0	
6	3	ω^2		ω^2	1	x_2
9	4		ω^2	ω^2	0	x_3
10	5		ω^4	ω^2	0	x_4

4. Conclusions

We have presented several linear array algorithms for signal processing applications that are amenable to wafer-scale integration. The idea was to design algorithms that execute on a logical linear array wherein adjacent processors may be separated by arbitrary number of buffers. Realization of unidirectional and bidirectional logical linear arrays on Diogenes and CHiP-like architectures were discussed. We now discuss the performance of the these algorithms on logical linear arrays vis-a-vis their performance on physical linear arrays.

The clock rate in unidirectional logical linear arrays is dependent on the time required to transfer data between adjacent buffers. As it is independent of the separation between adjacent processors, the throughput is the same as that on a fault-free linear array. The response time of the algorithms however, will be increased by the number of non-faulty processors that need to be bypassed to obtain an array of the desired size.

In the tree implementation of the Type-2 array, the clock rate is dependent on the time required to broadcast a token from a parent to its son, that is, by the length of the longest edge in the tree of processors around which the array is configured.

If the required number of fault-free processors lie within a connected set (wherein each processor in the set is connected by at most an edge to another processor in the set) then the maximum length of a broadcast edge is one and the throughput then is the same as that obtained in a fault-free linear array. Simulation results [15] show that a significant fraction (95%) of fault-free processors lie

VLSI Signal Processing

within a connected set only if the yield of individual processors is greater than 75%. However, by allowing unit wire-through (wherein two fault-free processors can be connected by bypassing at most one faulty processor) the yields of individual processors can decrease to as much as 35%. Unit wire-through increases the maximum length of a broadcast edge to be at most two. Hence the clock rate would now be half of that of a fault-free linear array. In this context, note that the linear array obtained by the structural approach [10] would have a throughput degraded by a factor of six over a fault-free linear array.

References

[1] R. Aubusson and I. Catt, "Wafer-Scale Integration - A Fault-Tolerant Procedure," *IEEE Journal of Solid-State Circuits*, SC-13, 3(1978).

[2] D.S. Fussell and P.J. Varman, "Fault-Tolerant Wafer-Scale Architectures for VLSI," *Proceedings of the Ninth Annual Symposium on Computer Architecture*, (April, 1982), pp. 190-198.

[3] J.W. Greene and A. Gamal, "Area and Delay Penalties in Restructurable Wafer-Scale Arrays," *Proceedings of the Third Caltech Conference on Very Large Scale Integration* (1983).

[4] I. Koren, "A Reconfigurable and Fault-Tolerant VLSI Multiprocessor Array," *Proceedings of the Eighth Annual Symposium on Computer Architecture*, (May, 1981).

[5] H.T. Kung, "Special-Purpose Devices for Signal and Image Processing: An Oppurtunity in VLSI," *Proceedings SPIE* (July, 1980), pp.76-84.

[6] H.T. Kung and M. Lam, "Wafer-Scale Integration and Two-Level Pipelined Implementations of Systolic Arrays," *Proceedings of the MIT Conference on Advanced Research in VLSI*, (January, 1984).

[7] H.T. Kung and C.E. Leiserson, "Systolic Arrays (for VLSI)," *Sparse Matrix Proceedings 1978*, SIAM, (1979), pp. 256-282.

[8] H.T. Kung and R.L. Picard, "Hardware Pipelines for Multi-Dimensional Convolution and Resampling," *Proceedinds of IEEE Workshop on Computer Architecture for Pattern Analysis and Image Database Management*, (November, 1981).

[9] S.Y. Kung, "VLSI Array Processors for Signal Processing," *Proceedings of the MIT Conference on Advanced Research in Integrated Circuits*, (January, 1980).

[10] F.T. Leighton and C.E. Leiserson, "Wafer-Scale Integration of Systolic Arrays,", *Proceedings of the Twentythird Annual Symposium on Foundations of Computer Science*, (November, 1982).

[11] F. Manning, "An Approach to Highly Integrated Computer-Maintained Cellular Arrays," *IEEE Transactions on Computers*, C-32,10(June, 1977).

[12] C.A. Mead and L.A. Conway, "Introduction to VLSI Systems," Addison-Wesley, Reading, Massachusetts, (1980).

[13] A. Rosenberg, "The Diogenes Approach to Testable Fault-Tolerant Networks of Processors," *IEEE Transaction on Computers*, C-32, 10(October, 1983).

[14] L. Snyder, "Introduction to the Configurable Highly Parallel Computer," IEEE Computer, Vol. 15, 1(January, 1982).

[15] P.J. Varman, "Wafer-Scale Integration of Linear Processor Arrays," Ph.D Dissertation, The University of Texas at Austin, (August, 1983).

[16] P.J. Varman and D.S. Fussell, "Design of Robust Systolic Algorithms," *Proceedings of the 1983 International Conference on Parallel Processing*, (August, 1983).

A PRACTICAL COMPARISON OF THE SYSTOLIC AND WAVEFRONT ARRAY PROCESSING ARCHITECTURES

D.S.Broomhead, J.G.Harp, J.G.M^{C}Whirter, K.J.Palmer, J.G.B.Roberts

Royal Signals and Radar Establishment,
St Andrews Road,
Great Malvern,
Worcestershire.

ABSTRACT

The development of a wavefront array processor for recursive least-squares minimization is described. It comprises a triangular array of processors each of which emulates the INMOS transputer chip. The array is programmed in occam – an associated high level language also developed by INMOS Ltd. Preliminary experiments have been carried out to compare the performance of the wavefront array with that of the associated systolic array and the results are presented.

1. INTRODUCTION

The application of VLSI technology to advanced signal processing systems has become increasingly dependent on the use of parallel processing. This is due to the very high data rates which are involved and the increasingly sophisticated algorithms which are required. For example, in adaptive digital beamforming it may be necessary to carry out a real-time, least-squares minimization in order to combine the complex, baseband signals from an array of ten or more sensors each of which is sampled at several megahertz.

In general the design of efficient multi-processor systems, particularly those involving multiple instruction as well as multiple data streams is extremely complicated. However as a result of the highly regular data flow and repetitive, data-independent operations which are involved in many signal processing applications, it is possible to achieve the necessary degree of parallel processing using highly regular arrays which are simple to control and much easier to design.

The systolic array architecture proposed by Kung and Leiserson (Ref. 1) is particularly suitable for signal processing. A systolic array is an array of individual processing nodes which are connected together in a nearest neighbour fashion to form a regular lattice. Most of the processors perform the same basic operation and the signal processing task is distributed across the entire processor array in a highly pipelined fashion. The processors operate synchronously and the only control required is a simple globally distributed clock. The need to distribute a common clock signal to every processor without incurring any appreciable clock skew is one possible disadvantage of the systolic array approach.

An important variant on the systolic array architecture is the wavefront array processor suggested by S.Y.Kung (Ref. 2). In this case the required processing function is distributed in exactly the same way over an identical array of processors but, unlike its systolic counterpart, the wavefront array does not operate synchronously. Instead the operation of each processor is controlled locally and depends on the necessary input data being available and on its previous output data having been delivered to the appropriate neighbouring processors. As a result the associated processing wavefront develops naturally

VLSI Signal Processing

within the array and the need to impose a temporal skew on the input data may be avoided.

Because of its local control properties the wavefront processor architecture is particularly attractive from the system design point of view. Furthermore, in situations where the time required for a given processor to perform its basic operations is variable (e.g. the number of shifts required for a floating point addition is data-dependent) it is possible that a wavefront array processor may achieve a higher throughput rate than the corresponding systolic array which must always be clocked at a rate consistent with the slowest processor operation.

So far, very little practical comparison seems to have been made between systolic and wavefront arrays, possibly due to the inherent difficulty associated with simulating asynchronous systems of this complexity. However a wavefront array processor for recursive least-squares minimization has recently been developed at the Royal Signals and Radar Establishment. It comprises a triangular array of processors each taking the form of a microprocessor circuit which emulates the INMOS transputer chip (Ref. 3). The entire array may be programmed in occam - an associated high level programming language which has also been developed by INMOS Ltd (Ref. 4). Using this array it has been possible to gain some experience in the application of wavefront processing techniques and to make a comparison with the systolic array approach. The purpose of this paper is to report some of the preliminary results which have been obtained.

In section 2 the basic triangular array architecture for recusive least-squares minimization is briefly reviewed. The INMOS transputer and occam programming language are described in section 3 and the development of a wavefront processor using an array of transputer emulator circuits is discussed in section 4. Finally, in section 5 we present some experimental results which were obtained when the apparatus was used to compare the performance of the wavefront processor with that of the associated systolic array.

2. THE SYSTOLIC/WAVEFRONT LEAST-SQUARES PROCESSOR

For convenience in this section we shall first consider how the process of recursive least-squares minimization may be carried out using a triangular systolic array and then describe how the array may be modified to operate as a wavefront processor.

The least-squares minimization problem may be stated as follows. Given any $N \times p$ matrix X with $N \geq p$ and an N-element vector $\underline{y}$ find the p-element vector of weights $\underline{w}$ which minimises $\|\underline{e}\|$ where

$$\underline{e} = X\underline{w} + \underline{y}$$

and $\|.\|$ denotes the usual Euclidean norm. The problem may be solved by the method of orthogonal triangularization (QR decomposition) which is numerically well-conditioned and may be described as follows (Ref. 5). An $N \times N$ orthogonal matrix Q is generated such that

$$QX = \begin{bmatrix} R \\ 0 \end{bmatrix} \quad \text{and} \quad Q\underline{y} = \begin{bmatrix} \underline{u} \\ \underline{v} \end{bmatrix}$$

where R is a $p \times p$ upper triangular matrix and $\underline{u}$ is a p-element vector. Then, since Q is orthogonal we have

$$\|\underline{e}\| = \|Q\underline{e}\| = \left\| \begin{bmatrix} R \\ 0 \end{bmatrix} \underline{w} + \begin{bmatrix} \underline{u} \\ \underline{v} \end{bmatrix} \right\|$$

It follows that the least squares weight vector $\underline{w}$ must satisfy the equation

$$R\underline{w} + \underline{u} = \underline{0}$$

which may readily be solved by a process of back-substitution

The orthogonal triangularization may be implemented using a sequence of Givens rotations (Ref. 6). A Givens rotation is an elementary orthogonal transformation of the form

$$\begin{bmatrix} c & s \\ -s & c \end{bmatrix} \begin{bmatrix} 0\ldots.0,r_i\ldots.r_k\ldots. \\ 0\ldots.0,x_i\ldots.x_k\ldots. \end{bmatrix} = \begin{bmatrix} 0\ldots.0,r'_i\ldots.r'_k\ldots. \\ 0\ldots.0,\;0\;\ldots.x'_k\ldots. \end{bmatrix}$$

where $c^2 + s^2 = 1$. The elements c and s may be regarded as the cosine and sine respectively of a rotation angle θ which is chosen to eliminate the leading element of the lower vector, ie such that

$$-sr_i + cx_i = 0$$

It follows that $c = r_i/r'_i$ and $s = x_i/r'_i$ where $r'_i = (r_i^2 + x_i^2)^{1/2}$.

A sequence of such elimination operations may be used to carry out orthogonal triangularization of the matrix X. The Givens rotation method is recursive in the sense that the data from the matrix X is introduced row by row and as soon as each row $\underline{x}_n^T$ has been absorbed into the computation the resulting triangular matrix R(n) represents an exact QR decomposition for all data processed up to that stage. A systolic array which performs recursive least-squares minimization using Givens rotations is illustrated in figure 1 for the case p = 4. It may be considered to comprise three distinct sections, (1) the basic triangular array labelled ABC, (2) the right hand column of cells labelled DE and (3) the final processing cell labelled F. The entire array is controlled by a single clock and contains three types of processing cells. Each cell receives its input data from the directions indicated on one clock cycle, performs the specified functions and delivers the appropriate output values to neighbouring cells as indicated on the next clock cycle.

Cells within the basic array store one element of the recursively evolving triangular matrix R(n) which is initialized to zero at the onset of the calculation and then updated every clock cycle. Cells in the right hand column store one element of the evolving vector $\underline{u}(n)$ which is also initialized to zero and updated every clock cycle.

The boundary cell in each row (indicated by a large circle in figure 1) computes the rotation parameters c and s appropriate to the internally stored components and vertically propagating data vector. These rotation parameters are then passed horizontally to the right on the next clock cycle. The internal cells (indicated by squares) are subsequently used to apply the same transformation to all other elements of the received data vector. Since a delay of one clock cycle per cell is incurred in passing the rotation parameters along the row, it is necessary to impose a corresponding time skew on the input data vectors as indicated in figure 1. This arrangement ensures that, as each row $\underline{x}_n^T$ of the matrix X moves down through the array, it interacts with the previously stored triangular matrix R(n-1) and is eliminated by the sequence of Givens rotations. As each element y_n moves down through the right hand column of processors it undergoes the same sequence of Givens rotations, interacting with the previously stored vector $\underline{u}(n-1)$ and generating the updated vector $\underline{u}(n)$ in the process. It follows that the exact least-squares solution could be derived at

VLSI Signal Processing

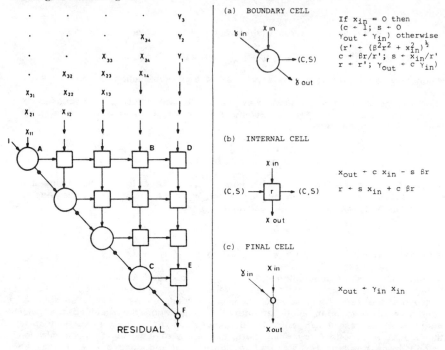

FIGURE 1 : Systolic Array for Recursive Least-squares Minimization

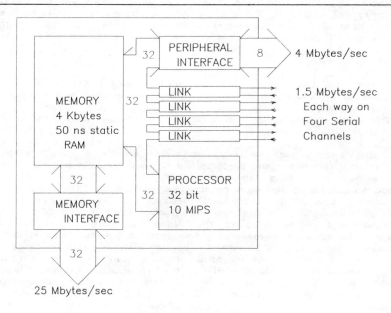

FIGURE 2 : Schematic of INMOS 32-bit Transputer

every stage of the process by solving the corresponding triangular linear system for $\underline{w}(n)$ (Ref. 7).

In many least-squares applications the primary objective is to compute the sequence of residuals

$$e_n = \underline{x}_n^T \underline{w}(n) + y_n$$

while the associated weight vectors $\underline{w}(n)$ are not of direct interest. It has recently been shown (Ref. 8) how the residual e_n at each stage of the computation may be generated quite simply using the systolic array in figure 1 without any need to solve the associated triangular linear system for $\underline{w}(n)$. The parameter x_{out} produced by the bottom processor in the right hand column of cells is simply multiplied by the parameter y_{out} which emerges from the final boundary processor and this produces the residual directly.

The triangular array illustrated in figure 1 may also be operated in the wavefront processor mode. In this case each cell retains the same basic function as before but its operation is no longer controlled by a globally distributed clock signal. Instead, the operation of each cell is controlled locally and depends on the necessary input being available and on its previous outputs having been accepted by the appropriate nearest neighbours. As a result it is not necesary to impose a temporal skew on the input data matrix and the associated processing wavefront develops naturally within the array. Consider, for example, the top processor in the right hand column. In the wavefront case this will not operate on its first input data sample y, until the required rotation parameters c and s are available from the neighbouring processor on its left. This occurs four cycles after the top boundary processor has operated on its first input data sample x_{11} and so the temporal data skew is imposed automatically.

3. THE INMOS TRANSPUTER AND OCCAM PROGRAMMING LANGUAGE

The transputer microprocessors and occam programming language have been developed by INMOS Ltd and are based on the process model of computation (Ref. 9) which is sufficiently general to include both sequential and concurrent processing in a natural manner.

A process is taken to mean an independent computation, autonomous in the sense that it has its own programme and data, but able to communicate with other current processes by message passing via explicitly defined channels. In general, a process may itself consist of a number of processes. These will be grouped into sets, called constructs. There are several types of construct according to whether the set is to be executed in sequence, in parallel, or cyclically, or whether a single member is to be chosen for execution. At the root of this hierachical structure are three primitive processes:
  (a) assignment, which changes the value of a variable
  (b) input, which receives a value from a channel
  (c) output, which sends a value to a channel.

Figure 2 is taken from the INMOS advanced information document for the IMS T424 transputer (Ref 3). The chip is clearly a complete computer. Indeed, it has been designed such that its external behaviour corresponds to the formal model of a process. It has a 32 bit processor capable of 10 MIPS, 4 Kbytes of 50 ns static RAM and, significantly, a variety of efficient communications interfaces. These make it possible to construct various process networks. In particular, the INMOS links are the hardware representation of the channels for process communication. They have a programmable data rate of up to 1.5 Mbytes s^{-1}. Data is transmitted as a sequence of bytes, each byte being

VLSI Signal Processing

ackowledged by the receiver before the next is transmitted. Since processes must be independent except while in communication, this protocol is essential for conformity with the computational model. It enables the processes to run asynchronously, only synchronizing when they need to communicate.

As well as the INMOS links there is a memory interface and an 8 bit bidirectional peripheral interface. The former can be used to extend the on-board memory to accommodate a large programme or large amount of data. It allows access of up to 6 Gbytes of external direct address memory space, with a maximum data rate of 25 Mbytes s^{-1}. The latter is to provide communications with standard external devices, and may be used to construct a broadcast facility.

The transputer should be seen then as a powerful building block from which new concurrent devices can be constructed. It does, of course, have significant limitations for many signal processing applications. For instance, it has no floating point processor and is only 4-coordinate. However, such objections may largely be overcome by the development of more specialised transputers. In any event, the ability to build, with standard components, a customised MIMD machine for prototype purposes will surely have a wide impact on the development of algorithms and architectures. The present work is an example of this type of application.

The occam language (Ref. 4) has been designed alongside the transputer and compiled occam will run on the transputer as efficiently as its own native code. In keeping with the philosophy underlying the development of the transputer, occam naturally describes a concurrent system as a set of independent processes which use locally defined variables, and can communicate only *via* declared channels. Channel communication in occam exactly mirrors communication *via* transputer links. In particular there is a handshake protocol that allows data to pass only when the sender and receiver have both signified their readiness to each other. This imposes order, although there is no concept of global time in occam and all concurrent processes run asynchronously.

Writing to and reading from a channel are represented in occam by :

```
chan1? x
chan2! x
```

Thus a simple process which reads from a channel and writes the result to a second channel using an intermediate variable x would be written in occam as:

```
VAR x :
SEQ
  chan.in? x
  chan.out! x
```

The letters "SEQ" signify a sequential construct in occam. All processes below which are indented to the right of SEQ are to be executed sequentially in the order that they appear. Examples of other constructors are: PAR, which indicates that the indented processes below are to be executed in parallel, and WHILE which cycles through its set of indented processes for as long as a specified expression remains true. Thus, the following "democratic" process continues to read, in parallel, the inputs from the channels For and Against until it finds a majority in the latter:

```
VAR for.vote,against.vote:
SEQ
  for.vote:= 0
  against.vote:= 0
  WHILE for.vote>= against.vote
    PAR
      For? for.vote
      Against? against.vote
```

Occam, in keeping with the hierachical structure of the process model, allows the user to define named processes which can be used in a programme as any other process. Named processes can be compiled as seperate entities, and downloaded onto specified nodes of a transputer array.

From the point of view of system design, occam is designed to run in the same logical fashion on a single processor as on several. Thus designs can be tested by running the occam programme on a conventional von Neumann machine - the occam compiler implements time sharing so that all concurrent processes gain access to the processor. After this initial development stage the programme can then be augmented with the necessary constructors to define the way in which it loads onto a given network and locations of the named processes in the local node memories.

Finally in this section, in order to make contact with the objectives of the paper, it is worth emphasising the direct relationship between channel communications defined in occam and implemented in transputer links and the communication protcol envisaged for wave front array processors. Transputer nets programmed in occam may, quite naturally, be regarded as wavefront processors, given that they have have a sufficiently simple data flow.

4. IMPLEMENTATION

The least-squares minimisation algorithm has been impemented as a wavefront processor using an array of microprocessors programmed in occam. As the transputer is not yet available, transputer emulator boards have been built to enable algorithms to be evaluated and performance assesments made.

Each emulator consists of a circuit board, as shown in Plate 1, containing an 8 Mhz Intel 8086 processor with transputer-like communication channels. Five bidirectional ports are implemented with Signetics S2661 USARTs. Four ports operate synchronously (ie handshaking) at 200Kbaud with system software ensuring that the channel facilities appear the same as transputer silicon. A fifth channel is programmed to operate as an asynchronous RS232 link to enable programmes to be loaded and for interfacing terminals etc. Each board has 8Kbytes of RAM for programme and data storage and 8Kbytes of EPROM for bootstrap loader and monitor. Additionally, each emulator has an Intel 8254 timer chip so that occam timing facilities can be implemented in hardware. The ratio of cpu execution time to I/O time is similar to that of the transputer with operation at approximately one-fifth of transputer speeds.

The array of emulator boards is organized as shown in figure 3. Processors P11 and P12 generate suitable input data which is diplayed on vdu #1. Processors P0-P9 compute the root-free least-squares minimisation as discussed in section 2, with P0, P2, P5, P9 being boundary cells; P4, P7, P8 being internal cells and P1, P3 and P6 being edge cells. Processor P12 receives the result and, after formatting, displays it on vdu #2. An option is available for displaying every 256th point for timing and performance estimation purposes. The complete array of processors is shown in plate 2.

VLSI Signal Processing

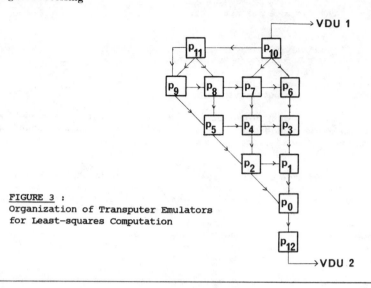

FIGURE 3 :
Organization of Transputer Emulators
for Least-squares Computation

```
PROC boundary (CHAN up, down, left, right)=
  VAR s1,s2,dprime,xin,din,dout,pout,qout:
  SEQ
    s1:=1
    WHILE TRUE
      SEQ
        left ? din
        up ? xin
        s2:=s1+(din*xin*xin)
        dprime:=din/s2
        dout:=s1*dprime
        qout:=xin*dprime
        s1:=s2
        right ! xin
        right ! qout
        down ! dout:

PROC internal(CHAN up, down, left, right)=
  VAR r,xin,pin,qin,xout,xin2,pin2,qin2:
  SEQ
    r:=1
    WHILE TRUE
      SEQ
        up ? xin
        left ? pin
        left ? qin
        xout:=xin-(pin*r)
        r:=r+(qin*xin)
        right ! pin
        right ! qin
        down ! xout:
```

FIGURE 4 : Occam Processes for the Boundary and Internal Cells

The emulators are programmed in occam with native 8086 code being generated in the occam programming system. Software development was carried out on a Sirius host computer. Occam programmes are written, executed and tested on the Sirius. After algorithm verification, occam processes are allocated to processors and native 8086 code generated. The occam programme specifies a load tree with one processor at the root through which the entire system is loaded. A configuration programme then produces a file containing system loaders for each processor together with routing information, code packets for each processor and, finally, start signals. The file is then down-loaded from the Sirius to the emulator array and executed. The occam processes "boundary", and "internal" are given in figure 4, showing the simplicity of communication between processors. The process "edge" is identical to "internal" except that the first two output statements are omitted. The least-squares algorithm was implemented using fixed point (16 bit) arithmetic (floating point procedures have been written but execution times have not been measured). Although this would not be sufficient for most practical applications it is adequate for our investigation the purpose of which is not to test the algorithm but to assess the performance of the wavefront array processor. 16 bit integer arithmetic is sufficient to check that the array is operating correctly.

5. AN EXPERIMENTAL COMPARISON OF SYSTOLIC AND WAVEFRONT PROCESSORS

It has already been noted that the handshake protocol implicit in occam channel communications and implemented in the transputer links naturally leads to wavefront architectures for systems having a simple data flow. Thus, in order to make a comparison between systolic and wavefront processors using the array of emulator boards described in the previous section, it is necessary to consider ways in which a wavefront array might appear to behave as a systolic array.

The point of distinction between the two is the in way in which communicating processes are synchronised. In the systolic case the whole array is entrained by the global clock which ensures that all the nodes have an identical cycle time. In contrast, there is no such constraint with the wavefront array which synchronises nodes locally when the data is to be passed. The local synchronisation is inherently more flexible and, whenever there is intrinsic variability in the processing time required by the nodes, this should enable the wavefront processor to out-perform its systolic counterpart. The current objective is to demonstrate this in a real system. Of course the two processors will behave identically if the timings of the nodes of the wavefront array are constrained to be uniform. In practice, this can be accomplished easily using a transputer/emulator board array such as the one described in the previous section.

The occam process:

                WAIT NOW AFTER restart.time

waits until the reading of the local clock, given by NOW, exceeds the value of the variable restart.time. At the heart of the occam process on each node of the least squares array is a WHILE construct which repeatedly cycles the process through the steps of the algorithm. This was modified by the inclusion of the above WAIT process together with the assignment:

                restart.time:= restart.time + cycle.time

VLSI Signal Processing

There are two possible effects of this modification. If the time taken to execute the WHILE construct without the WAIT is less the the value chosen for cycle.time then the WAIT process will ensure that the whole construct takes cycle.time to execute. On the other hand, should the rest of the WHILE construct take more than cycle time to execute, then the WAIT will have no appreciable effect.

If it can be assumed that the local clock on each transputer or emulator board runs at the same rate to within the accuracy of the experiment, then, by choosing the same, sufficiently large, value of cycle.time for each node process, the array can be made to run like its systolic counterpart. In this case the systolic clock period is cycle.time. A further modification to the WHILE construct was the addition of a process to generate random numbers uniformly distributed on a chosen interval. In order to avoid timing variations when making the comparison between systolic and wavefront architectures this process was always executed, its output was, however, used in two distinct ways. On the one hand it was used to generate different sequences of values for cycle.time on the nodes of the array. This allowed the study of a wavefront processor with variable node timings having well-understood statistical properties. On the other hand, using a fixed and uniform value of cycle.time while assigning the random numbers to a dummy variable resulted in the systolic-like behaviour as we have discussed above.

Figure 5 summarises the results of these experiments. There are three sorts of point plotted on the graph. The closed circles correspond to the systolic-like mode of the array. They give the throughput time per piece of data as a function of the value of cycle.time. The graph through these points has two parts. There is a portion over which the throughput time increases linearly with cycle.time. This is the region over which cycle.time dominates the other execution times of the node processes, and, from the structure of the WAIT process, gives a plot which extrapolates to the origin of the graph. As cycle.time is decreased, however, it eventually becomes less than the intrinsic execution time of the WHILE construct. As was discussed above this marks a cut-off beyond which the throughput time becomes constant. Actually, since the least squares array has three types of node there is not a single cut-off. This effect, however, is too small to be observed in the present, relatively crude experiment.

The other points plotted relate to the operation of the wave front array with random timings on the nodes. The error bars refer to plots of measured throughput time versus the mean value of the random delay. The mean is calculated using the distribution of the randomly generated values of cycle.time, allowance being made for the fact that below cut-off the delay is constant. The important point to be emphasised is that for a wavefront array processor it is the **mean** of the distribution of node execution times that appears to determine the throughput rate. In contrast the systolic array clock period must always exceed the node process execution time. The worst case times are known in this case and have been marked with crosses on the graph demonstrating clearly the advantage of the wavefront array processor. The broader the distribution of the node timings, of course, the greater this advantage will be. This feature of the wavefront architecture, although quite predictable, has not previously been demonstrated in practise. In this paper we have assumed that the node process times are uncorrelated and uniformly distributed. In situations where this is not the case the advantage of the wavefront architecture may not be so great. Further research is required to investigate these points.

VLSI Signal Processing

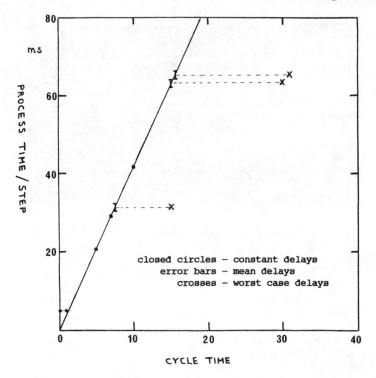

FIGURE 5 : Plot of Process Time per Unit Step (ms) *versus* Processor Cycle Time (arbitary units)

PLATE 1 : Transputer Emulator

VLSI Signal Processing

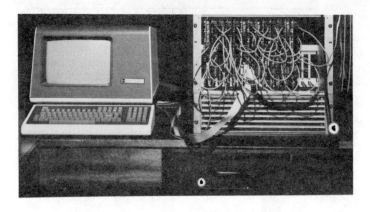

PLATE 2 : Wave Front Array Processor

REFERENCES

1. H. T. Kung and C. E. Leiserson, "Algorithms for VLSI processor arrays" in Introduction to VLSI systems eds. C. Mead and L. Conway, Addison-Wesley, 1980

2. S. Y. Kung, K. S. Arun, R. J. Gal-Ezer and D. V. Bhasakar Rao, "Wavefront array processor: Language, architecture and applications" IEEE trans. computers, special issue on parallel and distributed computers, Vol C-31, No 11, p 1054 (1982)

3. "IMS T424 transputer", Advanced information document, INMOS Ltd. (1983)

4. P. Wilson, "Occam architecture eases system design", Computer Design, Nov 1983, p 107 and Dec 1983, p 109.

5. G. Golub, "Numerical methods for solving linear least-squares problems", Numerische Mathematik, vol 7, p206 (1965)

6. W. M. Gentleman, "Least-squares computations by Givens rotations without square roots", J. Inst. Maths. Applics., vol 12, p329 (1973)

7. W. M. Gentleman and H. T. Kung, "Matrix triangularization by systolic arrays", Proc. SPIE, vol 298, "Real Time Signal Processing IV", p19 (1981)

8. J. G. M^{C}Whirter, "Recursive least-squares minimization using a systolic array", Proc. SPIE, vol 431, "Real Time Signal Processing VI", p105 (1983)

Some Experiments in VLSI Leaf-cell Optimization[†]

Kazuo Iwano

Kenneth Steiglitz

Department of Electrical Engineering and Computer Science
Princeton University
Princeton, N. J. 08544

Abstract

This paper describes a method for local optimization of VLSI leaf cells, using the parameterized procedural layout language ALLENDE [5]. Tradeoffs among delay time, power consumption, and area are illustrated. Three different implementations of the 1-bit full adder are compared: a random logic circuit, a data selector, and a PLA. The fastest random logic 1-bit full adder has a time-power product about 1/3 that of the fastest data selector, and about 1/4 that of the fastest PLA. The 4-bit parallel adder is used to illustrate the effect of loading when leaf cells are combined.

1. Introduction

In the design of a custom VLSI chips it often happens that there is one cell that is used many times, usually in an array or a recursive structure. The fact that a cell is used many times means that there is a large potential payoff in its optimization, and that the problem can be made small enough to be manageable. Arrays of cells are especially common in digital signal processing applications, where regular structures, like systolic arrays, lead to designs that are easy to lay out efficiently, and have high throughput. As examples, bit-parallel and bit-serial multipliers can be constructed from one- and two-dimensional arrays of one-bit full adders, as can a wide variety of pipelined FIR and IIR filters (see [1], for example). As another example, a processor for updating one-dimensional cellular automata has been designed at Princeton which consists of a one-dimensional array of 5-input/1-output PLA's [10]. In such cases the problem of making most efficient use of a given piece of silicon breaks down into two distinct problems: 1) choice of the global packing strategy (the method of laying out and interconnecting leaf cells, and connecting them to power and clocks), and 2) the design of the iterated structure itself (which we call the *leaf cell*). In this paper we study the second problem: the design of efficient leaf cells. The example used throughout is the most common in digital signal processing, the 1-bit full adder.

There are three important measures of how good a leaf cell is: its time delay T; its peak or average power dissipation P_{max} or P_{ave}; and its area A.

[†] This work was supported by National Science Foundation Grants ECS-8307955, U.S. Army Army Research Office, Durham, NC, under Grant DAAG29-82-K-0095, DARPA Contract N00014-82-K-0549, and ONR Grant N00014-83-K-0275.

Ideally, the designer should be able to trade off these measures, one against the other. For example, in one application the clock may be fixed at a known value T_0, and it would therefore be senseless to make the the cell faster. On the other hand, peak power may be a real constraint because of heat dissipation limitations, and at the same time it may be important to keep the area small so as to fit as many cells on one chip as possible. We might therefore try to minimize some measure of the peak power and area (the product, for example), while enforcing the constraint $T \leq T_0$. In other applications speed may be critical, and it may be important to minimize T while observing constraints on P_p and A, and so on. In general, we would like to have enough information about the tradeoffs among the measures T, P and A to make intelligent design decisions. As we will see, the $P-T$ tradeoff is often of most interest, since the area is often a less sensitive function of design parameters (at least for fixed topology).

2. Formulation

The basic approach we take will be to search for local improvements on random initial designs. The search strategy will be to consider all single or double changes in element size along the critical path. When only single changes are tried, we call the procedure "1-change", when double changes are tried, "2-change". The idea is that the critical path indicates which parameters are most important to performance at any given point in the analysis.

We will limit the optimization to choice of pulldown widths. The method can be extended to choice of layers, orientation, and topologies. We will, however, study three radically different topologies for the full adder: the PLA, data-selector, and random logic.

The main analysis tools used in these experiments are the timing simulator CRYSTAL, and the power-estimation program POWEST, together with the rest of the Berkeley tool package [2].

Another essential component of the work is a procedural, constraint-based layout language for specifying VLSI layouts; in this case, we used the new language ALLENDE being developed at Princeton, a successor to ALI2 and CLAY [3,4,5]. This allows us to specify circuit parameters and have a cifplot generated automatically.

3. The Critical-Path Optimization

Figure 1 shows how the optimization is performed in our experiments. In Figure 1 **faparm** is an input parameter vector to **ANALYSIS** which has diffusion widths of nodes as described in section 4. The initial **faparm** is generated at random by **RANDOM** according to its input file **pattern**. ANALYSIS takes **faparm** as its input and generates an appropriate layout and its resulting T, P, and A, as well as **the nodes on the critical path** (hereafter called the critical path nodes). Since every node on the critical path has an associated parameter in **faparm**, **CASEGEN** can generate **faparm**s as subcases by using the one-(two-)change method. Here the one-(two-)change method changes one(two) parameter(s) associated with the critical path nodes by one step. (From here on the 1-change method is denoted by 1-opt or Random 1-opt, and the 2-change method by 2-opt or Random 2-opt.)

The optimization strategy is shown in the flowchart of figure 2. When the first improvement occurs, this case is picked up for the next iteration. If no improvement occurs but there exists a case which has the same cost and has

VLSI Signal Processing

not yet been analyzed, this case is adopted next. Otherwise a new random **faparm** is generated for the next iteration, to search for other locally optimal points. We used two cost criteria for optimization: T, and $P_{max}T$ (hereafter denoted by PT). Figure 3 shows an outline of the main procedures used in the **ANALYSIS** loop. A short description of each follows below:

1) **ALLENDE** This procedural constraint-based VLSI layout language produces an integrated circuit layout in Caltech International Form (CIF) corresponding to the specified parameters [5].

2) **MEXTRA** MEXTRA reads CIF and extracts the nodes to create a circuit description for further analyses [2].

3) **CRYSTAL** CRYSTAL is used for finding the worst-case delay time of the circuit [2].

4) **POWEST** POWEST is used for finding the average and maximum power consumption of the circuit.

5) **CRITICAL** CRITICAL reports the critical path nodes by using the output of CRYSTAL.

6) **LIST** This command stores the vector of results (T,P,A) in the HISTORY file for further optimization.

In figure 3 the squares surrounded by dotted lines are files used for inputs or outputs of the above procedures.

1) **faparm** The faparm has parameters for layout generation; for example, the diffusion width of each node, the permutation of product terms in a PLA, etc.

2) **layout generating program** There are several ALLENDE programs implementing desired circuit topologies such as the PLA, random logic, etc. Each program requires parameters in its corresponding faparm.

3) **the critical path nodes** The critical path nodes are extracted from the output of CRYSTAL. Each node can be associated with parameters in faparm. This is done by looking up a table for each topology, which associates each node with its corresponding parameter.

4. Full-Adder Circuit Implementations

As mentioned in the Introduction, we adopted the 1-bit full-adder circuit as an example for experimentation, because it is relatively simple, but is a basic arithmetic logic circuit. The 1-bit full-adder circuit can be implemented in many ways. We chose three kinds of circuits: the **PLA, Data Selector**, and **Random logic**. Each layout has several parameters. We will use the vector representation of these parameters; that is $d = (d_1, d_2, \ldots, d_n)$ means that the diffusion width of node i is $d_i \lambda$. We also use the vector $k = (k_1, k_2, \ldots, k_n)$ to mean that the pullup to pulldown ratio of the inverter, NOR, or NAND circuit in which node i exists is k_i. The vector k is fixed for each circuit.

1) **PLA**

VLSI Signal Processing

Figure 4 shows the full-adder circuit diagram implemented by a programmable logic array (PLA) [7]. This layout has the following 17 parameters and 2 permutations.

$d = (d_{and_1}, \ldots, d_{and_7}, d_{or_1}, d_{or_2}, d_{in_{1,1}}, \ldots, d_{in_{3,2}}, d_{out_1}, d_{out_2}, \pi_1, \pi_2)$

$k = (4,4,4,4,4,4,4,4,4,4,4,4,4,4,4,4,4)$

- 7 pulldown diffusion widths of the AND plane.
- 2 pulldown diffusion widths of the OR plane.
- 6 pulldown diffusion widths for inputs.
- 2 pulldown diffusion widths for outputs.
- 1 permutation of product terms in the AND plane.
- 1 permutation of outputs.

In the optimization process, the two permutations are fixed for the sake of simplicity. However those two permutations are chosen in advance in order to give the best result before the optimization by doing experiments based on various random permutations as inputs.

2) Berkeley PLA

The PLA generated by using **mkpla** of the Berkeley VLSI tools [2,8] is used for the purpose of cost comparison with the PLA implemented in 1). This PLA is not optimized, but uses the following fixed parameter vector.

$d = (4,4,4,4,4,4,4,4,4,8,8,8,8,8,8,8)$

$k = (4,4,4,4,4,4,4,4,4,4,4,4,4,4,4,4)$

3) Data Selector

Figure 5 shows the full-adder circuit diagram of a Data Selector implementation [9]. The following truth table is used.

C_i	B	S	C_o
0	0	A	C_i (or B)
0	1	$\bar{A}$	A
1	0	$\bar{A}$	A
1	1	A	C_i (or B)

This circuit selects inputs (A, $\bar{A}$, or C_i) instead of calculating S and C_o. Here C_i is the input carry signal, C_o is the output carry signal, and S is the output sum signal. A and B denote the two other inputs. This layout has the following 8 parameters.

$d = (d_A, d_B, d_{C_i}, d_1, d_2, d_3, d_{C_o}, d_S)$

$k = (4,4,4,4,8,4,8,8)$

- 3 pulldown diffusion widths for input inverters.
- 3 pulldown diffusion widths for internal inverters.
- 2 pulldown diffusion widths for output inverters.

4) Random Logic

Figure 6 shows the circuit diagram of the Random Logic Implementation [6].

This layout has the following 4 parameters.
$d = (d_1, d_2, d_{C_o}, d_S)$
$k = (8, 12, 4, 4)$
- 2 pulldown diffusion widths for internal inverters.
- 2 pulldown diffusion widths for output inverters.

All the circuits above were verified by **ESIM** [2] or **SIMULATE** [5].

5. Parameterization

The diffusion width of the pullup in each stage is automatically determined and implemented by ALLENDE in the following way. Suppose that the current parameter vector is $d = (d_1, d_2, \ldots, d_n)$, and the pullup-to-pulldown ratio vector of the specified layout is $k = (k_1, k_2, \ldots, k_n)$. (The choice of pullup-to-pulldown ratio is discussed in [7].) For each node i, define the variables Z_{pu}, Z_{pd}, and a pullup-to-pulldown ratio K as follows.

$$Z_{pu} = \frac{L_{pu}}{W_{pu}}, \quad Z_{pd} = \frac{L_{pd}}{W_{pd}}, \quad K = \frac{Z_{pu}}{Z_{pd}}$$

where

L_{pu} (L_{pd}) is the length of pullup (pulldown).
W_{pu} (W_{pd}) is the width of pullup (pulldown).
$W_{pd} = d_i$, $K = k_i$ and $L_{pd} = 2$.

L_{pu} and W_{pu} are determined as follows.

If $W_{pd} \leq 2K$

$$W_{pu} = 2$$
$$K = \frac{L_{pu}/2}{2/W_{pd}} \quad \text{or} \quad L_{pu} = \frac{4K}{W_{pd}}.$$

If $W_{pd} > 2K$

$$W_{pu} = W_{pd} / K$$
$$K = \frac{L_{pu}/W_{pu}}{2/W_{pd}} \quad \text{or} \quad L_{pu} = \frac{2KW_{pu}}{W_{pd}}.$$

We adopted following choices.

1) $\lambda = 2 \mu$

2) The timing estimation program CRYSTAL uses an input pulse which is 1 $nsec$ wide.

6. Results

Table 1 shows a comparison of the performance of our implementations. Each row represents one locally optimal point using as criterion the item indicated by *. The units of A, P_{ave}, P_{max}, T, APT and PT are λ^2, $(10^{-6} * W)$, $(10^{-6} * W)$, ns, $(10^{-12} * \lambda^2 * W * ns)$ and $(10^{-8} * W * ns)$ respectively in all tables. Figure 7 shows P_{max} vs T curves for different topologies, while figure 8 shows several P_{max} vs T trajectories obtained during the process of optimization using the 1-change and 2-change methods for the Data Selector and the Random

VLSI Signal Processing

Table 1. performance comparison (1 bit full adder)

type	A	P_{ave}	P_{max}	T	APT	PT	parameter
PLA	21560	6472	10183	12.8*	2802	1303	1)
	21840	5678	9241	15.3*	3087	1413	2)
	21762	5503	8616	14.9*	2794	1284	3)
PLA(Berkeley)	22176	7314	11749	12.8*	3339	1504	4)
Data Selector	8100	3765	6117	15.8*	783	966	8 8 8 8 8 8 8 8
	8100	3529	5645	16.5*	754	931	8 8 8 4 8 8 8 8
	8190	3764	6116	15.9*	796	972	12 8 8 8 8 8 8 8
Random Logic	7742	1331	1957	16.5*	392	323	16 12 3 2
	9600	1683	2427	16.4*	382	398	16 24 2 3
	9800	1644	2329	16.4*	378	382	16 24 2 2
	9600	1723	2506	16.5*	397	413	16 24 3 3
	5194	705	1096	22.6	128	248*	6 8 2 2
	4704	626	1018	25.9	124	264*	4 6 3 2
	5136	744	1174	22.9	138	269*	6 8 2 3

1) d = (4,4,4,4,4,4,3,4,4,8,8,8,4,4,4,8,2)
2) d = (4,2,3,3,3,3,3,4,3,8,8,8,4,4,4,8,2)
3) d = (3,3,3,4,4,4,4,3,3,8,8,8,4,4,4,4,3)
4) d = (4,4,4,4,4,4,4,4,4,8,8,8,8,8,8,8,8)

Table 2 performance comparison (4 bit parallel adder)

type	A	P_{ave}	P_{max}	T	APT	PT	parameter
Data Selector	41310	16536	28218	75.3*	877761	212482	4 8 8 8 16 8 16 16
	44550	16536	28218	84.1*	1057230	237313	4 8 8 8 16 8 24 16
	45409	16534	28213	84.3*	1079990	287836	4 8 8 16 16 16 16 16
	44523	13248	21641	91.0*	876805	196933	4 8 8 8 16 4 8 4
	42845	12301	19748	92.5*	782645	182669	4 8 4 4 16 8 8 4
	43747	11362	17868	94.9*	741806	169567	4 8 4 4 16 4 8 4
	43605	12354	20692	98.0*	884229	202782	2 8 4 8 16 8 4 8
	45441	11885	19753	100.8*	904777	199110	2 8 4 8 16 8 4 4
	44523	12305	19755	101.1*	889227	199723	4 8 4 8 16 4 4 8
	44649	11831	18808	103.2*	866631	194099	4 8 4 4 16 8 4 4
	43747	11362	17868	103.6*	809812	185112	4 8 4 4 16 4 4 8
Random Logic	35552	6577	10335	41.1*	151014	42476	16 12 8 2
	34848	6734	10649	41.4*	153634	44087	16 12 8 3

VLSI Signal Processing

Logic circuit. Each point takes about 1.5 minutes of cpu time on a VAX 11/750. Many of the locally optimal solutions have identical parameter values on the critical path, but differ in other coordinates because of different random starting values.

7. Parallel Adder : The effect of loading factors

The preceding results did not take the loading on the output of the circuit into account. When these circuits are used in arrays, this may become important. To study this problem, we implemented two circuits for a 4-bit parallel adder, using the **Data Selector** and the **Random Logic** 1-bit full adders of the previous section. The results are shown in Table 2.

8. Discussion of Results

8.1. P_{max} vs T tradeoff

Figure 8 shows P_{max}-T trajectories followed by the critical path optimization process, when minimizing T for the **Random Logic** circuit. The dotted envelope shows the final tradeoff curve for P vs T. Notice that the locally optimal point obtained by using PT as the cost criterion lies very close to the trajectory obtained when minimizing T. (See point **a**, with $P = 12.5mW$, and $T = 22.4ns$.) For comparison, the optimization for PT gave us a locally optimal point **b** with $P = 10.9mW$ and $T = 22.6ns$, very close to point **a**. Thus, optimization using the two criteria is consistent.

8.2. Performance comparison among the PLA, Data selector, and Random logic.

Table 3 normalized performance comparison (1-bit full adder)

type	A	P_{ave}	P_{max}	T	APT	PT
Random Logic	100	100	100	100	100	100
Data Selector	105	283	313	96	200	299
PLA	278	486	520	78	715	403
PLA(Berkeley)	286	550	600	78	852	466

Table 3 shows a normalized performance comparison of the best locally optimal point for each layout, minimizing T. The **Random Logic** seems to be the best choice in all respects except T. However, it is the fastest among the 4-bit parallel adder implementations. The T of the 4-bit parallel adder using **Random Logic** is less than 4 times the T of the 1-bit full adder, while in the other layouts it is more than 4 times the T of the 1-bit full adder. The reason is that this **Random Logic** 1-bit full adder circuit calculates the carry signal and propagates it before the calculation of the sum signal, so the carry ripple propagates faster than the sum. As a result, the 4-bit parallel adder takes only 2.5 times as much time as the 1-bit full adder. Figure 7 shows the P-T tradeoff curve of each layout. The curve for the **Random Logic** circuit is below the one for the **Data Selector**, which is below that for the **PLA**. Hence we can order the layouts with **Random logic** best, **Data Selector** next, and **PLA** last. This result agrees with our intuition because this order is the same as the order of circuit specialization.

8.3. Comparison between our PLA and the Berkeley PLA

Both **PLA**'s have almost the same costs, except for P. The reason is that our locally optimal point occurs at the choice $d = (4,4,4,4,4,4,3,4,4,8,8,8,4,4,4,8,2)$,

VLSI Signal Processing

while the **Berkeley PLA** adopts $d = (4,4,4,4,4,4,4,4,4,8,8,8,8,8,8,8,8)$. The **Berkeley PLA** is therefore very close to locally optimum with respect to T.

8.4 Comparison with Myers' work

Myers did similar performance comparisons of various 1-bit full adder implementations [9], but did not use any optimization. His results, shown in Table 4 below, are quite different from ours, shown in Table 3. Our results show that an appropriate choice of layout and its optimization makes the **Random Logic** circuit better than the **Data Selector**, and that the **PLA** can be made very fast at the expense of Power.

Table 4. 1-bit full adder normalized performance comparison (Myers[9])

type	A	P_{max}	T	APT	PT
Random Logic	100	100	100	100	100
Data Selector	45	50	125	28	72.5
PLA	105	110	170	196	187

8.5. 4-bit Parallel Adder

Tables 1 and 2 show that the locally optimal point of the 1-bit full adder is attained with a pull-down diffusion width of the carry output stage $d_{C_o} = 2$ or 3, while the corresponding width for the 4-bit parallel adder is $d_{C_o} = 8$. The pullup width remains 2. This suggests that the critical path passes through the pull-down of the output carry stage, which is indeed the case.

On the hand, for the **Data Selector**, the critical path passes through the pullup of the output carry stage, and in fact it is the pullup width that expands during optimization of the 4-bit parallel adder.

8.6. Comparison of the 1-change and 2-change methods

Figure 8 and Table 5 show a comparison between the 1-change and the 2-change methods when applied to the **Random Logic** implementation. Table 5 is discussed in the next section. The slope of the 2-change method is steeper than that of the 1-change method, but the 2-change method reaches better locally optimal points. Hence in this case the 2-change method works better than the 1-change method does. However, the 2-change method does not work as well as the 1-change method for the **Data Selector**, which has many more parameters. The 2-change method took more iterations than the 1-change method and did not obtain better locally optimal points.

8.7. Effectiveness of our optimization: Cost Improvement ratio

Table 5 below shows the average initial delay times T_0 (obtained from random starts), the average locally optimal delay time T_{opt}, the average percent improvement of the delay time T, and the best locally optimal delay time T_{best}. We can see from this that **2-opt** performs much better than **1-opt**. We should note that it is very important to choose a good order in which to try improvements, because this saves unnecessary search time evaluating changes that are unlikely to be improvements. For example, we chose the diffusion widths of the 3-input NAND gate as the first parameters tried for the **Random Logic** circuit.

VLSI Signal Processing

Table 5 Cost improvement of our optimization methods

type	opt	criterion	$T_{initial}$	T_{opt}	% improvement	T_{best}
Random Logic	1-opt	T	29.7	19.2	33	19.1
Random Logic	2-opt	T	29.7	16.8	42	16.4
Data Selector	1-opt	T	24.3	17.7	25	15.8
Data Selector	2-opt	T	23.5	18.0	23	15.8
PLA	1-opt	T	19.3	16.3	16	12.8

9. References

[1] P. R. Cappello, K. Steiglitz, "Completely Pipelined Architectures for Digital Signal Processing," *IEEE Trans. on Acoustics, Speech, and Signal Proc.*, vol. ASSP-31, No.4, pp. 1016-22, Aug. 1983.

[2] R. N. Mayo, J. K. Ousterhout, W. S. Scott, "1983 VLSI Tools," Report No. UCB/CSD 83/115, Computer Science Division (EECS), University of California, Berkeley, Calif., March 1983.

[3] S. C. North, "Molding Clay: A Manual for the CLAY Layout Language," VLSI Memo #3, EECS Department, Princeton University, Princeton, N. J., July 1983.

[4] R. J. Lipton, S. C. North, R. Sedgewick, J. Valdes, G. Vijayan, "VLSI Layout as Programming," *ACM Trans. on Programming Languages and Systems*, July 1983.

[5] J. Mata, "ALLENDE User Manual," VLSI Memo #9, EECS Department, Princeton University, Princeton, N. J., May 1984.

[6] R. Rondell, P. C. Treleaven, *VLSI architecture*, Prentice-Hall Inc., Englewood Cliffs, N. J., 1983.

[7] C. Mead, L. Conway, *Introduction to VLSI Systems*, Addison-Wesley Publishing Co. Menlo Park, Ca., 1980.

[8] J. Mata, "A PLA Generator for the ALLENDE Layout System," EECS Department, Princeton University, Princeton, N. J., June 1984.

[9] D. J. Myers, "Multipliers for LSI and VLSI Signal Processing Applications," Masters Degree thesis, Edinburgh University, Edinburgh, England, Sept. 1981.

[10] R. R. Morita, "Pipelined Architecture for a Cellular Automaton," Senior Independent Project Report, EECS Department, Princeton University, May 1984.

VLSI Signal Processing

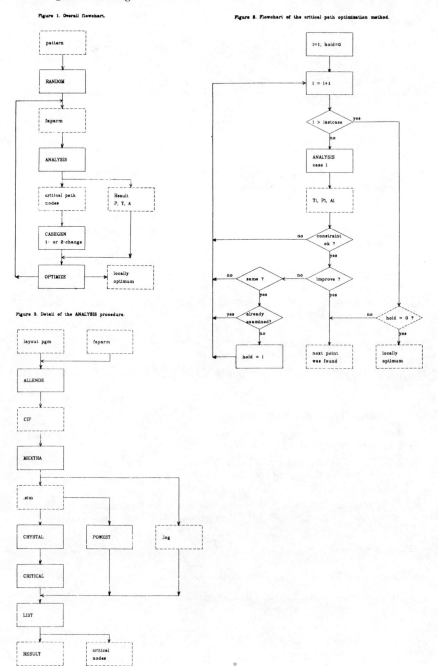

Figure 1. Overall flowchart.

Figure 2. Flowchart of the critical path optimization method.

Figure 3. Detail of the ANALYSIS procedure.

VLSI Signal Processing

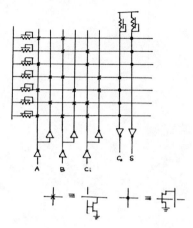

Figure 4. PLA

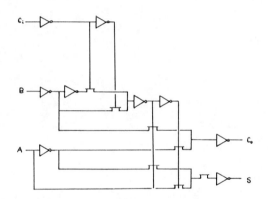

Figure 5. Data Selector

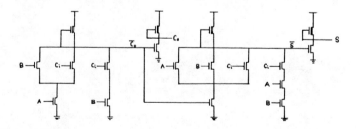

Figure 6. Random Logic

VLSI Signal Processing

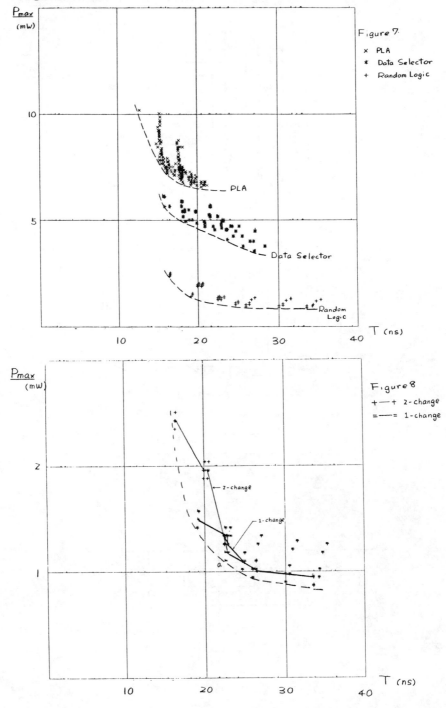

LATE PAPERS

A New Approach to Floating Point DSP

Robert Perlman
senior engineer, Product Planning
Advanced Micro Devices
Sunnyvale, CA 94088

ABSTRACT

A new high-speed, single-chip floating point processor, the Am29325, is introduced; this processor incorporates features of interest to those implementing high-performance digital signal processing systems. Processor architecture is described, and the advantages of the architecture for DSP and array processing applications are discussed. Typical small- and large-system designs are presented.

INTRODUCTION

Floating point arithmetic engines are natural candidates for very-large-scale integration, due to the popularity of the function, and to the large amounts of design time and circuit board space needed to implement such a function in SSI and MSI. Early efforts to integrate floating point operators in a single chip or chip set usually resulted in serial-parallel designs which, while considerably faster than software floating point implementations, did not approach the speeds of fully parallel, dedicated hardware designs.

Recent improvements in process technology have made possible, for the first time, the joining of combinatorial floating point addition/subtraction and multiplication functions in a single VLSI device. The Advanced Micro Devices Am29325 Floating Point Processor contains all hardware necessary to perform high-speed, 32-bit floating point addition, subtraction, multiplication, and format conversion operations, in either IEEE or DEC floating point formats. The device also contains a flexible 32-bit data path, with facilities for local operand storage.

The integration of three elements - a combinatorial adder/subtractor, combinatorial multiplier, and data path - marks the fundamental difference between the Am29325 and previous floating point implementations. By combining these functions, the design addresses not only the problem of implementing fast floating point operators, but also the equally important problem of efficiently transferring operands from one operation to the next. The data path architecture is optimized for performing often-used arithmetic sequences, such as multiplication- accumulation and Newton-Raphson division.

The Am29325 is fabricated with the IMOX-S (for Ion-iMplantation, OXide isolation with Scaling) process, a refinement of earlier AMD bipolar processes. IMOX-S has a feature size of 1.5 microns; three layers of metal are used for interconnects. The Am29325 die contains 48,000 devices on 129,000 square mils of silicon, and is packaged in a 144-lead pin-grid-array. Standard cell techniques were used to reduce design time and simplify chip layout. Improvements in turn-around time were significant: custom design of the Am29116, a 16-bit bipolar microprocessor, took 51 months, while design of the Am29325, a device three times as large, took only 31 months.

The floating point processor is the first in a series of general-purpose, microprogrammable devices primarily intended for 32-bit systems. Other family members include the Am29331 microprogram sequencer, the Am29332 ALU, the Am29323 32-by-32-bit fixed-point multiplier, and the Am29334 register file.

Am29325 ARCHITECTURE

The Am29325 comprises a high speed floating point arithmetic unit, a status flag generator, and a 32-bit data path (fig. 1).

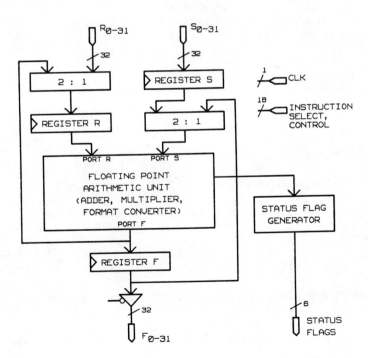

Fig. 1: Am29325 Floating Point Processor block diagram.

VLSI Signal Processing

Arithmetic unit - The three-port, combinatorial arithmetic unit contains a high speed adder/subtractor, a 24-by-24-bit multiplier, an exponent processor, and other logic needed to implement floating point operations. Two input ports, R and S, provide operands for the instruction to be performed. The result of an operation appears on output port F.

The arithmetic unit executes one of eight instructions (table 1). Three of the instructions - R PLUS S, R MINUS S, and R TIMES S - add, subtract, and multiply 32-bit floating point numbers. A fourth instruction, 2 MINUS S, subtracts 32-bit floating point operand S from 2. The 2 MINUS S instruction is used to perform Newton-Raphson division, a means of calculating the quotient A/B. Unlike conventional division, in which quotients are calculated with a series of subtractions and shifts, the Newton-Raphson division algorithm first calculates the reciprocal (1/B) using an iterative equation, then computes the quotient by post-multiplying the reciprocal by A.

The remaining four instructions perform data format conversions. Instructions INT-TO-FP and FP-TO-INT convert between floating point and 32-bit, 2's complement integer formats. The IEEE-TO-DEC and DEC-TO-IEEE instructions convert between IEEE and DEC floating point formats.

Instructions may be performed in either of two single-precision floating point formats - the IEEE format, as specified in proposed standard P754, draft 10.0 (ref. 1), or the DEC F format (ref. 2). These formats are similar, each having an 8-bit biased exponent, a 24-bit significand comprising a 23-bit

MNEMONIC	OPERATION
R PLUS S	Add floating point operands R and S
R MINUS S	Subtract floating point operand S from floating point operand R
R TIMES S	Multiply floating point operands R and S
2 MINUS S	Constant floating point subtraction for Newton-Raphson division (see text)
INT-TO-FP	Convert floating point operand R to integer
FP-TO-INT	Convert integer operand R to floating point
IEEE-TO-DEC	convert IEEE floating point operand R to DEC floating point format
DEC-TO-IEEE	convert DEC floating point operand R to IEEE floating point format

Table 1: Floating point arithmetic unit operations

mantissa appended to an implied or "hidden" MSB, and a sign bit. There are, however, a number of differences between IEEE and DEC floating point conventions, in both the format and the manner in which operands are handled during the course of an operation. These differences are automatically accounted for when the desired format is selected.

The arithmetic unit implements four IEEE-mandated rounding modes that map the infinitely precise result of a calculation to a representable floating point value. An additional VAX-compatible rounding mode is provided for users of the DEC floating point format.

Status flag generator - The status flag generator produces six flags that report operation status. Four of the flags report exception conditions stipulated in IEEE standard P754. The first of these, the INVALID flag, indicates that an operation does not have a sensible answer; multiplying infinity by zero is one example of an invalid operation. Operations producing results either too large or too small to be represented in the selected floating point format are identified by the second and third exception flags, UNDERFLOW and OVERFLOW. The fourth exception flag, INEXACT, indicates that the result of an operation is not infinitely precise. Two additional flags not called for in the IEEE standard, ZERO and NAN, identify zero-valued or non-numerical results.

Data path - The integrated data path comprises two input buses, a three-state output bus, and two data feedback buses, all 32 bits wide. Operands enter the Am29325 through input buses R_{0-31} and S_{0-31}; results exit through three-state output bus F_{0-31}. Each of the R, S, and F buses has a 32-bit edge-triggered register for data storage. An independent clock enable is provided for each register, so that new data can be clocked in or old data held. Input registers R and S, and output register F can be made transparent independently. When all three registers are made transparent, the Am29325 operates in a purely combinatorial "flow-through" mode.

The two feedback data paths transport processor output operands back to the inputs. The first feedback path routes data from the output of the arithmetic unit to a 32-bit multiplexer at the input of register R; the multiplexer selects the operation result or R_{0-31}. The other feedback path carries the output of register F to a second 32-bit multiplexer, which selects either register S or register F as the input for port S of the arithmetic unit.

To allow easy interface with a variety of 16- and 32-bit systems, buses R, S, and F can be programmed to operate in one of three I/O modes. The first and most straightforward of these is the 32-bit, 2-input-bus mode; in this mode, the R and S buses are configured as independent 32-bit input buses, the F bus as a

VLSI Signal Processing

32-bit output bus. The second I/O option is a 32-bit, single-input-bus mode, in which the R and S operands are taken from a single 32-bit input bus on alternate clock edges. For the third option, a 16-bit, two-input-bus mode, the R, S, and F buses are 16 bits wide. Thirty-two-bit operands are placed on these 16-bit buses by time-multiplexing the 16 MSBs and LSBs of each data word during alternate halves of the clock cycle. Internal data paths and registers remain 32 bits wide when this 16-bit I/O mode is selected.

ARCHITECTURAL ADVANTAGES FOR DSP APPLICATIONS

The architecture of the Am29325 offers several advantages to the implementor of DSP and array processing systems:

Efficient data routing - Three aspects of the Am29325 architecture contribute to efficient data routing. First, placing the adder/subtractor and multiplier on the same die eliminates the shuffling of data between separate adder/subtractor and multiplier chips. Minimizing chip-to-chip communication is an important consideration in high-performance system design, since, in VLSI-based systems, the time needed to transfer data between chips can often limit maximum operating speed.

Second, the on-board data paths allow the intermediate result of a calculation to be routed to the input of the floating point arithmetic unit, for use as an input operand in the next phase of the calculation. This feature not only keeps data on-chip, but also makes an external implementation of a similar data path unnecessary. Such an external data path would be expensive, both in components and circuit-board real estate; implementing the two 32-bit multiplexers alone would consume over a dozen MSI devices.

Third, the absence of pipeline delays in the floating point arithmetic unit makes it possible to use the result of one calculation as the input operand for the very next calculation, a crucial feature when implementing algorithms with tight data feedback loops. Users of floating point processors with pipeline delays have one of two choices when implementing such an algorithm - they can either halt the operation while waiting for the desired result to drop out of the pipeline, thus reducing computational efficiency, or can interleave different sets of calculations to keep the arithmetic unit busy, at the cost of more complicated programming. Using a zero-pipeline-delay arithmetic unit avoids both of these unappealing choices.

Am29325 data routing efficiency is best appreciated by considering the manner in which multiplication-accumulation is performed. In a typical multiplication-accumulation calculation,

N input terms x_i are multiplied by coefficients k_i. These products are then added, producing the weighted sum

$$s = \sum_{i=0}^{N-1} k_i x_i$$

Multiplication-accumulation is performed in a two-step loop, with two additional steps for initialization (fig. 2a-d). To initialize the process, data and coefficient values x_0 and k_0 are clocked into registers R and S (fig. 2a). Next, the values x_0 and k_0 are multiplied, and the product placed in register F; at the same time, data and coefficient values x_1 and k_1 are clocked into registers R and S (fig. 2b). In the first step of the multiplication-accumulation loop, values x_1 and k_1 are multiplied, and the product placed in register R (fig. 2c). In the second step, products x_1*k_1 and x_0*k_0 are added, and their sum placed in register F; x_2 and k_2 are placed in registers R and S (fig. 2d).

The two loop steps are then repeated for as many iterations as needed to complete the calculation. The internal data paths wrap back products and accumulations, thus keeping the arithmetic unit busy with a multiplication or addition every clock cycle; a new multiplication-accumulation is performed every two clock cycles. Partial results remain on-chip until the multiplication-accumulation is completed.

High I/O bandwidth - The three 32-bit I/O buses provide high-bandwidth access to the floating point arithmetic unit. When the device is operated in the 32-bit, two-input-bus I/O mode, no multiplexing of I/O buses is required, thus improving system speed and easing critical timing constraints.

Transparent operation - In many applications, the R, S, and F registers will be used to store an operation's inputs and outputs; it is in this register-to-register mode that the Am29325 operates the fastest. In some applications, however, it may be desirable to bypass the internal registers, either because system requirements dictate a data path structure substantially different from that provided, or because the floating point operations must be concatenated with other combinatorial functions. These situations can be accommodated by making all three registers transparent, turning the floating point processor into a purely combinatorial device; this "flow-through" mode of operation would not be possible if the Am29325 used multiplexed I/O.

VLSI Signal Processing

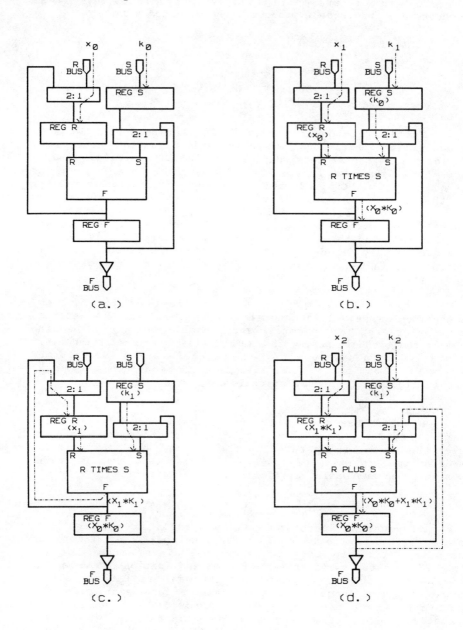

Fig. 2(a-d): Performing floating point multiplication-accumulation with the Am29325.

VLSI Signal Processing

SYSTEM DESIGN

A block diagram for a typical small system design is shown in fig. 3. The system consists of an Am29325, an Am29334 four-port register file, data memory, coefficient memory, microprogrammed controller, clock generator, and host system interface. Although small enough to fit on a single circuit board, this system contains all the elements needed for floating point digital-signal and array processing.

Because of its three-bus I/O structure and internal feedback paths, the Am29325 can be used to advantage in both cascade and parallel configurations. Fig. 4 illustrates a simple cascade system, a variation on the previous architecture. In this system, the output port of one floating point processor feeds the input port of another. This arrangement is particularly advantageous when performing high-speed

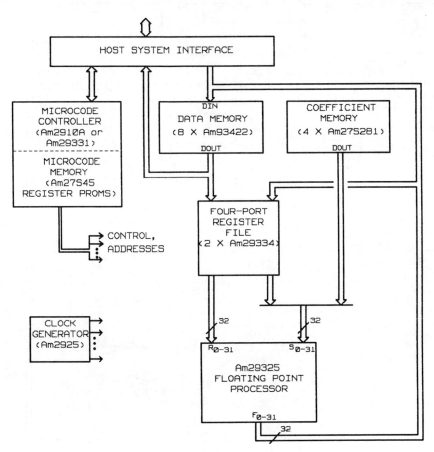

Fig. 3: Typical small-system design.

VLSI Signal Processing

multiplication-accumulation; the first Am29325 forms products, while the second computes the accumulation in parallel. The accumulation is performed using a feedback data path in the second part - no external feedback path is necessary. By doing the multiplications and additions in parallel, the effective throughput rate is one clock per multiplication-accumulation, twice that of the system shown in fig. 3.

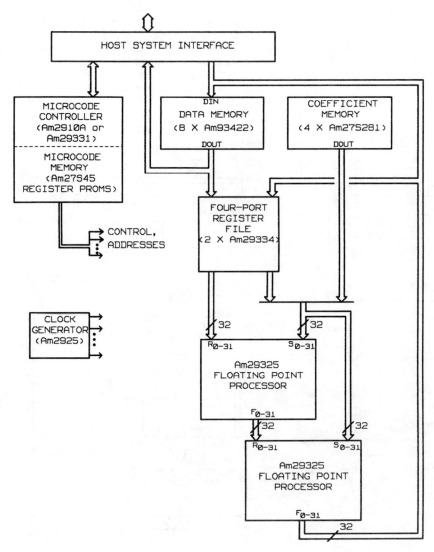

Fig. 4: System using two floating point processors in cascade.

VLSI Signal Processing

Parallel configurations are also useful, and are easily implemented. In one such configuration, the Am29325 is used with other members of the Am29300 family to create a 32-bit floating-point/integer processor (fig. 5). In the system shown, the Am29332 ALU and the Am29323 32-by-32-bit parallel multiplier share three 32-bit buses with the Am29325; data can be passed from one processor to another through the Am29334 register file. Combining these parts produces a system that can perform high-speed floating point, integer, and logical operations. The user can further expand the system by adding 32-bit operators of his own devising to the three-bus architecture.

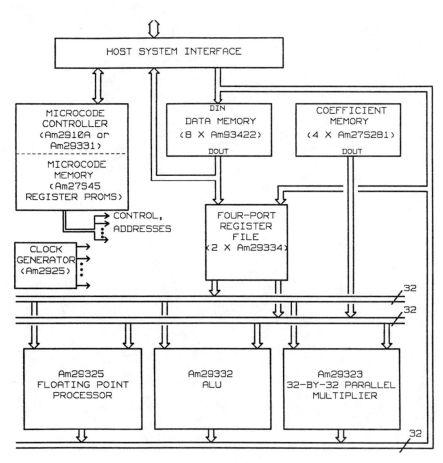

Fig. 5: Thirty-two-bit floating point/integer processor.

REFERENCES

1. **A Proposed Standard for Binary Floating-Point Arithmetic**, IEEE Floating-Point Working Group, draft 10.0, December 2, 1982.

2. **VAX Architecture Handbook**, Digital Equipment Corporation, 1981.

Signal to Symbols: Unblocking the Vision Communications/Control Bottleneck

Steven P. Levitan, Charles C. Weems and Edward M. Riseman *

Department of Computer and Information Science
University of Massachusetts, Amherst MA 01003

I. Introduction

In this paper, we propose the construction of a Content Addressable Array Parallel Processor (CAAPP) for machine vision processing [WEE82, 84]. The CAAPP is a new "processor per pixel" parallel image architecture which represents the synthesis of both content addressable processors (such as STARAN or ASPRO [BAT82]) and mesh connected parallel array processors (such as ILLIAC IV [BAR68] or CLIP-4 [DUF78]). The resulting architecture can be used for both associative and array type operations encountered in image processing and computer vision tasks to produce simple solutions that are difficult for parallel machines which provide only one of these capabilities.

In our work, we have taken a pragmatic view of VLSI technology because we wish to actually construct this machine. Hence, we have approached the design in a very conservative manner, making use only of existing technology (3μ NMOS), to insure rapid and successful development. None the less, the architecture represents a genuine increase in processing power over the best machines available today (Figure 1).

Figure 1
Overview of the CAAPP System

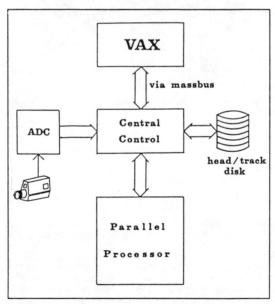

The CAAPP will be capable of performing all low level image processing tasks, but more importantly it will provide a mechanism for transforming low level image data into higher level symbolic data directly — without mediation (or serial processing) by symbolic "host" processors. Thus, it allows for a new style of high level algorithms where processing decisions can be based on direct global feedback information from the processing elements. We have closed the feedback loop between low level and high level processing by providing a control interface. The key to this capability is the provision of fast global feedback operations from the array to the controller (Figure 2).

---

* Work supported in part by DARPA grant N00014- 82-k-0464

VLSI Signal Processing

Figure 2
Associative Processing Closes the Algorithm "Control Loop"

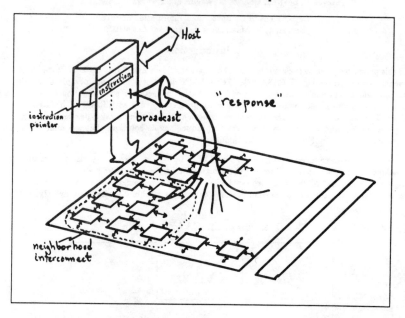

II. The Computer Vision Problem: Image Understanding

The processing requirements needed to solve the vision problem are not well understood. The difficulty is that general computer vision is far from being solved, and is currently a rapidly evolving area of research. At this point in time, no one can give a detailed algorithmic specification for a general vision interpretation system. However, it is possible to give a list of features that must be present in *any* machine that is to be used to significantly advance the vision problem. We believe that if such machines are built, they will greatly facilitate research and clarify many issues in machine vision development.

By computer vision, or image understanding, we mean much more than image processing which, usually, refers to the enhancement and classification of images. The goal of computer vision is the automatic transformation of an image to a symbolic form that represents a description and an understanding of the content of the image. In general, the computer vision problem subsumes the tasks performed in normal image processing.

The image understanding process can be thought of as an iconic to symbolic (or signal to symbol) transformation. The input image data is essentially an array of signal data and forms an iconic representation of the real world. To perform image interpretation the machine must transform this data into a symbolic form. The transformation is from low level information (e.g., the pixel at coordinates [112,47] has a blue intensity value of 17) to symbolic representations of objects in the scene, in terms of predefined knowledge about objects in the world (e.g., region 75 in the image is an instance of the object class HOUSE–DOOR). This task involves monocular static image interpretation as well as integrating information from multiple sensory sources including stereo input, motion sequences, and laser ranging information.

From our perspective, the computer vision problem will be described as involving three levels of processing. These are referred to as the low, intermediate and high levels (Figure 3). The low level consists mainly of operations on pixels and local neighborhoods of pixels. The result of what we call low level processing is a transformed image with labeled regions and line segments. However, we are assuming that no operations

Figure 3
Image Interpretation

```
Communications and Control Across Multiple Levels of Representation
High Level - Schema - Symbolic Descriptions of Objects - Control Strategies
              Rule-Based              ↑    Object Matching and Inference:
              Object Hypothesis       ↓    Grouping, Splitting and Adding
                                           Regions, Lines and Surfaces
Intermediate Level - Symbolic Description of Regions, Lines, Surfaces

              Segmentation            ↑    Goal-Oriented Resegmentation:
              Feature Extraction      ↓    Additional Features, Finer Resolution
Low-Level - Pixels  - Arrays of Intensity, RGB, Depth
                     (Static monocular, stereo, motion)
```

relating different image events have been performed nor have there been any inferences on the object identity of these events.

The *intermediate level* of representation provides an interface between the low and high levels of representation, that is, between pixel-based representation and symbolic elements representing visual knowledge stored in a database. In the UMASS VISIONS system [HAN78, 83], which is the environment in which most of this research was conducted, the intermediate level consists of a symbolic description of the two dimensional image in terms of regions and line segments (that are still in registration with the raw image data) as well as their associated attributes which can be used in the interpretation process. In some systems this level would consist of representations of surfaces, or more generally, "intrinsic" features of the physical environment [BAR78], [MAR82].

Intermediate processing includes several kinds of activities. First is the set of bottom-up tasks which are needed to complete the intermediate level of representation. This includes the extraction of the features for regions, lines, and vertices as well as the relations between these entities. The results of this processing are representations of image entities stored along with the normal "pixel" information in the processing elements which hold the respective image objects. Note that the relationships between the objects in the image and objects in the world are not yet elucidated.

The second group of intermediate processing activities involve grouping, splitting, and labelling processes, in either data-directed or knowledge directed modes (i.e., bottom-up or top-down) to form intermediate events which more naturally match stored object descriptions. Some operations in this class are:
- Labelling points of high curvature on the perimeter of a region.
- Merging co-linear line segments based on the properties of their adjacent regions,
- Merging adjacent regions based on their relationship to shared line segments.

The *high level processing* controls the intermediate level of processing where the symbolic two-dimensional representations of the intermediate level must be related to object descriptions stored in a knowledge base. The result of high level processing is a symbolic representation of the content of a specific image in terms of the general stored knowledge of the object classes and the physical environment.

Communication between these levels is by no means unidirectional. In most cases, recognition of an object or part of a scene at the high level will establish a strategy for further processing and probing the low and intermediate levels, in order to pull out additional features under the guidance of a partial interpretation. This might involve joining together region, line, and surface information to form a symbolic representation which would more easily and naturally match a stored object description. Another example would be to verify a region is an object, say a tree. Assuming the original hypothesis was based on simple texture measures, for verification we might want to use more expensive feature extraction techniques to reduce uncertainty.

Thus, *the key to vision processing is a flow of communication and control both up and down through all representation levels.* In the upward direction, the communication consists of segmentation results from multiple algorithms, and possibly from multiple sensory sources. It also involves the computation of a set of attributes of each extracted image event to be stored in a symbolic representation. The summary information and statistics allow processes at the higher levels to evaluate the success of lower level operations. It is also

VLSI Signal Processing

Table 1
Performance of Parallel Architectures

Machines	Algorithms						
	Broad	Report	Extrema	Pack	Sort	MST	Conv
Serial	1	1	N	N	$N \log N$	$E \log E, N^2$	N
Linear	N	N	N	N	N	N^2	$\sqrt{N}$
Star	N	N	N	N	$N \log N$	N^2	N
Tree	$\log N$	$\log N$	$\log N$	$\log N$	N	$N \log N$	?
Shuffle	$\log N$	$\log N$	$\log N$	$\log N$	$(\log N)^2$	$N \log N$	?
BPM	1	1	$\log Max$	N	N	$N \log N, E$	N
CAPP	1	1	$\log Max$	N	$N \log Max$	E	N
Full	$\log N$	$\log N$	$\log N$	$\log N$	$(\log N)^2$	$N \log N$	a
Mesh	$\sqrt{N}$	$\sqrt{N}$	$\sqrt{N}$	$\sqrt{N}$	$\sqrt{N}$	?	a
CAAPP	1	1	$\log Max$	$\sqrt{N}$	$\sqrt{N}$	?	a

the mechanism for the passing of actual symbols. In the downward direction the communication consists of commands for selecting subsets of the image, for specifying further processing in particular portions of the image, and requests for additional information in terms of the intermediate representation.

To summarize — a key issue in achieving an effective architecture is the ability to maintain the low and intermediate representations, pixels and symbolic region, line and surface representations simultaneously in the same machine. The necessity of dumping an image out, for evaluation by a sequential program, must be avoided at all cost. It is too time consuming to transfer the volume of information contained in an image. Even if it took no time to dump the information, the time required for serial evaluation would still be too great. Dumping an image for outside evaluation defeats the entire purpose of having a special parallel processor for computer vision. Instead, the computer vision machine must be able to provide enough feedback to the controlling processor to allow all of the operations to take place within the vision machine itself.

III. The Need for Communication in Parallel Processing Systems

One of our working hypotheses has been that the communication structures of parallel architectures are one of the most important features to be considered in any new machine design. Levitan's [LEV84] recent doctoral dissertation documents our research over the last several years to identify and quantify the important aspects of communication in parallel algorithms and architectures. We will briefly present a few key findings.

During the initial design phase of the CAAPP, we evaluated several different architectures' abilities to support the communication needs of parallel algorithms. We needed a way to evaluate the abilites of our machine designs. There are generally three ways to do this: build the machines and see how well they perform, simulate the machines, or use *a priori* measures or metrics on the designs. Our work showed that many generally accepted measures of machine design are not good predictors of algorithm performance. Instead, we chose to evaluate machines using several "kernel" tasks as a "Performance Suite" of algorithms. We chose broadcasting data, reporting results, selecting or extrema finding, packing vectors, sorting, and contagion (propagating multiply dependent pointer updates in parallel as in the Minimum Spanning Tree problem) as a basis for evalutation of machine performance.

In our research we simulated these six tasks on each of seven parallel computer architectures. We have recently extended the set of tasks to include convolution with a mask of area a. This operation is typical of many low level nearest neighbor image operations needed in vision algorithms. The seven machine designs were chosen as a representative sample of both current and proposed designs for new computer architectures.

One of the results of this investigation was to identify the strengths and weaknesses of hardware broadcast and report mechanisms versus local and distant neighbor mechanisms in parallel architectures. We illustrate some of these results in Table 1.

For distributing global information, and collecting summary information global mechanisms are essential. The Broadcast Protocol Machine (BPM) and the (linear) CAPP the CAAPP perform these operations best. The Tree, Shuffle, and Full interconnect machines which have distant neighbor connections can perform these operations in logarithmic time. The Mesh has local 2-d neighborhood operations which gives $\sqrt{N}$ time. The other machines with simple local interconnections are no faster than a serial processor.

On the other hand, for algorithms that primarily use unique messages and local neighbor communication, the global mechanisms are not sufficient. The key is the need for unique messages rather than common data replicated to (or collected from) all processors. For these operations (ie., packing and convolution) nearest neighbor connections are required. For the special case of convolution on a square data set with a 2-d mask, the Mesh, the CAAPP and Full interconnect are of course best.

In summary, our results show that parallel machines need both neighbor communication (2-d for image operations) and global communication abilities. The CAAPP incorporates both these local and global communication structures as high speed primitive operations of the hardware.

IV. An Architecture for Machine Vision

The CAAPP has been designed to support vision processing research. It is also sufficiently general that new approaches, to the various aspects of vision, can be easily implemented on it. It is quite simple to build special purpose machines that implement particular image processing algorithms with great speed. However, as mentioned above, computer vision research is a dynamic, rapidly changing area. New algorithms are constantly under development and experimentation. A vision machine must therefore be sufficiently fast and general to allow complex experimentation up to the interpretation level.

The basic architectural issues to be addressed for vision stem from the requirements of the problem:
- The ability to process both pixel and symbol data
- A fast processing rate
- The ability to select particular subsets of the pixels for special processing
- Feedback mechanisms that allow focussing of attention and data-directed processing, without having to dump the image for external evaluation.
- The ability to transform an image into a set of meaningful symbols that describe it.

The general solution that we have developed is a machine that is a fusion of mesh connected cellular array processors with associative or content addressable parallel processing capability.

Previous research has shown that a mesh connected cellular array is a structure that is extremely well suited to performing basic image processing tasks. With one processing element per pixel, such a machine can perform many of the basic image processing operations, including both the pixel and local neighborhood classes of operations, very quickly. The problem with the cellular arrays that have been proposed is that they generally do not provide for selective processing of pixel subsets (such as collections of regions or line segments), nor do they supply feedback to the controller. In other words, they do not provide the necessary bidirectional communication between symbolic processing and pixel processing. An image is simply loaded, some operations are applied to it, and then the image is returned for external sequential processing or human presentation.

Research on content addressable parallel processors (CAPPs) has always emphasized selecting and processing arbitrary subsets of the data elements, providing feedback to the controller and doing whatever is necessary to keep from having to move data in and out of the processor. This is because the time required for loading the data, which is roughly equivalent to the time to serially process the data with one operation, must be included in the total processing time. In order to claim any significant speed increase over a serial processor, a CAPP must be able to average the data load time with a large number of parallel operations. One way of achieving this is to reduce the number of times that the data must be transferred in and out, by eliminating the need to externally evaluate the results of processing. This can be done by providing global summary mechanisms that feed back to the controlling processor, thereby allowing it to perform the evaluation of the processing without removing the data from the processor.

V. Description of the CAAPP

The machine we are proposing to build is constructed as a square grid of 128×128 processing elements (or cells); if the research is successful we are intending to extend this prototype to 512×512, which corresponds to the usual number of pixels in a digitized image. Each cell contains 128 bits of storage, five register bits, and a one-bit ALU for bit serial arithmetic and logic functions. Information in each cell can be moved North, South, East, or West on the array so that neighboring cells can communicate with each other.

The 128×128 memory array is controlled by a microprogrammed controller capable of issuing an array command every 100 nanoseconds. If the controller is interfaced to a VAX, it will be able to receive macro instructions from the VAX as fast as the VAX can issue them.

Figure 4
Communication Network
for 64 Processing Element Chip

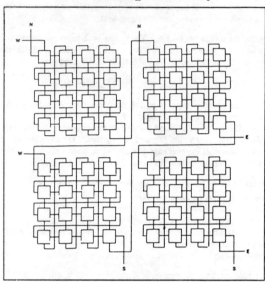

The machine allows global broadcast from the controller to all cells, an activity bit set by each cell for its response, and the global response from the array of cells to the controller in terms of a count or some/none functions. In particular, a comparand may be broadcast from central control and cells whose contents fail to match the broadcast comparand will be turned off so that exact match to comparand, greater than (less than) comparand, maximum, and minimum searches may be performed in parallel on all cells of the memory.

The individual chips we are designing will contain 64 cells in an 8×8 array in a 45-pin package (Figure 4). The 64 cells on a chip are arranged into four quadrants of 16 cells each. Each cell is a long, narrow strip running horizontally across the width of a quadrant. This was done to allow the cells to share the control lines which run vertically down from the instruction and address decoders. The control lines are actually shared among 32 cells in pairs of vertically stacked quadrants with buffering between the quadrants. To provide control lines for both halves of the chip, the instruction decoder is replicated, mirror imaged on both sides of the address decoder. Each PC board will contain 64 chips in an 8×8 array (64 cells $\times$ 64 cells) and the memory as a whole will contain 4 such cards (expandable to 64 in an 8×8 array) giving the overall 128 cell $\times$ 128 cell memory design. Through a clever organization of the architecture we have managed to reduce the number of off-card connections to only 146 lines per card, thereby eliminating what has been a major source of unreliability in other parallel processors.

The chips and the PC cards are designed to be independent of the overall size of the memory array. All array size dependent functions are implemented in a single column and row of "edge cards" that lie conceptually to the left and below the leftmost column and the bottom row of the array. Thus we will be developing not only a large parallel processor, but also a set of building blocks that can be easily assembled to make other special purpose machines tailored for specific applications.

Images are loaded into the CAAPP in a parallel/serial scheme, one scan line at a time. This takes 1.64 milliseconds, or about 1/20th of of a frame time for a 16 bit (color) image. In addition to the array movement operations and the usual content addressable functions, three important global functions are included. These are: (1) report whether any of the cells is a "responder" (has 1 in its X register), (2) count number of "responders", and (3) find the "first" responder. Together these provide the key to adaptive processing techniques. For example, an image enhancement algorithm could adapt automatically to different light levels in different parts of the image in order to extract the same amount of detail from all parts of the image.

VLSI Signal Processing

Figure 5
Response Count Circuitry

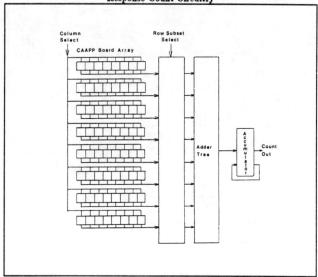

The response count scheme uses the idea of adder trees within the chips and on the circuit boards to develop a board level response count. The response count register of each board is connected via tri-state bus drivers to a backplane bus. Unlike the other bus lines on the backplane, however, this bus is broken at the end of each row of circuit boards. The busses are connected to the controller through a full selector and an adder tree which sums all of the selected rows to a single value. The value is then input to a high speed adder which has its own output as its second input. This is diagrammed in Figure 5.

A response count then proceeds as follows: The response count register on each of the chips latches the count of the current response pattern (0.1 us) and then shifts this bit serially to an external register (0.7 us). Note that as soon as the count is latched, the array can go back to work – shifting control is independent of array processing. The output of the external register feeds another adder tree which is connected to the 13 bit board response count register. After another instruction time (0.1 us) this value is latched. At this point the response count register on the chip can begin feeding its next count out to the external register. Meanwhile the controller begins polling the columns of board response count registers by selecting them one after the other to be output onto the busses. Each column of response counts passes through the selector, then the adder tree and finally to the summing register (0.1 us for the select, followed by 0.1 us to develop the sum) at which point it is accumulated into the previous sum.

The total count time for a 64 board array is thus 2.5 microseconds of which 1.6 microseconds is required for the polling operation. As this can be carried on independently of the rest of the operation, successive counts can be obtained at the maximum polling rate.

Each processing element, or cell (Figure 6) is a bit serial processor consisting of a bit serial ALU, 128 bits of memory (M), local (and global) interconnection hardware, and five single bit registers:

X   The primary accumulator bit, which is also used for communications.
Y   The second accumulator bit.
Z   The carry bit, used for arithmetic operations.
A   The activity bit, used for enabling and disabling this cell on any given operation.
B   The secondary activity bit, used as a temporary storage for activity "flags".

Figure 7 shows the basic micro-operations performed by each cell. instructions are of the form: "select two sources, perform some function on them, and store the result in some destination." Instructions involving the 128 bit memory must use the same location for read-modify-write operations.

VLSI Signal Processing

Figure 6
Functional Diagram of One Processing Element

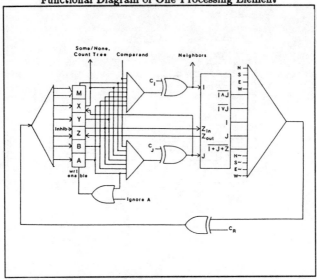

Figure 7
Processing Element Micro-Instruction Set

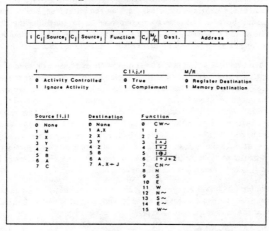

VI. Image Interpretation Applications on the CAAPP

In this section we present some examples of ways that the CAAPP could be used to solve machine vision tasks. We are certainly not claiming that these are the only ways the CAAPP could used, or even that these are the best ways. But rather we wish to give a general list and several specific examples of both low and intermediate level processing to give the "flavor" or style of processing that can be performed with the CAAPP.

VLSI Signal Processing

Convolution

As our first example we discuss one of the most common low level image processing and computer vision operations: discrete, two dimensional, convolution. To perform convolution in the CAAPP each cell distributes its own data to every cell in the neighborhood. Because every other cell is also doing this, the end result is that the central cell (and hence all cells) gets the data it needs from all of the cells in the neighborhood. It should be noted that the time required to perform a convolution using the CAAPP is independent of the size of the image (assuming the image is no larger than the array) and only dependent upon the area of the convolution mask. Since the CAAPP does cell level arithmetic bit-serially, the size of the data values also affects the speed of the algorithm. For 8 bit pixel values, the algorithm would take: 0.7, 2.1, 4.0, and 9.9 milliseconds for convolutions with square masks of 3×3, 5×5, 7×7 and 11×11 pixels respectively.

Histogramming

Our histogramming algorithm uses "Select Less Than" and "Count Responders" sub-routines. The algorithm simply selects ranges of values (buckets) starting with the lowest range and working up to the highest range. For simplicity, the range of each bucket is taken to be the maximum range divided by the number of buckets. The time for the algorithm is dominated by the time to perform the count responders operation for each bucket. The time is: (Number of Buckets) $\times 1.6 + 0.9$ microseconds; A 256 bucket 8-bit histogram would take 410.5 microseconds.

Histogram-Based Segmentation

Histogram based segmentation techniques have been studied by [OHL78], [NAG82], [KOH83] and more recently by [REY84]. On the CAAPP we can first perform a histogram, then decide what region labels we want to associate with the peaks and valleys of the histogram, finally we broadcast back to the array the labels that each pixel should assume. However, this simple technique does not give us a robust segmentation. We need to perform a connected components region labeling to produce regions with unique labels.

The connected components operation is done by locally propagating labels in parallel from seed points which are put in the array and iteratively merged with surrounding pixels of the same region. This operation would take about 10 milliseconds for all regions less than 128 pixels in diameter, in parallel. For larger regions we would use a different technique, serialize by the (small) number of large regions left in the image. A segmentation can now be defined in terms of regions of connected pixels, each with a unique label.

To proceed with building the intermediate symbolic representation, we form a "name cell", a distinguished pixel in each region. This would just be the upper right hand pixel of the region. We would now proceed to extract attributes of the region and add them to (symbolic) data fields in the name cell for each region, as we discussed in the section on Image Understanding.

High Level Querys

Once a symbolic intermediate level representation has been formed, high level processing can be initiated. For example, object hypothesis can be formed via rule-oriented techniques [WEY84], [RIS84]. This could involve broadcasting, sequentially, the range of a set of expected feature values for a particular object and setting weighted responses for regions which match these feature values. While the broadcasting of object features proceeds sequentially, each will take place very quickly so that the strongest responders for several objects can be found in a single frame time.

Motion

We use Motion Processing as a final example of an application developed for the CAAPP. This is a robust algorithm for quickly and accurately decomposing a flow field into its rotational and translational components to recover the parameters of sensor motion [STE83], [LAW84]. It makes use of both the associativity and array processing capabilities of the CAAPP. The algorithm is an exhaustive search procedure via a top-down parallel correlation of a set of rotational and translational flow field templates to find a component pair which most closely accounts for the motion depicted in a given flow field. Timing calculations and simulations reveal that the CAAPP could perform the rotational-translational decomposition using two 512×512 frames in slightly more than 1/4 second.

The preceeding algorithms illustrate that the CAAPP is within an order of magnitude of being what we call an Image Operand Class machine (a processor capable of performing a meaningful whole image operation in roughly the same time that a normal uniprocessor performs a floating point multiply). The implications for close to real time vision and motion analyses are clear.

VII. Conclusions

The architecture presented here is novel, buildable, and opens up a whole range of very high speed computation at the more difficult levels of machine vision. Both characteristics of the CAAPP, array processing and associative processing, seem to be fundamental to parallel vision algorithms. These capabilities allow communication which supports both global and local processing, which in turn supports transformations between representations of the image in a highly flexible manner.

VIII. Acknowledgments

We greatfully acknowledge the pioneering work and contributions of Caxton C. Foster on Content Addressable Processors. We would also like to thank Allen Hanson and Daryl Lawton for their contributions to the research reported here.

IX. References

[BAR68] G. H. Barnes, et al, *The ILLIAC IV Computer*, IEEE Transactions on Computers Vol. C-17, No. 8 (August 1968), 746-757.

[BAR78] H. Barrow and J. Tenenbaum, *Recovering Intrinsic Scene Characteristics from Images*, in "Computer Vision Systems", A. Hanson and E. Riseman, eds., Academic Press, 1978, pp. 3-26.

[BAT82] K.E. Batcher, *Bit Serial Parallel Processing Systems*, IEEE Transactions on Computers Vol. C-31, No. 5 (May 1982), 377-384.

[DUF78] M.J.B. Duff, *Review of the CLIP Image Processing System*, Proceedings of the National Computer Conference (1978), 1055-1060, AFIPS.

[HAN78] A.R. Hanson and E.M. Riseman, *VISIONS: A Computer System for Interpreting Scenes*, in "Computer Vision Systems", A. Hanson and E. Riseman, eds., Academic Press, 1978, pp. 303-333.

[HAN83] A.R. Hanson and E.M. Riseman, *A Summary of Image Understanding Research at the University of Massachusetts*, COINS Technical Report 83-35 (October 1983), University of Massachusetts at Amherst.

[KOH83] R.R. Kohler, *Integrating Non-Semantic Knowledge into Image Segmentation Processes*, Ph.D. Dissertation and COINS Technical Report 84-04 (September 1983), University of Massachusetts at Amherst.

[LAW84] D.T. Lawton, *Processing Dynamic Image Sequences from a Moving Sensor*, Ph.D. Dissertation and COINS Technical Report 84-05 (February 1984), University of Massachusetts at Amherst.

[LEV84] S.P. Levitan, *Parallel Algorithms and Architectures: A Programmer's Perspective*, Ph.D. Dissertation and COINS Technical Report 84-11 (May 1984), University of Massachusetts at Amherst.

[MAR82] D. Marr, "Vision", W.H. Freeman, San Francisco, 1982.

[NAG82] P.A. Nagin, A.R. Hanson and E.M. Riseman, *Studies in Global and Local Histogram-Guided Relaxation Algorithms*, IEEE Transactions on Pattern Analysis and Machine Intelligence PAMI-4 (May 1982), 263-277.

[OHL78] R. Ohlander, K. Price, and R. Reddy, *Picture Segmentation Using a Recursive Region Splitting Method*, Computer Graphics and Image Processing Vol. 8 (1978), 313-333.

[REY84] G. Reynolds, N. Irwin, A. Hanson and E. Riseman, *Hierarchical Knowledge-Directed Object Extraction Using a Combined Region and Line Representation*, Proceedings of the Workshop on Computer Vision: Representation and Control (April 30–May 2, 1984), 238-247, Annapolis, Maryland.

[RIS84] E.M. Riseman and A.R. Hanson, *A Methodology for the Development of General Knowledge-Based Vision Systems*, to appear, Proceedings of the IEEE Workshop on Principles of Knowledge-Based Systems (3-4 December 1984), Denver, Colorado.

[STE83] M. Steenstrup, D. Lawton and C. Weems, *Determination of the Rotational and Translational Components of a Flow Field Using a Content Addressable Parallel Processor*, Proceedings of the International Conference on Parallel Processing (August 1983), 492-495.

[WEE82] C. Weems, S. Levitan and C. Foster, *Titanic: A VLSI-Based Content Addressable Parallel Array Processor*, Proc. of IEEE International Conference on Circuits and Computers (September 1982), 236-239.

[WEE84] C. Weems, *Image Processing with a Content Addressable Array Parallel Processor*, Ph.D. Dissertation (1984), Computer and Information Science Department, University of Massachusetts at Amherst.

[WEY84] T.E. Weymouth, *Using Object Descriptions in a Schema Network for Machine Vision*, Ph.D. Dissertation (1984), Computer and Information Science Department, University of Massachusetts at Amherst.

EMSP: A DATA-FLOW COMPUTER FOR SIGNAL PROCESSING APPLICATIONS

J. D. Seals
R. R. Shively

AT&T Bell Laboratories
Whippany, New Jersey 07981

ABSTRACT

The Navy's signal processing requirements over the next decade will far exceed the capabilities of current signal processing computers. To meet these needs, AT&T Bell Laboratories is currently designing and building the Navy's AN/UYS-2 Enhanced Modular Signal Processor (EMSP). [1] [2] [3] Key elements in the EMSP concept are: 1) a hybrid data-flow and control-flow program organization, 2) a modular structure with well defined interfaces to enable evolutionary technology infusion over the life of the system, and 3) decentralized control of task scheduling and data movement.

The multi-processor type EMSP architecture employs a hybrid of data-flow and control-flow control. The data-flow control for task scheduling, together with a directed graph methodology for transcribing it, provide a means for exploiting the parallelism inherent in most signal processing applications. Within functional elements responsible for executing the tasks, a more conventional control-flow approach is used.

This paper will provide an overview of the EMSP architecture, the distributed control and task scheduling strategies, and the commonality in the control module implementation across functional element types that results.

INTRODUCTION

The computational throughput requirements for the Navy's current and projected signal processing requirements far exceed the capabilities of any existing computer. Current requirements range from ten million to hundreds of millions of multiply-add operations per second. Projected processing estimates show that the these requirements will increase tenfold in the next ten years. Even if it were technically feasible to build a single-thread computer capable of meeting the high end of this performance range under a mil spec environment, a single large processor would not offer the reliability, nor the modularity required to span the range of volume, power, and performance required by the Navy. Modular, multi-processor signal processors which exploit the parallelism inherent in signal processing applications are needed to meet these requirements.

There are three major approaches for the allocation and scheduling of parallel tasks: control-flow, data-flow, and reduction. [4] Of these three approaches, data-flow and control-flow are the most mature. Control-flow can achieve parallel processing via a single central sequence of instructions which are carried on simultaneously on many processors and data streams (Single Instruction, Multiple Data Processing) and pipelining. Control-flow is a highly efficient means of executing concurrently on multiple processors if the process can be described in a do-loop like statement. Unfortunately, not all processes exhibit this characteristic and it is difficult to write programs as simplified operations that can be applied to multiple data streams.

Data-flow [5] [6] [7] [8] allows any task to be executed if its input data is available. In data-flow control, sequencing is performed by the flow of data in an asynchronous manner. There are no program counters or central control. Since any task may be executed when its inputs are available, concurrent processing is easily and naturally supported. There are two major problems which detract from the simplicity of data-flow. First, the cost of communication and bookkeeping operations are significant, and secondly, it is difficult to schedule tasks such that the work load is evenly distributed over all processors.

While these two approaches are seemingly different and conflicting, it is becoming recognized [9] [10] [11] that the attributes of data-flow and control-flow are complementary. If one uses data-flow at the task or functional level rather than at the elementary level used in traditional data flow, the costs associated with communications and bookkeeping are minimal compared with the gain in concurrent processing. Control processing within the task provide efficient execution at the fine grain or elementary operation level. Modularity in hardware and efficient support of technology evolution requires an application software interface that isolates programming from configuration change or upgrade, as well as supporting data flow execution. The graph-language methodology [12] [13] that has evolved as the logical means of expressing data flow programs appears to provide such an interface. The Navy sponsored ECOS (EMSP Common Operational Software Methodology) graph notation is used for EMSP.

EMSP DESIGN PHILOSOPHY

EMSP incorporates a hybrid data-flow and control-flow program organization to realize high throughput and utilization of computational resources. Major concepts are:

- to use data-flow at the task level to exploit the parallelism inherent in signal processing applications,
- to use control-flow processing within the processing elements to eliminate the communication and bookkeeping costs that data-flow would incur at the fine grain (elemental operation) level,
- to provide decentralized scheduling and control, and
- to use intelligent control within each processing element to monitor and control work loads.

EMSP data-flow control scheduling extends to the level of tasks or functions. Unlike traditional data-flow architectures which schedule elemental (add, multiply) operations on a single operand or operand pair, EMSP uses data-flow to schedule macro functions such as a matrix multiply or FFT. Since these functions may operate on arrays of hundreds or thousands of operands, the scheduling and bookkeeping is reduced by several orders of magnitude over data-flow at a fine grain level. For example, a 1024 point vector multiply would require 1024 node schedulings at the micro level, but only one scheduling at the functional (macro) level.

A bonus of the task level data-flow control is the direct mapping between a signal processing graph and a data-flow graph. In the ECOS data-flow program description, signal processing applications are defined as a directed graph, with the nodes representing signal processing functions such as FFT and FIR, and the arcs representing the flow of data between nodes. Conversion of this graph into an executable data-flow graph involves little more than defining read, produce, and consume amounts, defining the threshold amount of data in each queue required to dispatch the target node, and adding any needed synchronization nodes.

Most signal processing nodes are highly repetitive loops operating on one or more arrays of data. The parallelism in these operations can be described for the most part by simple loop constructs, amenable to control-flow. EMSP uses control-flow techniques (SIMD and pipeline processing) to exploit this parallelism, thereby eliminating the burden of communication and bookkeeping that data-flow would incur at the micro level.

Each programmable element in EMSP contains a control processor, denoted Control Unit (CU), to act as a local task dispatcher. At any given instant, several nodes may be in various stages of processing within the element. For example, the CU may have initiated the output of one node, while the data from a second node is being processed, and data from a third node is being read. Once the output of the first node and the computation of the second node is completed, the CU will request a new task. The overlapping of setup and breakdown leads to a balanced system in which all processing elements share equitably in the processing of a graph.

DATA-FLOW SCHEDULING AND CONTROL

The data flow scheduling and control of nodes is distributed between a Scheduler, the Global Memories, and the Processing Elements.

The Global Memories (GMs) maintain data queues (graph arcs) and report on their status. Each data queue is a dynamic structure with an associated threshold and capacity. The queue threshold indicates the minimum number of data elements needed to satisfy one of the conditions for a node firing. When the number of elements in a given queue reach or exceeds threshold, a queue-over-threshold message is sent to the Scheduler. No further threshold messages are sent until the queue has been consumed and its contents again exceed threshold. Each queue has a capacity above which a queue-over-capacity message is sent to the Scheduler. This inhibits the input node from firing until the queue falls below capacity and a queue-under-capacity message sent to the Scheduler.

The Scheduler's primary function is to determine when a node is ready-to-execute and to match it to a free processing element (PE). A node is ready-to-execute when all of its input queues are at, or above threshold and all synchronization events are satisfied. The Scheduler performs these tasks via four tables: the queue-to-node table, the node status table, the ready-node list, and the free PE list.

The queue-to-node table is a connectivity map which identifies the input and output nodes associated with a given queue. This map points to entries in the the node status table. Each node entry in the node status table contains the node's id, priority, firing counter, instruction stream location, and graph instance. The firing counter indicates the number of conditions (queues over threshold, synchronization events) that remain to be satisfied before the node can fire. As each of these conditions is satisfied, the firing counter is decremented. When the counter reaches zero, it is matched to a processor on the free processor list or placed on the ready node list if no free processor is available. Nodes on the ready node list are matched to free processors, as they become available.

Once the Scheduler has matched a node to a free PE, it obtains the node's instruction stream id from the node status table and sends a message to the Global Memory containing the instruction stream. It then increments the firing counter by the number of conditions needed to fire that node again.

At this point, the Scheduler has essentially completed its tasks. The GM receives the message, locates the instruction stream, and forwards it the designated PE. The instruction stream contains information on what data is needed to execute the node, where is it stored, and what operations are to be performed. For example, it might instruct the PE to fetch 1024 data elements each from X and Y (stored in GM #i and #j respectively), to store these inputs in specific locations in operand memory, to perform a vector multiply, and store the results in queue Z in GM #k. It would then instruct the GM to consume X and Y.

The instruction stream is decoded by the PE. It forwards requests for the data needed to execute this node (to the GMs) and stores the data in an operand memory as it is received. After a node has executed, a message is sent to the Scheduler, causing the sending PE to be placed on the free PE list. Since a PE overlaps setup and breakdown with execution, a new node is typically setup and awaiting execution.

It should be noted that a second instance of the a node cannot fire until the first has completed. This prevents data from getting out of sequence and simplifies error recovery. Also, once a queue sends a queue over threshold message, it cannot send another until it has been consumed (zero consume is permitted). This insures that a queue cannot send multiple queue-over-threshold messages and prematurely fire a node.

Any node can be suspended or inhibited by sending a message which increments its firing counter. For example, if a given queue is nearing its maximum capacity, a queue-over-capacity message inhibits the input node for that queue. When the queue falls below capacity, a queue-under-capacity message is sent and the firing counter is decremented. Similarly, messages to suspend and start a given node or node sequence will increment and decrement the firing counter respectively.

THE EMSP ARCHITECTURE

The EMSP architecture (see Figure 1) encompasses a diverse family of machine configurations which are tailored to meet the specific processing and packaging requirements of a given application. This versatility is realized by defining a set of autonomous and asynchronous functional elements (FEs) which form the basic system building blocks. Each functional element uses the same protocols and electrical interfaces. The functional elements are, in turn, constructed from a set of standard electronic modules (SEMs).

Each FE is a functionally complete architectural component supporting hardware and software functions necessary to performs its assigned tasks. The FE types were carefully selected to provide a balanced distribution of work load and control. Currently, five FE types are defined: the Arithmetic Processors (APs), the Global Memories (GMs), the Scheduler (SCH), the Command Program Processor (CPP), and the Input/Output Processors (IOPs). Support of technology infusion is provided by formal management of FE communication interfaces. A new FE, or alternate realization of an existing one, can be integrated provided it uses the EMSP protocol and electrical interface.

VLSI Signal Processing

Communication between FEs is supported by the control busses (CBUSs) and the the data transfer networks (DTNs). The control busses are used to communicate control data, data requests, and test functions. The DTN is a dynamically reconfigurable, nonblocking matrix switch for the movement of data queues between FEs.

The Arithmetic Processor implementation (Fig. 2) is a variant on what has been termed the decoupled access/execute processor. [14] [15] [16] Traditionally, von Neumann type processors are accelerated by including a sub-processor dedicated to the pre-access and post-storage of operands and results, respectively, while another processor interconnected via hardware queues executes the operations (the Address Generator, AG, and Arithmetic Unit, AU, respectively in Fig. 2). This concept is extended once more in the AP structure with another processor dedicated to concurrent pre-access and post-storage of *arrays* of data to/from the GM, as well as executing the system level commands related to node execution (the CU in Fig. 2). The memory-access bottleneck that is characteristic of von Neumann processors is relaxed by providing distinct microprogram and data memories for each of the three subprocessors, as well as a third memory (the Coefficient Memory) for each AU. Thus, with two AU's in the SIMD AP structure, a total of 10 concurrent memory references can take place each machine cycle in each AP, not including the direct-memory-access references occurring in support of DTN transfers.

Each AU consists of two mulipliers and four register arithmetic logic units (RALUs). Each dual AU Arithmetic Processor is capable of up to 28 million multiplies and 56 million RALU operations (adds, subtracts, conditional tests, etc.) per second. The 2:1 ratio of adders to multipliers enhances algorithms such as the radix 4 FFT.

The Global Memory (GM) is an intelligent memory processor which provides storage and memory management of queues, node scheduling assistance, and execution of data management functions. The GM realizes data queues as dynamically allocated arrays which are addressed by name. The Global Memory monitors the threshold and capacity of each queue it contains and sends appropriate scheduling messages to the Scheduler. The basic GM is sized for one million 32 bit words and can be upgraded in one million 32 bit word increments to four million words. The memories provide single bit error correction, and double bit error detection. A GM is capable of transferring five million 32 bit words/sec at burst rate.

The Scheduler determines the readiness of a node to fire and makes the assignment of ready nodes to available processors, as described above. The Scheduler is capable of matching in excess of 7,000 and 28,000 nodes per second in sustained and burst rate respectively. It can handle 40,000 (sustained) and 160,000 (burst) events (threshold messages, etc.) per second and contains table memory for 16,000 nodes and 65,000 data queues. A doubly redundant version provides high reliability.

The Command Program Processor (CPP) provides graph management, system control, performance monitoring and fault control, and operating system functions for the EMSP. The CPP is a single or dual embedded AN/UYK-44 militarized and reconfigurable minicomputer. It is not typically involved in high speed signal processing, but does provide six I/O interface channels to any existing host, displays, and/or terminals via a standard digital format.

The Input/Output Processor (IOP) provides versatile control of I/O data. It can accept any mix of up to 32 standard (NTDS, RS-232) or high speed (RS-422A, 1553B, etc.) channels at an aggregate rate of 5 million 16 bit words per second. Each IOP can contain up to a megaword (in 128K increments) of internal buffer memory and an arithmetic processor for simple multiplexing and packing operations. The IOP can be enhanced by the addition of up to 32 I/O channel processors (IOCP) each of which is capable of 5 million multiply-add operations per second. This permits front end processing (FIRs, Complex Demodulation, Decimation) to be performed on the data before it enters a Global Memory. This can substantially reduce data bandwidth, GM capacity, and AP requirements.

The Data Transfer Network (DTN) is used to move data and instruction streams between FEs. Each DTN is a unidirectional, asynchronous, full-access nonblocking switch with up to 16x16 ports, each 32 bits wide. Each switch port can be time division multiplexed by up to four functional elements through the use of concentrator and distributor modules. VLSI realization of the switch enables two bits of the 16x16 port switch with each device. A novel mode control permits configuration of a switch as a 4 bit 8x8 switch, and 8 bit 4x4 switch, or a 16 bit 2x2 switch with the the same device. Transfers are packetized to minimize acknowledgement delays, an important requirement for future technology enhancement.

ACKNOWLEDGEMENTS

The EMSP design is the result of the efforts of many talented and dedicated individuals at AT&T Bell Laboratories. While it is impossible to list all of these contributors, we would like to acknowledge the major contributions of N. Brown, R. Morris, P. Gloudemans, P. Punguliya, R. Jordan, C. Stanziola, C. Jordan, and D. Long.

REFERENCES

1. LCDR William L. Hatcher, III, "EMSP: The Navy's Next Generation Real-Time Signal Processor," SOUTHCON, January 1984.

2. N.H. Brown, "The EMSP Data Flow Computer," HICSS-17, International Conference on Systems Sciences, January 1984.

3. C.B. Robbins, "Navy Real-time Signal Processing Development - Second Generation Planned Service Standard,"SPIE, Vol. 298, Real Time Signal Processing IV, August 1984.

4. Eric J. Lerner, "Data-flow Architecture," IEEE Spectrum, April 1984.

5. Arvind, Lecture Notes, Tutorial: "Data Flow Languages and Architecture," The 8th Annual Symposium on Computer Architecture, May 11,1981.

6. T. Agerwala, Arvind, "Data Flow Systems," Computer pp10-13, Feburary 1982.

7. A.L. Davis, R.M. Keller, "Data Flow Program Graphs", Computer pp26-41, Feburary 1982.

8. J.B. Dennis and D.P.Mesunas, "A Preliminary Architecture for a Basic Data Flow Processor," Proc. Second Annual Symposium Computer Architecture, January 1975, pp 126-132.

9. R.P. Hopkins et al.,"A Computer supporting Data Flow, Control Flow, and Updateable Memory, Technical Report 144, Computing Laboratory," The University of Newcastle upon Tyne, September 1979.

10. P.C. Treleaven et al., "Combining Data Flow and Control Computing," The Computer Journal, Vol.25, No.2, 1982.

11. R.G. Babb II, "Parallel Processing with Large Grain Data Flow Techniques," IEEE Computer, July 1984.

12. "ECOS/ACOS Common Operational Support Software Methodology Specification," June 30, 1983, Analytic Disciplines Inc., Navy Contract NRL, PMS408

13. Y.S. Wu et al.,"Architectural approach to alternate low-level primitive structures (ALPS) for acoustic signal processing," IEE Proceedings, Vol.131, June 1984.

14. J.E. Smith, "Decoupled Access/Execute Computer Architectures," Conference Proceedings of the 9th Sumposium on Computer Architecture, Austin, Texas, 1984.

15. Y.S. Wu, "Microprogramming Applications for Signal Processing," Microarchitecture of Computing Systems, R.W. Herenstein and R. Zuks, North Holland Co., 1975.

16. R.R. Shively, "Architecture of a Programmable Digital Signal Processor," IEEE Transactions on Computers,Vol.C-31, No.1, Januaray 1982.

VLSI Signal Processing

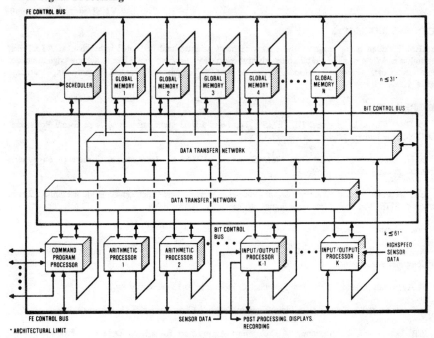

FIGURE 1 - THE EMSP MODULAR DATA FLOW ARCHITECTURE

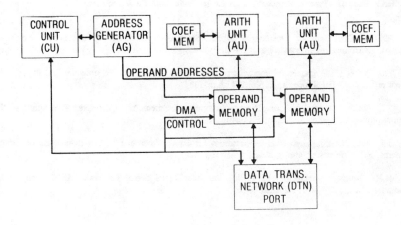

FIGURE 2 - THE EMSP'S ARITHMETIC PROCESSOR

13 VLSI Signal Processing

An Integrated Coherent Demodulation Technique

Kerry Hanson
Texas Instruments, Inc.
MS 7845
P.O. Box 1444
Houston, Tx 77251

A circuit is described that provides coherent demodulation of quadrature phase shift keyed signals that are used in voice band modems. Specifically an IC is described that uses switched capacitor circuits to provide full duplex communications over dial up telephone lines at 1200 bits per second. The circuits include Hilbert filters, phase locked loops, voltage controlled oscillators and carrier detection functions. A brief description of an adaptive equalization circuit used in this application is also included.

The TMS99542 is an integrated modem that communicates at 1200 BPS an is compatible with the Bell 212 modem, the Racal-Vadic modem and the CCITT V.22 specification[1]. These modems use quadrature phase shift keyed (QPSK) modulation and are synchronous in nature. All of the options, modes and alternatives of these modems are also supported by the IC. A block diagram of the modem IC is shown in figure one.

These modems are full duplex which means that they can transmit and receive simultaneously. The transmit path is shown along the top of figure one and the receive path is shown along the bottom. The data to transmitted is first buffered. Two events occur here. The first is that proper synchronous timing is added to the data stream. The data is synchronized by inserting or deleting stop bits in the character data. If the data is coming into the modem synchronously in the first place, then this step is bypassed. In addition to synchronizing the data, the buffer also scrambles it by multiplying it with the output of a pseudo random shift register. The resulting bit stream looks random even though it isn't. This is done in order to insure that the modem meets certain tariff requirements.

The next step is the modulation of the synchronous transmitted bit stream. The data is phase encoded into quadrature phase assignments and then multiplied by the appropriate carrier frequency. The two resulting biphase modulated signals are then summed and the result is a QPSK modulated carrier signal. Some additional filtering is required in the transmitter to prepare the signal for transmission on a telephone line.

Essentially the opposite occurs in the received carrier signal path. The incoming signal from the telephone line is first filtered with an anti-alias filter to remove any high frequency components. The received signal is then filtered with a high order band pass filter. This filter provides three different processing functions. First, it filters out any out of band noise added by the telephone channel. Since the modem is full duplex the receive band pass filters must also filter out the transmitted signal. This is a fairly stringent requirement since the transmitted signal is almost always much stronger than the received signal. Normally,

427

VLSI Signal Processing

the receive filter provides at least sixty dB of transmitter rejection. The third function of the receive filter is that it provide both amplitude and group delay equalization to account for some the distortion introduced by the telephone line.

A characteristic of the demodulator used in the TMS99542 is that the received signal must have a fixed amplitude. A limiter cannot be used because it introduces odd harmonics that then degrade the performance of the demodulator. The amplitude must be fixed by an automatic gain control circuit (AGC). The AGC is part of the last two stages of the band pass filter. Over the entire dynamic range of the modem the output of the receive filter is held to within plus or minus 1.5 dB. This insures that the demodulator will work on only the phase and not the amplitude of the signal. It is easy to change the gain of a circuit using switched capacitor circuits since the amplification is set by the ratio of two capacitors [2].

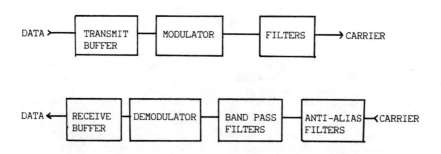

Figure one - TMS99542 Block Diagram

The actual demodulator uses a coherent technique based on a Costas phase locked loop. It will be described in more detail in the next section. The demodulated data stream then enters the receive buffer. This circuit is just the opposite of the transmit buffer. First, it unscrambles the bit stream and extracts the original transmitted data. This is done by simply multiplying the bit stream by the out put of another pseudo random shift register. This shift register automatically synchronizes itself with the transmitter shift register. The second function of the receive buffer is to reinsert any character stop bits that were deleted by the transmit buffer. This is all done automatically and transparently to the user of the modem [3].

The basic modulation scheme used by these modems is called differential quadrature phase shift keyed modulation. The phase of a base band signal is measured at specific points in time. This phase is stored for one sample time and then compared with the next phase sample. The result is a measurement of the phase shift rather than the absolute phase of the signal. The phase difference of the signal is much easier to detect rather than the absolute phase and the result is only a very small degradation in bit error rate performance. As implemented, the base band

VLSI Signal Processing

actually consists of two biphase modulated signals. The modulation occurs by multiplying an analog signal by plus or minus one. The base band is then mixed with two carrier frequency signals that are ninety degrees out of phase. The two signals are summed and the result is a four phase differential PSK signal. Since there are four phases each shift actually represents two bits of data.

When there are only four possible phases then a QPSK signal is exactly the same as a quadrature amplitude modulated signal (QAM). Another way of looking at the carrier then is as two amplitude modulated signals that are ninety degrees apart. This is a key characteristic of the coherent demodulation technique used on the TMS99542.

The demodulation technique is called a Costas phase locked loop. This technique is the most popular one used in the demodulation of QAM signals. A block diagram of this technique is shown in figure two [4].

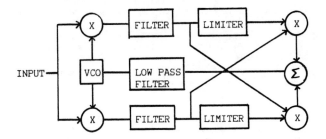

Figure two - Costas Phase Locked Loop

A Costas PLL is similar to an ordinary phase locked loop in that it contains the same basic signal processing blocks. The difference is in the number of blocks. Figure two shows that the input signal is actually referenced against two outputs of the voltage controlled oscillator which are ninety degrees apart. This performs the function of separating the two biphase modulated signals and allowing independent demodulation of each. The other effect of these phase comparators is to mix the carrier frequency back down to the base band signal. The double frequency components are filtered out and then each signal is limited. The data is pulled off at this point. Another difference with a normal phase locked loop is that both signals are then mixed again. The base band data is filtered out and the result is just the error signal as shown in figure two. The error signal drives the voltage controlled oscillator which, when everything is working properly, results in demodulated phase assignments that are then converted into the original data stream. The entire demodulator is implemented in switched capacitor circuits.

VLSI Signal Processing

There are two enhancements to the Costas phase locked loop as implemented on the TMS99542. The first is a Hilbert filter at the input to the Costas PLL. As shown in figure two the two arm filters must completely reject the double frequency component at the output of the phase comparators. This requires either high order low pass filters or filters with a high Q. In one case a lot of chip real estate is required and in the other case significant group delay is introduced into the signal path resulting in a reduction in bit error rate performance. A method of avoiding this problem is to separate the input signal into two identical signals that are exactly ninety degrees out of phase. The filters that do this are called Hilbert filters are the phase shift is completely frequency independent. These two signals are added and subtracted as appropriate at the two phase comparators resulting in just the difference product rather than both the sum and difference products. This means the output of the phase comparators is just the base band signal and the arm filters can be eliminated. As actually implemented, the filters are still required but are simple first order low pass filters.

The second enhancement to the basic Costas PLL demodulator is the addition of an adaptive equalizer. This is a circuit that adaptively cancels both the amplitude and delay distortion of the received signal that was introduced by the telephone line. This circuit constantly readapts and matches itself to the received signal in real time. The specific circuit used on the TMS99542 is a decision feedback adaptive equalizer. This circuit stores the sign of the base band signal for the most recent sample and the two previous samples. This is multiplied by the base band signal according to a simple algebraic relation and the result is that some of the distortion in the base band is canceled. The new, equalized base band signal is then used to generate new values for the stored samples. The equalizer is updated every sample time. When combined with the fixed equalization in the receive filter the result is good bit error rate performance over a wide range of telephone channels[5].

The voltage controlled oscillator (VCO) was one of the most difficult circuits on the chip to design. The technique finally chosen was a digital counter with the count range set by a digital equivalent of the analog error voltage. The digital signal is generated in a six bit successive approximation A/D converter. Since a fixed master clock frequency is used for the VCO any variation in the error voltage will cause internally generated phase jitter in the phase locked loop. Careful loop design was required to account for this effect and still provide good performance.

To provide complete demodulation of the data two separate phase locked loops are required. The first one is the Costas PLL and it has been described. Another phase locked loop is required to recover the embedded clock signal from the demodulated data. A block diagram of this phase locked loop is shown in figure three. The input to this loop is the two demodulated base bands from the Costas loop. These signals are full wave rectified and then mixed with the output of a VCO running at the clock rate. These signals are summed and filtered to remove the data and retain only the clock information. Normally the output of the low pass filter just drives the VCO. Note that this PLL has an additional parallel path through an integrator. Without the integrator this loop is called a first order phase locked loop. This means the error voltage is proportional to the frequency difference between the VCO and the input signal. If an integrator is inserted in this path then the loop is called a second order loop. The integrator reduces the average error signal to zero. The only time the signal is not zero is when the VCO is slewing to match the input.

The advantage of a first order loop is that it has a wide lock range However, it won't sample the base band data at the proper time if there is a frequency offset. The data rate may vary over a fairly wide range resulting in a frequency offset. A second order loop operates properly with frequency offsets but with adequate PLL parameters there are conditions when it may not lock onto the incoming signal. The solution to these problems is to provide both kinds of loops in parallel. That way the first order loop can acquire the signal and, when locked on, the second order loop can reduce the phase error to zero. Care must be taken in setting the gain through both paths but the result is good clock recovery performance.

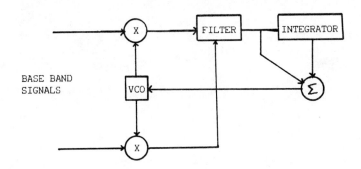

Figure three - Clock Recovery Loop

In conclusion, an integrated circuit is described that provides all the modulation, demodulation, filtering and data buffering functions of a full duplex modem. Special emphasis was placed on the demodulator design to insure optimum performance. The demodulator is a coherent PLL demodulator and includes Hilbert filters and adaptive equalizers. All signal processing functions are implemented in switched capacitor circuits. This allows good performance while minimizing chip real estate.

References

[1] K. Hanson et al, "A 1200 BPS PSK Full Duplex Modem", ISSCC conf., 1984.

[2] P. Allen, E. Sanchez-Sinencio, <u>Switched Capacitor Circuits</u>, Van Nostrand Reinhold, 1984.

[3] D. Tugal, O. Tugal, <u>Data Transmission Analysis, Design, Applications,</u> McGraw-Hill, 1982.

[4] F. Gardner, <u>Phaselock Techniques,</u> John Wiley & Sons, 1966.

[5] S. Qureshi, "Adaptive Equalization", IEEE Communications Magazine, March, 1982.